Mechatronics

**Electronic control systems
in mechanical and electrical engineering**

Third Edition

W. Bolton

We work with leading authors to develop the
strongest educational materials in engineering,
bringing together cutting-edge thinking and best learning
practice to a global market.

Under a range of well-known imprints, including
Prentice Hall, we craft high-quality print and
electronic publications which help readers to
understand and apply their content,
whether studying or at work.

To find out more about the complete range of our
publishing, please visit us on the World Wide Web at:
www.pearsoned.co.uk

Mechatronics

Electronic control systems
in mechanical and electrical engineering

Third Edition

W. Bolton

Harlow, England • London • New York • Boston • San Francisco • Toronto
Sydney • Tokyo • Singapore • Hong Kong • Seoul • Taipei • New Delhi
Cape Town • Madrid • Mexico City • Amsterdam • Munich • Paris • Milan

Pearson Education Limited
Edinburgh Gate
Harlow
Essex CM20 2JE
England

and Associated Companies throughout the world

Visit us on the World Wide Web at:
www.pearsoned.co.uk

First published 1995
Second edition 1999
Third edition 2003

ISBN-10: 0-13-121633-3
ISBN-13: 978-0-13-121633-4

British Library Cataloguing-in-Publication Data
A catalogue record for this book is available from the British Library

Library of Congress Cataloging-in-Publication Data
A catalog record for this book is available from the Library of Congress

10 9 8 7

08 07 06

Printed by Bell & Bain Ltd., Glasgow

Contents

Preface

Overview

The integration of electronic engineering, electrical engineering, computer technology and control engineering with mechanical engineering is increasingly forming a crucial part in the design, manufacture and maintenance of a wide range of engineering products and processes. A consequence of this is the need for engineers and technicians to adopt an interdisciplinary and integrated approach to engineering. The term *mechatronics* is used to describe this integrated approach. A consequence of this approach is that engineers and technicians need skills and knowledge that are not confined to a single subject area. They need to be capable of operating and communicating across a range of engineering disciplines and linking with those having more specialised skills. This book is an attempt to provide a basic background to mechatronics and provide links through to more specialised skills.

Readership

The *first edition* was designed to cover the Business and Technology Education Council (BTEC) Mechatronics A and B Units (1413G and 1414G) for higher technicians, these units being the core units of their Higher National Certificate/Diploma courses and designed to fit alongside more specialist units such as those for design, manufacture and maintenance determined by the application area of the course. The book was widely used for such courses and has also found use in undergraduate courses in both Britain and in the United States. Following feedback from lecturers in both Britain and the United States, the *second edition* was considerably extended and with its extra depth it was not only still relevant for its original readership but also suitable for undergraduate courses. The *third edition* involves refinements of some explanations, more discussion of microcontrollers and programming, increased use of models for mechatronics systems, and the grouping together of key facts in the Appendices.

Aims

The overall aim of the book is to give a comprehensive coverage of mechatronics which can be used with courses for both technicians and undergraduates in engineering, and hence, to help the reader:

1 Acquire a mix of skills in mechanical engineering, electronics and computing which is necessary if he/she is to be able to comprehend and design mechatronics systems.

2 Become capable of operating and communicating across the range of engineering disciplines necessary in mechatronics.

In the various chapters, the book aims to help the reader:

Chapter 1: *Mechatronics*
Appreciate what mechatronics is about.
Comprehend the various forms and elements of control systems: open-loop, closed-loop and sequential.
Recognise the need for models of systems in order to predict their behaviour.

Chapter 2: *Sensors and transducers*
Describe the performance of commonly used sensors.
Evaluate sensors used in the measurement of: displacement, position and proximity; velocity and motion; force; fluid pressure; liquid flow; liquid level; temperature; light.
Explain the problem of bouncing when mechanical switches are used for inputting data.

Chapter 3: *Signal conditioning*
Explain the requirements for signal conditioning.
Explain how operational amplifiers can be used, the requirements for protection and filtering, the principle of the Wheatstone bridge and, in particular, how it is used with strain gauges, the principles and main methods of analogue-to-digital and digital-to-analogue converters, multiplexers and data acquisition using DAQ boards.
Explain the principle of digital signal processing.
Explain the principle of pulse-modulation.

Chapter 4: *Data presentation systems*
Explain the problem of loading.
Describe the basic principles of use of commonly used data presentation elements: meters, analogue chart recorders, oscilloscopes, visual display units, printers.
Explain the principles of magnetic recording on floppy and hard discs.
Explain the principles of displays and, in particular, the use of LED seven-segment and dot matrix displays and the use of driver circuits.

Explain how data presentation can occur with the use of DAQ boards.

Design measurement systems.

Chapter 5: *Pneumatic and hydraulic actuation systems*

Interpret system drawings, and design simple systems, for sequential control systems involving valves and cylinders.

Explain the principle of process control valves, their characteristics and sizing.

Chapter 6: *Mechanical actuation systems*

Evaluate mechanical systems involving linkages, cams, gears, ratchet and pawl, belt and chain drives, and bearings.

Chapter 7: *Electrical actuation systems*

Evaluate the operational characteristics of electrical actuation systems: relays, solid-state switches (thyristors, bipolar transistors and MOSFETs, solenoid actuated systems, d.c. motors, a.c. motors and steppers).

Chapter 8: *Basic system models*

Devise models from basic building blocks for mechanical, electrical, fluid and thermal systems.

Chapter 9. *System models*

Devise models for rotational–translational, electro-mechanical and hydraulic–mechanical systems.

Chapter 10: *Dynamic responses of systems*

Model dynamic systems by means of differential equations.

Determine the response of first- and second-order systems to simple inputs.

Chapter 11: *System transfer functions*

Define the transfer function and determine the responses of systems to simple inputs by its means, using Laplace transforms.

Identify the effect of pole location on transient response.

Use MATLAB and SIMULINK to model systems.

Chapter 12: *Frequency response*

Analyse the frequency response of systems subject to sinusoidal inputs.

Plot and interpret Bode plots, using such plots for system identification.

Chapter 13: *Closed-loop controllers*

Predict the behaviour of systems with proportional, integral, derivative, proportional plus integral, proportional plus derivative and PID control.

Explain how such modes of control can be realised with operational amplifiers and digital controllers and controller settings determined.

Explain what is meant by velocity feedback and adaptive control.

Chapter 14: *Digital logic*

Use the binary, octal, hexadecimal and binary coded decimal number systems; explain how numbers can be signed and the twos complement method of handling negative numbers.

Explain the advantages of the Gray code.

Describe parity methods of error detection.

Recognise the symbols and Boolean representation of, write truth tables for and use in applications, the logic gates of AND, OR, NOT, NAND, NOR and XOR.

Use Boolean algebra to simplify Boolean expressions and present them in the form of sums of products or product of sums.

Use Karnaugh maps to determine the Boolean expressions to represent truth tables.

Explain the operation of decoders.

Explain how SR, JK and D flip-flops can be used in control systems.

Chapter 15: *Microprocessors*

Describe the basic structure of a microprocessor system.

Describe the architecture of common microprocessors and how they can be incorporated in microprocessor systems.

Describe the basic structure of microcontrollers and the architecture of commonly encountered microcontrollers and how their registers are used to carry out tasks.

Explain how programs can be developed using flow charts or pseudocode.

Chapter 16: *Assembly language*

Use assembly language to write programs.

Chapter 17: *C language*

Use C to write programs.

Chapter 18: *Input/output systems*

Identify interface requirements and how they can be realised; in particular buffers, handshaking, polling and serial interfacing.

Explain how interrupts are used with microcontrollers.

Explain the function of peripheral interface adapters and program them for particular situations.

Explain the function of asynchronous communications interface adapters.

Chapter 19: **Programmable logic controllers**
Describe the basic structure of PLCs.
Program a PLC, recognising how the logic functions, latching and sequencing can be realised.
Develop programs involving timers, internal relays, counters, shift registers, master relays, jumps and data handling.

Chapter 20: **Communication systems**
Describe centralised, hierarchical and distributed control systems, network configurations and methods of transmitting data, describing protocols used in the transmission of data.
Describe the Open Systems Interconnection communication model.
Describe commonly used communication interfaces: RS-232, Centronics, IEEE-488, personal computer buses, VXIbus, and I²C bus.

Chapter 21: **Fault finding**
Recognise the techniques used to identify faults in microprocessor-based systems, including both hardware and software.
Explain the use of emulation and simulation.
Explain how fault finding can be developed with PLC systems.

Chapter 22: **Mechatronics systems**
Compare and contrast possible solutions to design problems when considered from the traditional and the mechatronic points of view, recognising the widespread use of embedded systems.
Analyse case studies of mechatronics solutions.
Design mechatronics solutions to problems.

Structure of the book

Each chapter of the book is copiously illustrated and contains problems, answers to which are supplied at the end of the book. With Chapter 22 research and design assignments are also included, clues as to their possible answers also being given.

The diagram on the next page indicates the overall structure of the book. Chapter 1 is a general introduction to mechatronics. Chapters 2 to 7 form a coherent block on basic systems hardware, Chapters 8 to 13 are concerned with developing system models, Chapters 14 to 21 are concerned with digital and microprocessor systems with Chapter 22 providing an overall conclusion in considering the design of mechatronics systems.

An introduction

Chapter 1 Mechatronics

Basic system elements	**Actuation**	**System models**	**Basic digital principles**	**Microprocessor systems**
Chapter 2 Sensors and transducers	Chapter 5 Pneumatic and hydraulic actuation systems	Chapter 8 Basic system models	Chapter 14 Digital logic	Chapter 15 Microprocessors
Chapter 3 Signal conditioning	Chapter 6 Mechanical actuation systems	Chapter 9 System models		Chapter 16 Assembly language
Chapter 4 Data presentation systems	Chapter 7 Electrical actuation systems	Chapter 10 Dynamic responses of systems		Chapter 17 C language
		Chapter 11 System transfer functions		Chapter 18 Input/output systems
		Chapter 12 Frequency response		Chapter 19 Programmable logic controllers
		Chapter 13 Closed-loop controllers		Chapter 20 Communication systems

A conclusion

Chapter 22 Mechatronics systems

Chapter 21 Fault finding

Support materials

An Instructor's Guide and OHTs are available for lecturers to download at:

www.booksites.net/bolton

Acknowledgements

A large debt is owed to the publications of the manufacturers of the equipment referred to in the text. I would also like to thank those reviewers in both Britain and the United States who painstakingly read through the first edition and made suggestions for improvements and those lecturers who have since written to me.

W. Bolton

1 Mechatronics

1.1 What is mechatronics?

Consider the modern auto-focus, auto-exposure camera. To use the camera all you need to do is point it at the subject and press the button to take the picture. The camera automatically adjusts the focus so that the subject is in focus and automatically adjusts the aperture and shutter speed so that the correct exposure is given. Consider a truck smart suspension. Such a suspension adjusts to uneven loading to maintain a level platform, adjusts to cornering, moving across rough ground, etc. to maintain a smooth ride. Consider an automated production line. Such a line may involve a number of production processes which are all automatically carried out in the correct sequence and in the correct way. The automatic camera, the truck suspension, and the automatic production line are examples of a marriage between electronic control systems and mechanical engineering.

Such control systems generally use microprocessors as controllers and have electrical sensors extracting information from the mechanical inputs and outputs via electrical actuators to mechanical systems. The term *mechatronics* is used for this integration of microprocessor control systems, electrical systems and mechanical systems. A mechatronic system is not just a marriage of electrical and mechanical systems and is more than just a control system; it is a complete integration of all of them.

In the design of cars, robots, machine tools, washing machines, cameras, and very many other machines, such an integrated and interdisciplinary approach to engineering design is increasingly being adopted. The integration across the traditional boundaries of mechanical engineering, electrical engineering, electronics and control engineering has to occur at the earliest stages of the design process if cheaper, more reliable, more flexible systems are to be developed. Mechatronics has to involve a concurrent approach to these disciplines rather than a sequential approach of developing, say, a mechanical system then designing the electrical part and the microprocessor part.

Mechatronics brings together areas of technology involving sensors and measurement systems, drive and actuation systems, analysis of the behaviour of systems, control systems, and microprocessor systems. That essentially is a summary of this book. This chapter is an introduction to the topic, developing some of the basic concepts in order to give a framework for the rest of the book in which the details will be developed.

1.2 Systems

Mechatronics involves, what are termed, systems. A *system* can be thought of as a box which has an input and an output and where we are not concerned with what goes on inside the box but only the relationship between the output and the input. Thus, for example, a motor may be thought of as a system which has as its input electric power and as output the rotation of a shaft. Figure 1.1 shows a representation of such a system.

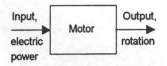

Fig. 1.1   An example of a system

A *measurement system* can be thought of as a black box which is used for making measurements. It has as its input the quantity being measured and its output the value of that quantity. For example, a temperature measurement system, i.e. a thermometer, has an input of temperature and an output of a number on a scale. Figure 1.2 shows a representation of such a system.

Fig. 1.2   An example of a measurement system

A *control system* can be thought of as a black box which is used to control its output to some particular value or particular sequence of values. For example, a domestic central heating control system has as its input the temperature required in the house and as its output the house at that temperature, i.e. you set the required temperature on the thermostat or controller and the heating furnace adjusts itself to pump water through radiators and so produce the required temperature in the house. Figure 1.3 shows a representation of such a system.

Fig. 1.3   An example of a control system

1.3 Measurement systems

Measurement systems can, in general, be considered to be made up of three elements (as illustrated in Fig. 1.4):

1   A *sensor* which responds to the quantity being measured by giving as its output a signal which is related to the quantity. For example, a thermocouple is a temperature sensor. The input to the sensor is a temperature and the output is an e.m.f. which is related to the temperature value.

Fig. 1.4   A measurement system and its constituent elements

2   A *signal conditioner* takes the signal from the sensor and manipulates it into a condition which is suitable for either display, or, in the case of a control system, for use to exercise control. Thus, for example, the output from a thermocouple is a rather small e.m.f. and might be fed through an amplifier to obtain a bigger signal. The amplifier is the signal conditioner.

3   A *display system* where the output from the signal conditioner is displayed. This might, for example, be a pointer moving across a scale or a digital readout.

As an example, consider a digital thermometer. This has an input of temperature to a sensor, probably a semiconductor diode. The potential difference across the sensor is, at constant current, a measure of the temperature. This potential difference is then amplified by an operational amplifier to give a voltage which can directly drive a display. The sensor and operational amplifier may be incorporated on the same silicon chip.

Sensors are discussed in Chapter 2 and signal conditioners in Chapter 3. Measurement systems involving all elements are discussed in Chapter 4. For further details of measurement systems, readers are referred to texts more specifically concerned with measurement, e.g. *Instrumentation Reference Book* edited by B.E. Noltingk (Butterworth-Heinemann 1995), *Measurement and Instrumentation Principles* by A.S. Morris (Newnes 2001) or *Newnes Instrumentation and Measurement* by W. Bolton (Newnes 1991, 1996, 2000).

1.4 Control systems

Your body temperature, unless you are ill, remains almost constant regardless of whether you are in a cold or hot environment. To maintain this constancy your body has a temperature control system. If your temperature begins to increase above the normal you sweat, if it decreases you shiver. Both these are mechanisms which are used to restore the body temperature back to its normal value. The control system is maintaining constancy of temperature. The system has an input from sensors which tell it what the temperature is and then compares this data with what the temperature should be and provides the appropriate response in order to obtain the required temperature. This is an example of *feedback control*; signals are fed back from the output, i.e. the actual temperature, in order to modify the reaction of the body to enable it to restore the temperature to the 'normal' value. *Feedback control* is exercised by the control system comparing the fed back actual output of the system with what is required and adjusting its output accordingly. Figure 1.5 illustrates this feedback control system.

One way to control the temperature of a centrally heated house is for a human to stand near the furnace on/off switch with a

Fig. 1.5   Feedback control for human body temperature

Fig. 1.6   Feedback control for room temperature

Fig. 1.7   Feedback control for picking up a pencil

thermometer and switch the furnace on or off according to the thermometer reading. That is a crude form of feedback control using a human as a control element. The term feedback is used because signals are fed back from the output in order to modify the input. The more usual feedback control system has a thermostat or controller which automatically switches the furnace on or off according to the difference between the set temperature and the actual temperature (Fig. 1.6). This control system is maintaining constancy of temperature.

If you go to pick up a pencil from a bench there is a need for you to use a control system to ensure that your hand actually ends up at the pencil. This is done by you observing the position of your hand relative to the pencil and making adjustments in its position as it moves towards the pencil. There is a feedback of information about your actual hand position so that you can modify your reactions to give the required hand position and movement (Fig. 1.7). This control system is controlling the positioning and movement of your hand.

Feedback control systems are widespread, not only in nature and the home but also in industry. There are many industrial process and machines where control, whether by humans or automatically, is required. For example, there is process control where such things as temperature, liquid level, fluid flow, pressure, etc. are maintained constant. Thus in a chemical process there may be a need to maintain the level of a liquid in a tank to a particular level or to a particular temperature. There are also control systems which involve consistently and accurately positioning a moving part or maintaining a constant speed. This might be, for example, a motor designed to run at a constant speed or perhaps a machining operation in which the position, speed and operation of a tool is automatically controlled.

1.4.1 Open- and closed-loop systems

There are two basic forms of control system, one being called *open loop* and the other *closed loop*. The difference between these can be illustrated by a simple example. Consider an electric fire which has a selection switch which allows a 1 kW or a 2 kW heating element to be selected. If a person used the heating element to heat a room, he or she might just switch on the 1 kW element if the room is not required to be at too high a temperature. The room will heat up and reach a temperature which is only determined by the fact the 1 kW element was switched on and not the 2 kW element. If there are changes in the conditions, perhaps someone opening a window, there is no way the heat output is adjusted to compensate. This is an example of open-loop control in that there is no information fed back to the element to adjust it and maintain a constant temperature. The heating system with the heating element could be made a closed-loop system if the person has a thermometer and switches the

1 kW and 2 kW elements on or off, according to the difference between the actual temperature and the required temperature, to maintain the temperature of the room constant. In this situation there is feedback, the input to the system being adjusted according to whether its output is the required temperature. This means that the input to the switch depends on the deviation of the actual temperature from the required temperature, the difference between them determined by a comparison element – the person in this case. Figure 1.8 illustrates these two types of system.

To illustrate further the differences between open- and closed-loop systems, consider a motor. With an open-loop system the speed of rotation of the shaft might be determined solely by the initial setting of a knob which affects the voltage applied to the motor. Any changes in the supply voltage, the characteristics of the motor as a result of temperature changes, or the shaft load will change the shaft speed but not be compensated for. There is no feedback loop. With a closed-loop system, however, the initial setting of the control knob will be for a particular shaft speed and this will be maintained by feedback, regardless of any changes in supply voltage, motor characteristics or load. In an open-loop control system the output from the system has no effect on the input signal. In a closed-loop control system the output does have an effect on the input signal, modifying it to maintain an output signal at the required value.

Open-loop systems have the advantage of being relatively simple and consequently low cost with generally good reliability. However, they are often inaccurate since there is no correction for error. Closed-loop systems have the advantage of being relatively accurate in matching the actual to the required values. They are, however, more complex and so more costly with a greater chance of breakdown as a consequence of the greater number of components.

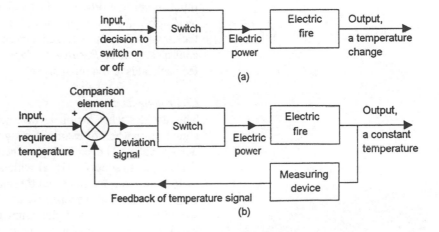

Fig. 1.8   Heating a room:
(a) an open-loop system,
(b) a closed-loop system

Fig. 1.9   The elements of a
closed-loop control system

1.4.2 Basic elements of a closed-loop system

Figure 1.9 shows the general form of a basic closed-loop system.
It consists of the following elements:

1   *Comparison element*
    This compares the required or reference value of the variable
    condition being controlled with the measured value of what is
    being achieved and produces an error signal. It can be
    regarded as adding the reference signal, which is positive, to
    the measured value signal, which is negative in this case:

    Error signal = reference value signal
                       − measured value signal

    The symbol used, in general, for an element at which signals
    are summed is a segmented circle, inputs going into
    segments. The inputs are all added, hence the feedback input
    is marked as negative and the reference signal positive so
    that the sum gives the difference between the signals. A
    feedback loop is a means whereby a signal related to the
    actual condition being achieved is fed back to modify the
    input signal to a process. The feedback is said to be *negative
    feedback* when the signal which is fed back subtracts from
    the input value. It is negative feedback that is required to
    control a system. *Positive feedback* occurs when the signal
    fed back adds to the input signal.

2   *Control element*
    This decides what action to take when it receives an error
    signal. It may be, for example, a signal to operate a switch or
    open a valve. The control plan being used by the element
    may be just to supply a signal which switches on or off when
    there is an error, as in a room thermostat, or perhaps a signal
    which proportionally opens or closes a valve according to the
    size of the error. Control plans may be *hard-wired systems* in
    which the control plan is permanently fixed by the way the
    elements are connected together or *programmable systems*

where the control plan is stored within a memory unit and may be altered by reprogramming it. Controllers are discussed in Chapter 11.

3   *Correction element*
The correction element produces a change in the process to correct or change the controlled condition. Thus it might be a switch which switches on a heater and so increases the temperature of the process or a valve which opens and allows more liquid to enter the process. The term *actuator* is used for the element of a correction unit that provides the power to carry out the control action. Correction units are discussed in Chapters 5 and 6.

4   *Process element*
The process is what is being controlled. It could be a room in a house with its temperature being controlled or a tank of water with its level being controlled.

5   *Measurement element*
The measurement element produces a signal related to the variable condition of the process that is being controlled. It might be, for example, a switch which is switched on when a particular position is reached or a thermocouple which gives an e.m.f. related to the temperature.

With the closed-loop system illustrated in Figure 1.8 for a person controlling the temperature of a room, the various elements are:

Controlled variable	–	the room temperature
Reference value	–	the required room temperature
Comparison element	–	the person comparing the measured value with the required value of temperature
Error signal	–	the difference between the measured and required temperatures
Control unit	–	the person
Correction unit	–	the switch on the fire
Process	–	the heating by the fire
Measuring device	–	a thermometer

An automatic control system for the control of the room temperature could involve a temperature sensor, after suitable signal conditioning, feeding an electrical signal to the input of a computer where it is compared with the set value and an error signal generated. This is then acted on by the computer to give at

Fig. 1.10 The automatic control of water level

its output a signal, which, after suitable signal conditioning, might be used to control a heater and hence the room temperature. Such a system can readily be programmed to give different temperatures at different times of the day.

Figure 1.10 shows an example of a simple control system used to maintain a constant water level in a tank. The reference value is the initial setting of the lever arm arrangement so that it just cuts off the water supply at the required level. When water is drawn from the tank the float moves downwards with the water level. This causes the lever arrangement to rotate and so allows water to enter the tank. This flow continues until the ball has risen to such a height that it has moved the lever arrangement to cut off the water supply. It is a closed-loop control system with the elements being:

Controlled variable	–	water level in tank
Reference value	–	initial setting of the float and lever position
Comparison element	–	the lever
Error signal	–	the difference between the actual and initial settings of the lever positions
Control unit	–	the pivoted lever
Correction unit	–	the flap opening or closing the water supply
Process	–	the water level in the tank
Measuring device	–	the floating ball and lever

The above is an example of a closed-loop control system involving just mechanical elements. We could, however, have controlled the liquid level by means of an electronic control system. We thus might have had a level sensor supplying an electrical signal which is used, after suitable signal conditioning, as an input to a computer where it is compared with a set value signal and the difference between them, the error signal, then used to give an appropriate response from the computer output. This is then, after suitable signal conditioning, used to control the movement of an actuator in a flow control valve and so determine the amount of water fed into the tank.

Figure 1.11 shows a simple automatic control system for the speed of rotation of a shaft. A potentiometer is used to set the reference value, i.e. what voltage is supplied to the differential amplifier as the reference value for the required speed of rotation. The differential amplifier is used to both compare and amplify the difference between the reference and feedback values, i.e. it amplifies the error signal. The amplified error signal is then fed to a motor which in turn adjusts the speed of the rotating shaft.

Fig. 1.11   Shaft speed control

The speed of the rotating shaft is measured using a tachogenerator, connected to the rotating shaft by means of a pair of bevel gears. The signal from the tachogenerator is then fed back to the differential amplifier.

1.4.3 Sequential controllers

There are many situations where control is exercised by items being switched on or off at particular preset times or values in order to control processes and give a step sequence of operations. For example, after step 1 is complete then step 2 starts. When step 2 is complete then step 3 starts, etc.

The term *sequential control* is used when control is such that actions are strictly ordered in a time or event driven sequence. Such control could be obtained by an electrical circuit with sets of relays or cam-operated switches which are wired up in such a way as to give the required sequence. Such hard-wired circuits are now more likely to have been replaced by a microprocessor-controlled system, with the sequencing being controlled by means of a software program.

As an illustration of sequential control, consider the domestic washing machine. A number of operations have to be carried out in the correct sequence. These may involve a pre-wash cycle when the clothes in the drum are given a wash in cold water, followed by a main wash cycle when they are washed in hot water, then a rinse cycle when the clothes are rinsed with cold water a number of times, followed by spinning to remove water from the clothes. Each of these operations involves a number of steps. For example, a pre-wash cycle involves opening a valve to fill the machine drum to the required level, closing the valve, switching on the drum motor to rotate the drum for a specific time, and operating the pump to empty the water from the drum. The operating sequence is called a *program*, the sequence of instructions in each program being predefined and 'built' into the controller used.

Figure 1.12 shows the basic washing machine system and gives a rough idea of its constituent elements. The system that used to be used for the washing machine controller was a mechanical system which involved a set of cam-operated switches, i.e. mechanical switches. Figure 1.13 shows the basic principle of one such switch. When the machine is switched on, a small electric motor slowly rotates its shaft, giving an amount of rotation proportional to time. Its rotation turns the controller cams so that each in turn operates electrical switches and so switches on circuits in the correct sequence. The contour of a cam determines the time at which it operates a switch. Thus the contours of the cams are the means by which the program is specified and stored in the machine. The sequence of instructions and the instructions used in a particular washing program are determined by the set of cams chosen. With modern washing machines the controller is a microprocessor and the program is not supplied by the mechanical arrangement of cams but by a software program.

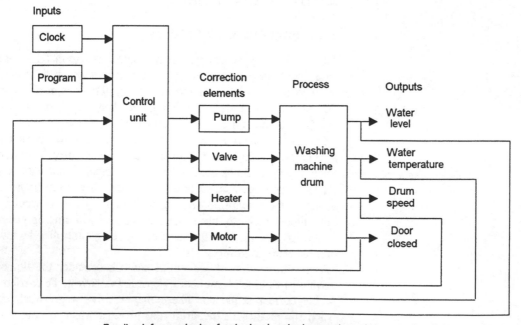

Feedback from outputs of water level, water temperature, drum speed and door closed

Fig. 1.12 Washing machine system

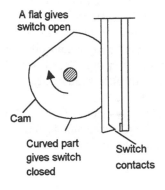

A flat gives
switch open

Cam

Curved part
gives switch
closed

Switch
contacts

Rotation of the cam closing
the switch contacts

Fig. 1.13   Cam-operated switch

For the pre-wash cycle an electrically operated valve is opened when a current is supplied and switched off when it ceases. This valve allows cold water into the drum for a period of time determined by the profile of the cam or the output from the microprocessor used to operate its switch. However, since the requirement is a specific level of water in the washing machine drum, there needs to be another mechanism which will stop the water going into the tank, during the permitted time, when it reaches the required level. A sensor is used to give a signal when the water level has reached the preset level and give an output from the microprocessor which is used to switch off the current to the valve. In the case of a cam-controlled valve, the sensor actuates a switch which closes the valve admitting water to the washing machine drum. When this event is completed the microprocessor, or the rotation of the cams, initiates a pump to empty the drum.

For the main wash cycle, the microprocessor gives an output which starts when the pre-wash part of the program is completed; in the case of the cam-operated system the cam has a profile such that it starts in operation when the pre-wash cycle is completed. It switches a current into a circuit to open a valve to allow cold water into the drum. This level is sensed and the water shut off when the required level is reached. The microprocessor or cams then supply a current to activate a switch which applies a larger current to an electric heater to heat the water. A temperature sensor is used to switch off the current when the water temperature reaches the preset value. The microprocessor or cams then switch on the drum motor to rotate the drum. This will continue for the time determined by the microprocessor or cam profile before switching off. Then the microprocessor or a cam switches on the current to a discharge pump to empty the water from the drum.

The rinse part of the operation is now switched as a sequence of signals to open valves which allow cold water into the machine, switch it off, operate the motor to rotate the drum, operate a pump to empty the water from the drum, and repeat this sequence a number of times.

The final part of the operation is when the microprocessor or a cam switches on just the motor, at a higher speed than for the rinsing, to spin the clothes.

1.5 Microprocessor-based controllers

Microprocessors are now rapidly replacing the mechanical cam-operated controllers and being used in general to carry out control functions. They have the great advantage that a greater variety of programs become feasible. In many simple systems there might be just an embedded microcontroller, this being a microprocessor with memory all integrated on one chip, which has been specifically programmed for the task concerned. A more adaptable form is the *programmable logic controller*. This is a

Fig. 1.14 Programmable logic controller

microprocessor-based controller which uses programmable memory to store instructions and to implement functions such as logic, sequence, timing counting and arithmetic to control events and can be readily reprogrammed for different tasks. Figure 1.14 shows the control action of a programmable logic controller, the inputs being signals from, say, switches being closed and the program used to determine how the controller should respond to the inputs and the output it should then give.

The following examples of control systems illustrate how microprocessor-based systems have not only been able to carry out tasks that previously were done 'mechanically' but also able to do tasks that were not easily automated before.

1.5.1 The automatic camera

The modern camera is likely to have automatic focusing and exposure. Figure 1.15 illustrate the basic aspects of a microprocessor-based system that can be used to control the focusing and exposure.

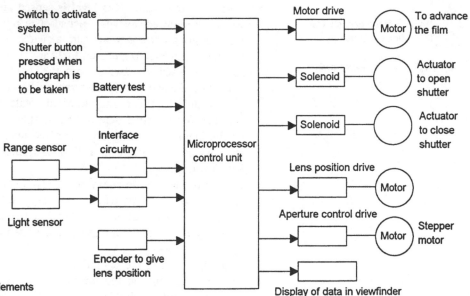

Fig. 1.15 Basic elements of the control system for an automatic camera

When the switch is operated to activate the system and the camera pointed at the object being photographed, the microprocessor takes in the input from the range sensor and sends an output to the lens position drive to move the lens to achieve focusing. The lens position is fed back to the microprocessor so that the feedback signal can be used to modify the lens position according to the input from the range sensor. The light sensor gives an input to the microprocessor which then gives an output to determine, if the photographer has selected the shutter controlled rather than aperture controlled mode, the time for which the shutter will be opened. When the photograph has been taken, the microprocessor gives an output to the motor drive to advance the film ready for the next photograph.

The program for the microprocessor is a number of steps where the microprocessor is making simple decisions of the form: is there an input signal on a particular input line or not and if there is output a signal on a particular output line. The decisions are logic decisions with the input and output signals either being low or high to give on-off states. A few steps of the program for the automatic camera might be of the form:

```
begin
        if battery check input OK
                then continue
                otherwise stop
loop
        read input from range sensor
        calculate lens movement
        output signal to lens position drive
        input data from lens position encoder
        compare calculated output with actual output
        stop output when lens in correct position
        send in-focus signal to viewfinder display
        etc.
```

1.5.2 The engine management system

The engine management system of a car is responsible for managing the ignition and fuelling requirements of the engine. With a four-stroke internal combustion engine there are several cylinders, each of which has a piston connected to a common crankshaft and each of which carries out a four-stroke sequence of operations (Fig. 1.16).

When the piston moves down a valve opens and the air–fuel mixture is drawn into the cylinder. When the piston moves up again the valve closes and the air–fuel mixture is compressed. When the piston is near the top of the cylinder the spark plug ignites the mixture with a resulting expansion of the hot gases. This expansion causes the piston to move back down again and so the cycle is repeated. The pistons of each cylinder are connected to a common crankshaft and their power strokes occur at different times so that there is continuous power for rotating the crankshaft.

Intake stroke Compression stroke Power stroke Exhaust stroke

Fig. 1.16 Four-stroke sequence

Fig. 1.17 Elements of an engine management system

The power and speed of the engine are controlled by varying the ignition timing and the air–fuel mixture. With modern car engines this is done by a microprocessor. Figure 1.17 shows the basic elements of a microprocessor control system. For ignition timing, the crankshaft drives a distributor which makes electrical contacts for each spark plug in turn and a timing wheel. This timing wheel generates pulses to indicate the crankshaft position. The microprocessor then adjusts the timing at which high voltage pulses are sent to the distributor so they occur at the 'right' moments of time. To control the amount of air–fuel mixture entering a cylinder during the intake strokes, the microprocessor varies the time for which a solenoid is activated to open the intake valve on the basis of inputs received of the engine temperature and the throttle position. The amount of fuel to be injected into the air stream can be determined by an input from a sensor of the mass rate of air flow, or computed from other measurements, and the microprocessor then gives an output to control a fuel injection valve.

The above is a very simplistic indication of engine management; for more detail the reader is referred to texts such as *Automobile Electrical and Electronic Systems* by T. Denton (Arnold 1995) or manufacturers' data sheets.

1.6 Response of systems

The response of any system to an input is not instantaneous. For example, if you switch on a kettle it takes some time for the water in the kettle to reach boiling point (Fig. 1.18). When a microprocessor controller gives a signal to, say, move the lens for focusing in an automatic camera then it takes time before the lens reaches its position for correct focusing. When you step onto the bathroom scales, the system does not immediately indicate your weight but gives a response which oscillates before settling down to indicate the weight value (Fig. 1.19). The responses of systems are functions of time. Thus, in order to know how systems behave when there are inputs to them, we need to devise models for systems which relate the output to the input so that we can work out, for a given input, how the output will vary with time and what it will settle down to.

1.7 The mechatronics approach

Fig. 1.18   The response to an input for a kettle system

Fig. 1.18   The response to an input for a bathroom scales system

The domestic washing machine (referred to earlier in this chapter) that used cam-operated switches in order to control the washing cycle is now out-of-date. Such mechanical switches are being replaced by microprocessors. A microprocessor may be considered as being essentially a collection of logic gates and memory elements that are not wired up as individual components but whose logical functions are implemented by means of software. The microprocessor-controlled washing machine can be considered an example of a mechatronics approach in that a mechanical system has become integrated with electronic controls. As a consequence, a bulky mechanical system is replaced by a much more compact microprocessor system which is readily adjustable to give a greater variety of programs.

1.7.1 In conclusion

Mechatronics involves the bringing together of a number of technologies: mechanical engineering, electronic engineering, electrical engineering, computer technology, and control engineering. This can be considered to be the application of computer-based digital control techniques, through electronic and electric interfaces, to mechanical engineering problems. Mechatronics provides an opportunity to take a new look at problems, with mechanical engineers not just seeing a problem in terms of mechanical principles but having to see it in terms of a range of technologies. The electronics, etc., should not be seen as a bolt-on item to existing mechanical hardware. A mechatronics approach needs to be adopted right from the design phase. There needs to be a complete rethink of the requirements in terms of what an item is required to do.

There are many applications of mechatronics in the mass-produced products used in the home. Microprocessor-based controllers are to be found in domestic washing machines, dish washers, microwave ovens, cameras, camcorders, watches, hi-fi and video recorder systems, central heating controls, sewing machines, etc. They are to be found in cars in the active suspension, antiskid brakes, engine control, speedometer display, transmission, etc.

A larger scale application of mechatronics is a flexible manufacturing engineering system (FMS) involving computer-controlled machines, robots, automatic material conveying and overall supervisory control.

Problems

1   Identify the sensor, signal conditioner and display elements in the measurement systems of (a) a mercury-in-glass thermometer, (b) a Bourdon pressure gauge.

2   Explain the difference between open- and closed-loop control.

3   Identify the various elements that might be present in a control system involving a thermostatically controlled electric heater.

4   The automatic control system for the temperature of a bath of liquid consists of a reference voltage fed into a differential amplifier. This is connected to a relay which then switches on or off the electrical power to a heater in the liquid. Negative feedback is provided by a measurement system which feeds a voltage into the differential amplifier. Sketch a block diagram of the system and explain how the error signal is produced.

5   Explain the function of a programmable logic controller.

6   Explain what is meant by sequential control and illustrate your answer by an example.

7   State steps that might be present in the sequential control of a dishwasher.

8   Compare and contrast the traditional design of a watch with that of the mechatronics-designed product involving a microprocessor.

9   Compare and contrast the control system for the domestic central heating system involving a bimetallic thermostat and that involving a microprocessor.

2   Sensors and transducers

2.1 Sensors and transducers

The term *sensor* is used for an element which produces a signal relating to the quantity being measured. Thus in the case of, say, an electrical resistance temperature element, the quantity being measured is temperature and the sensor transforms an input of temperature into a change in resistance. The term *transducer* is often used in place of the term sensor. Transducers are defined as elements that when subject to some physical change experience a related change. Thus sensors are transducers. However, a measurement system may use transducers, in addition to the sensor, in other parts of the system to convert signals in one form to another form.

This chapter is about transducers and in particular those used as sensors. The terminology that is used to specify the performance characteristics of transducers is defined and examples of transducers commonly used in engineering are discussed.

2.2 Performance terminology

The following terms are used to define the performance of transducers, and often measurement systems as a whole.

1   *Range and span*   The range of a transducer defines the limits between which the input can vary. The span is the maximum value of the input minus the minimum value. Thus, for example, a load cell for the measurement of forces might have a range of 0 to 50 kN and a span of 50 kN.

2   *Error*   Error is the difference between the result of the measurement and the true value of the quantity being measured.

Error = measured value − true value

Thus if a measurement system gives a temperature reading of 25°C when the actual temperature is 24°C, then the error is +1°C. If the actual temperature had been 26°C then the error would have been –1°C. A sensor might give a resistance change of 10.2 Ω when the true change should have been 10.5 Ω. The error is –0.3 Ω.

3   *Accuracy* Accuracy is the extent to which the value indicated by a measurement system might be wrong. It is thus the summation of all the possible errors that are likely to occur, as well as the accuracy to which the transducer has been calibrated. A temperature-measuring instrument might, for example, be specified as having an accuracy of ±2°C. This would mean that the reading given by the instrument can be expected to lie within + or –2°C of the true value. Accuracy is often expressed as a percentage of the full range output or full-scale deflection. The percentage of full-scale deflection term results from when the outputs of measuring systems were displayed almost exclusively on a circular or linear scale. A sensor might, for example, be specified as having an accuracy of ±5% of full range output. Thus if the range of the sensor was, say, 0 to 200°C, then the reading given can be expected to be within + or –10°C of the true reading.

4   *Sensitivity* The sensitivity is the relationship indicating how much output you get per unit input, i.e. ouput/input. For example, a resistance thermometer may have a sensitivity of 0.5 Ω/°C. This term is also frequently used to indicate the sensitivity to inputs other than that being measured, i.e. environmental changes. Thus there can be the sensitivity of the transducer to temperature changes in the environment or perhaps fluctuations in the mains voltage supply. A transducer for the measurement of pressure might be quoted as having a temperature sensitivity of ±0.1% of the reading per °C change in temperature.

5   *Hysteresis error* Transducers can give different outputs from the same value of quantity being measured according to whether that value has been reached by a continuously increasing change or a continuously decreasing change. This effect is called hysteresis. Figure 2.1 shows such an output with the hysteresis error as the maximum difference in output for increasing and decreasing values.

6   *Non-linearity error* For many transducers a linear relationship between the input and output is assumed over the working range, i.e. a graph of output plotted against input is assumed to give a straight line. Few transducers, however, have a truly linear relationship and thus errors occur as a result of the assumption of linearity. The error is defined as the maximum difference from the straight line. Various methods are used for the numerical expression of the non-linearity error. The differences occur in determining the

Fig. 2.1   Hysteresis

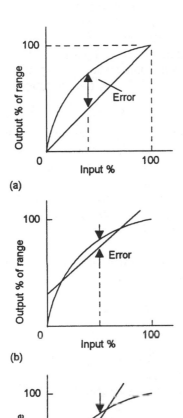

Fig. 2.2   Non-linearity error using:
(a) end-range values,
(b) best straight line for all values,
(c) best straight line through zero point

straight line relationship against which the error is specified. One method is to draw the straight line joining the output values at the end points of the range; another is to find the straight line by using the method of least squares to determine the best fit line when all data values are considered equally likely to be in error; another is to find the straight line by using the method of least squares to determine the best fit line which passes through the zero point. Figure 2.2 illustrates these three methods and how they can affect the non-linearity error quoted. The error is generally quoted as a percentage of the full range output. For example, a transducer for the measurement of pressure might be quoted as having a non-linearity error of ±0.5% of the full range.

7 *Repeatability/reproducibility*   The terms repeatability and reproducibility of a transducer are used to describe its ability to give the same output for repeated applications of the same input value. The error resulting from the same output not being given with repeated applications is usually expressed as a percentage of the full range output.

$$\text{Repeatability} = \frac{\text{max.} - \text{min. values given}}{\text{full range}} \times 100$$

A transducer for the measurement of angular velocity typically might be quoted as having a repeatability of ±0.01% of the full range at a particular angular velocity.

8 *Stability*   The stability of a transducer is its ability to give the same output when used to measure a constant input over a period of time. The term *drift* is often used to describe the change in output that occurs over time. The drift may be expressed as a percentage of the full range output. The term *zero drift* is used for the changes that occur in output when there is zero input.

9 *Dead band/time*   The dead band or dead space of a transducer is the range of input values for which there is no output. For example, bearing friction in a flow meter using a rotor might mean that there is no output until the input has reached a particular velocity threshold. The dead time is the length of time from the application of an input until the output begins to respond and change.

10 *Resolution*   When the input varies continuously over the range, the output signals for some sensors may change in small steps. A wire-wound potentiometer is an example of such a sensor, the output going up in steps as the potentiometer slider moves from one wire turn to the next. The resolution is the smallest change in the input value that

will produce an observable change in the output. For a wire-wound potentiometer the resolution might be specified as, say, 0.5° or perhaps a percentage of the full-scale deflection. For a sensor giving a digital output the smallest change in output signal is 1 bit. Thus for a sensor giving a data word of N bits, i.e. a total of 2^N bits, the resolution is generally expressed as $1/2^N$.

11 *Output impedance* When a sensor giving an electrical output is interfaced with an electronic circuit it is necessary to know the output impedance since this impedance is being connected in either series or parallel with that circuit. The inclusion of the sensor can thus significantly modify the behaviour of the system to which it is connected. See Section 4.1.1 for a discussion of loading.

To illustrate the above, consider the significance of the terms in the following specification of a strain gauge pressure transducer:

> Ranges: 70 to 1000 kPa, 2000 to 70 000 kPa
> Supply voltage: 10 V d.c. or a.c. r.m.s.
> Full range output: 40 mV
> Non-linearity and hysteresis: ±0.5% full range output
> Temperature range: −54°C to +120°C when operating
> Thermal zero shift: 0.030% full range output/°C

The range indicates that the transducer can be used to measure pressures between 70 and 1000 kPa or 2000 and 70 000 kPa. It requires a supply of 10 V d.c. or a.c. r.m.s. for its operation and will give an output of 40 mV when the pressure on the lower range is 1000 kPa and on the upper range 70 000 kPa. Non-linearity and hysteresis will lead to errors of ±0.5% of 1000, i.e. ±5 kPa on the lower range and ±0.5% of 70 000, i.e. ±350 kPa on the upper range. The transducer can be used between the temperatures of −54 and +120°C. When the temperature changes by 1°C the output of the transducer for zero input will change by 0.030% of 1000 = 0.3 kPa on the lower range and 0.030% of 70 000 = 21 kPa on the upper range.

2.2.1 Static and dynamic characteristics

The *static characteristics* are the values given when steady-state conditions occur, i.e. the values given when the transducer has settled down after having received some input. The terminology defined above refers to such a state. The *dynamic characteristics* refer to the behaviour between the time that the input value changes and the time that the value given by the transducer settles down to the steady-state value. Dynamic characteristics are stated in terms of the response of the transducer to inputs in particular forms. For example, this might be a step input when the input is suddenly changed from 0 to a constant value, or a ramp input when the input is changed at a steady rate, or a sinusoidal input of a specified frequency. Thus we might find the following terms (see Chapter 8 for a more detailed discussion of dynamic systems):

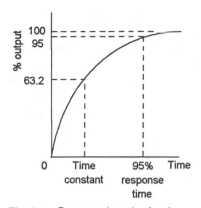

Fig. 2.3   Response to a step input

1   *Response time*   This is the time which elapses after a constant input, a step input, is applied to the transducer up to the point at which the transducer gives an output corresponding to some specified percentage, e.g. 95%, of the value of the input (Fig. 2.3). For example, if a mercury-in-glass thermometer is put into a hot liquid there can be quite an appreciable time lapse, perhaps as much as 100 s or more, before the thermometer indicates 95% of the actual temperature of the liquid.

2   *Time constant* This is the 63.2% response time. A thermocouple in air might have a time constant of perhaps 40 to 100 s. The time constant is a measure of the inertia of the sensor and so how fast it will react to changes in its input; the bigger the time constant the slower will be its reaction to a changing input signal. See Section 10.2.3 for a mathematical discussion of the time constant in terms of the behaviour of a system when subject to a step input.

3   *Rise time* This is the time taken for the output to rise to some specified percentage of the steady-state output. Often the rise time refers to the time taken for the output to rise from 10% of the steady-state value to 90 or 95% of the steady-state value.

4   *Settling time* This is the time taken for the output to settle to within some percentage, e.g. 2%, of the steady-state value.

To illustrate the above, consider the following data which indicates how an instrument reading changed with time, being obtained from a thermometer plunged into a liquid at time $t = 0$. The 95% response time is required.

Time (s)	0	30	60	90	120	150	180
Temp. (°C)	20	28	34	39	43	46	49

Time (s)	210	240	270	300	330	360
Temp. (°C)	51	53	54	55	55	55

Fig. 2.4   Thermometer in liquid

Figure 2.4 shows the graph of how the temperature indicated by the thermometer varies with time. The steady-state value is 55°C and so, since 95% of 55 is 52.25°C, the 95% response time is about 228 s.

The following sections give examples of transducers grouped according to what they are being used to measure. The measurements considered are those frequently encountered in mechanical engineering, namely: displacement, proximity, velocity, force, pressure, fluid flow, liquid level, temperature, and light intensity. For a more comprehensive coverage of transducers the reader is referred to more specialist texts, e.g. *Instrumentation Reference Book* edited by B.E. Noltingk (Butterworth 1988, 1995), *Measurement and Instrumentation Systems* by W. Bolton (Newnes 1996) and the concise *Newnes Instrumentation and Measurement Pocket Book* by W. Bolton (Newnes 1991, 1996, 2000). Examples of the types of specifications that might be expected from electrical transducers are given in *Transducer Handbook* by H.B. Boyle (Newnes 1992).

2.3 Displacement, position and proximity

Displacement sensors are concerned with the measurement of the amount by which some object has been moved; position sensors are concerned with the determination of the position of some object with reference to some reference point. Proximity sensors are a form of position sensor and are used to determine when an object has moved to within some particular critical distance of the sensor. They are essentially devices which give on–off outputs.

In selecting a displacement, position or proximity sensor, consideration has to be given to:

1   The size of the displacement; are we talking of fractions of a millimetre, many millimetres or perhaps metres? For a proximity sensor; how close should the object be before it is detected?
2   Whether the displacement is linear or angular; linear displacement sensors might be used to monitor the thickness or other dimensions of sheet materials, the separation of rollers, the position or presence of a part, the size of a part, etc. while angular displacement methods might be used to monitor the angular displacement of shafts.
3   The resolution required.
4   The accuracy required.
5   What material the measured object is made of; some sensors will only work with ferromagnetic materials, some with only metals, some with only insulators.
6   The cost.

Displacement and position sensors can be grouped into two basic types: contact sensors in which the measured object comes into mechanical contact with the sensor or non-contacting where there is no physical contact between the measured object and the sensor. For those linear displacement methods involving contact, there is usually a sensing shaft which is in direct contact with the object being monitored. The displacement of this shaft is then monitored by a sensor. The movement of the shaft may be used to cause changes in electrical voltage, resistance, capacitance, or mutual inductance. For angular displacement methods involving mechanical connection the rotation of a shaft might directly drive, through gears, the rotation of the transducer element. Non-contacting sensors might involve the presence in the vicinity of the measured object causing a change in the air pressure in the sensor, or perhaps a change in inductance or capacitance. The following are examples of commonly used displacement sensors.

2.3.1 Potentiometer sensor

A *potentiometer* consists of a resistance element with a sliding contact which can be moved over the length of the element. Such elements can be used for linear or rotary displacements, the displacement being converted into a potential difference. The rotary potentiometer consists of a circular wire-wound track or a film of conductive plastic over which a rotatable sliding contact can be rotated (Fig. 2.5). The track may be a single turn or helical. With a constant input voltage V_s, between terminals 1 and 3, the output voltage V_o between terminals 2 and 3 is a fraction of the input voltage, the fraction depending on the ratio of the resistance R_{23} between terminals 2 and 3 compared with the total resistance R_{13} between terminals 1 and 3, i.e. $V_o/V_s = R_{23}/R_{13}$. If the track has a constant resistance per unit length, i.e. per unit angle, then the output is proportional to the angle through which the slider has rotated. Hence an angular displacement can be converted into a potential difference.

With a wire-wound track the slider in moving from one turn to the other will change the voltage output in steps, each step being a movement of one turn. If the potentiometer has N turns then the resolution, as a percentage, is $100/N$. Thus the resolution of a wire track is limited by the diameter of the wire used and typically ranges from about 1.5 mm for a coarsely wound track to 0.5 mm for a finely wound one. Errors due to non-linearity of the track tend to range from less than 0.1% to about 1%. The track resistance tends to range from about 20 Ω to 200 kΩ. Conductive plastic has ideally infinite resolution, errors due to non-linearity of track of the order of 0.05% and resistance values from about 500 Ω to 80 kΩ. The conductive plastic has a higher temperature coefficient of resistance than the wire and so temperature changes have a greater effect on accuracy.

An important effect to be considered with a potentiometer is the effect of a load R_L connected across the output. The potential difference across the load V_L is only directly proportional to V_o if the load resistance is infinite. For finite loads, however, the effect of the load is to transform what was a linear relationship between output voltage and angle into a non-linear relationship. The resistance R_L is in parallel with the fraction x of the potentiometer

A rotary potentiometer

The circuit when connected to a load

R_L in parallel
with xR_p

The circuit as a potential divider

Fig. 2.5   Rotary potentiometer

resistance R_p. This combined resistance is $R_L x R_p / (R_L + x R_p)$. The total resistance across the source voltage is thus:

$$\text{total resistance} = R_p(1-x) + R_L x R_p / (R_L + x R_p)$$

The circuit is a potential divider circuit and thus the voltage across the load is the fraction the resistance across the load is of the total resistance across which the applied voltage is connected:

$$\frac{V_L}{V_s} = \frac{x R_L R_p / (R_L + x R_p)}{R_p(1-x) + x R_L R_p / (R_L + x R_p)}$$

$$= \frac{x}{(R_p/R_L)x(1-x) + 1}$$

If the load is of infinite resistance then we have $V_L = x V_s$. Thus the error introduced by the load having a finite resistance is:

$$\text{error} = x V_s - V_L = x V_s - \frac{x V_s}{(R_p/R_L)x(1-x) + 1}$$

$$= V_s \frac{R_p}{R_L}(x^2 - x^3)$$

To illustrate the above, consider the non-linearity error with a potentiometer of resistance 500 Ω, when at a displacement of half its maximum slider travel, which results from there being a load of resistance 10 kΩ. The supply voltage is 4 V. Using the equation derived above:

$$\text{error} = 4 \times \frac{500}{10\,000}(0.5^2 - 0.5^3) = 0.025$$

As a percentage of the full range reading, this is 0.625%.

2.3.2 Strain-gauged element

The electrical resistance strain gauge (Fig. 2.6) is a metal wire, metal foil strip, or a strip of semiconductor material which is wafer-like and can be stuck onto surfaces like a postage stamp. When subject to strain, its resistance R changes, the fractional change in resistance $\Delta R/R$ being proportional to the strain ε, i.e.

$$\frac{\Delta R}{R} = G\varepsilon$$

where G, the constant of proportionality, is termed the gauge factor. Since strain is the ratio (change in length/original length) then the resistance change of a strain gauge is a measurement of the change in length of the element to which the strain gauge is attached. The gauge factor of metal wire or foil strain gauges with the metals generally used is about 2.0. Silicon p- and n-type

(a)

Wire

Connection leads

(b)

Metal foil

Connection leads

(c)

Semiconductor

Connection leads

Fig. 2.6 Strain gauges: (a) metal wire, (b) metal foil, (c) semiconductor

(a)

(b)

(c)

Fig. 2.7   Strain-gauged elements

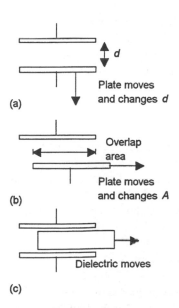

(a)

(b)

(c)

Fig. 2.8   Forms of capacitive sensing element

semiconductor strain gauges have gauge factors of about +100 or more for p-type silicon and –100 or more for n-type silicon. The gauge factor is normally supplied by the manufacturer of the strain gauges from a calibration made of a sample of strain gauges taken from a batch. The calibration involves subjecting the sample gauges to known strains and measuring their changes in resistance. A problem with all strain gauges is that their resistance not only changes with strain but also with temperature. Ways of eliminating the temperature effect have to be used and are discussed in Chapter 3. Semiconductor strain gauges have a much greater sensitivity to temperature than metal strain gauges.

To illustrate the above, consider an electrical resistance strain gauge with a resistance of 100 Ω and a gauge factor of 2.0. What is the change in resistance of the gauge when it is subject to a strain of 0.001? The fractional change in resistance is equal to the gauge factor multiplied by the strain, thus:

Change in resistance = $2.0 \times 0.001 \times 100 = 0.2 \ \Omega$

One form of displacement sensor has strain gauges attached to flexible elements in the form of cantilevers, rings or U-shapes (Fig. 2.7). When the flexible element is bent or deformed as a result of forces being applied by a contact point being displaced, then the electrical resistance strain gauges mounted on the element are strained and so give a resistance change which can be monitored. The change in resistance is thus a measure of the displacement or deformation of the flexible element. Such arrangements are typically used for linear displacements of the order of 1 mm to 30 mm and have a non-linearity error of about ± 1% of full range.

2.3.3 Capacitive element

The capacitance C of a parallel plate capacitor is given by

$$C = \frac{\varepsilon_r \varepsilon_0 A}{d}$$

where ε_r is the relative permittivity of the dielectric between the plates, ε_0 a constant called the permittivity of free space, A the area of overlap between the two plates and d the plate separation. Capacitive sensors for the monitoring of linear displacements might thus take the forms shown in Figure 2.8. In (a) one of the plates is moved by the displacement so that the plate separation changes; in (b) the displacement causes the area of overlap to change; in (c) the displacement causes the dielectric between the plates to change.

For the displacement changing the plate separation (Fig. 2.8(a)), if the separation d is increased by a displacement x then the capacitance becomes:

$$C - \Delta C = \frac{\varepsilon_0 \varepsilon_r A}{d + x}$$

Hence the change in capacitance ΔC as a fraction of the initial capacitance is given by:

$$\frac{\Delta C}{C} = -\frac{d}{d + x} - 1 = -\frac{x/d}{1 + (x/d)}$$

There is thus a non-linear relationship between the change in capacitance ΔC and the displacement x. This non-linearity can be overcome by using what is termed a *push-pull displacement sensor* (Fig. 2.9). This has three plates with the upper pair forming one capacitor and the lower pair another capacitor. The displacement moves the central plate between the two other plates. The result of, for example, the central plate moving downwards is to increase the plate separation of the upper capacitor and decrease the separation of the lower capacitor. We thus have:

$$C_1 = \frac{\varepsilon_0 \varepsilon_r A}{d + x}$$

$$C_2 = \frac{\varepsilon_0 \varepsilon_r A}{d - x}$$

Fig. 2.9   Push-pull sensor

When C_1 is in one arm of an a.c. bridge and C_2 in the other, then the resulting out-of-balance voltage is proportional to x. Such a sensor is typically used for monitoring displacements from a few millimetres to hundreds of millimetres. Non-linearity and hysteresis are about $\pm\,0.01\%$ of full range.

One form of capacitive proximity sensor consists of a single capacitor plate probe with the other plate being formed by the object, which has to be metallic and earthed (Fig. 2.10). As the object approaches so the 'plate separation' of the capacitor changes, becoming significant and detectable when the object is close to the probe.

Fig. 2.10   Capacitive proximity sensor

2.3.4 Differential transformers

The linear variable differential transformer, generally referred to by the abbreviation LVDT, consists of three coils symmetrically spaced along an insulated tube (Fig. 2.11). The central coil is the primary coil and the other two are identical secondary coils which are connected in series in such a way that their outputs oppose each other. A magnetic core is moved through the central tube as a result of the displacement being monitored.

When there is an alternating voltage input to the primary coil, alternating e.m.f.s are induced in the secondary coils. With the magnetic core central, the amount of magnetic material in each of the secondary coils is the same. Thus the e.m.f.s induced in each coil are the same. Since they are so connected that their outputs oppose each other, the net result is zero output.

Fig. 2.11   LVDT

However, when the core is displaced from the central position there is a greater amount of magnetic core in one coil than the other, e.g. more in secondary coil 2 than coil 1. The result is that a greater e.m.f. is induced in one coil than the other. There is then a net output from the two coils. Since a greater displacement means even more core in one coil than the other, the output, the difference between the two e.m.f.s increases the greater the displacement being monitored (Fig. 2.12).

The e.m.f. induced in a secondary coil by a changing current i in the primary coil is given by:

$$e = M\frac{\mathrm{d}i}{\mathrm{d}t}$$

where M is the mutual inductance, its value depending on the number of turns on the coils and the ferromagnetic core. Thus, for a sinusoidal input current of $i = I \sin \omega t$ to the primary coil, the e.m.f.s induced in the two secondary coils 1 and 2 can be represented by:

$$v_1 = k_1 \sin(\omega t - \phi) \text{ and } v_2 = k_2 \sin(\omega t - \phi)$$

where the values of k_1, k_2 and ϕ depend on the degree of coupling between the primary and secondary coils for a particular core position. ϕ is the phase difference between the primary alternating voltage and the secondary alternating voltages. Because the two outputs are in series, their difference is the output:

$$\text{output voltage} = v_1 - v_2 = (k_1 - k_2) \sin(\omega t - \phi)$$

When the core is equally in both coils, k_1 equals k_2 and so the output voltage is zero. When the core is more in 1 than in 2 we have $k_1 > k_2$ and:

$$\text{output voltage} = (k_1 - k_2) \sin(\omega t - \phi)$$

Fig. 2.12   LVDT output

Fig. 2.13   LVDT d.c. output

When the core is more in 2 than in 1 we have $k_1 < k_2$. A consequence of k_1 being less than k_2 is that there is a phase change of 180° in the output when the core moves from more in 1 to more in 2. Thus:

$$\text{output voltage}$$
$$= -(k_1 - k_2)\sin(\omega t - \phi) = (k_2 - k_1)\sin[\omega t + (\pi - \phi)]$$

Figure 2.12 shows how the size and phase of the output changes with the displacement of the core.

With this form of output, the same amplitude output voltage is produced for two different displacements. To give an output voltage which is unique to each value of displacement we need to distinguish between where the amplitudes are the same but there is a phase difference of 180°. A phase sensitive demodulator, with a low pass filter, is used to convert the output into a d.c. voltage which gives a unique value for each displacement (Fig. 2.13). Such circuits are available as integrated circuits.

Typically, LVDTs have operating ranges from about ±2 mm to ±400 mm with non-linearity errors of about ±0.25%. LVDTs are very widely used as primary transducers for monitoring displacements. The free end of the core may be spring loaded for contact with the surface being monitored, or threaded for mechanical connection. They are also used as secondary transducers in the measurement of force, weight and pressure; these variables are transformed into displacements which can then be monitored by LVDTs.

A rotary variable differential transformer (RVDT) can be used for the measurement of rotation (Fig. 2.14); it operates on the same principle as the LVDT. The core is a cardioid-shaped piece of magnetic material and rotation causes more of it to pass into one secondary coil than the other. The range of operation is typically ±40° with a linearity error of about ±0.5% of the range.

Fig. 2.14   RVDT

2.3.5 Eddy current proximity sensors

If a coil is supplied with an alternating current, an alternating magnetic field is produced. If there is a metal object in close proximity to this alternating magnetic field, then eddy currents are induced in it. The eddy currents themselves produce a

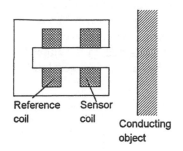

Reference coil   Sensor coil   Conducting object

Fig. 2.15   Eddy current sensor

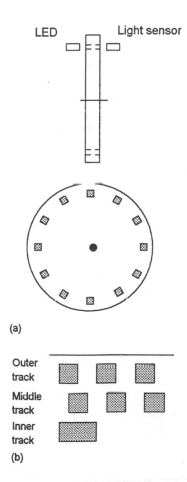

LED   Light sensor

(a)

Outer track
Middle track
Inner track
(b)

Fig. 2.16   Incremental encoder:
(a) the basic principle, (b) concentric tracks

magnetic field. This distorts the magnetic field responsible for their production. As a result, the impedance of the coil changes and so the amplitude of the alternating current. At some preset level, this change can be used to trigger a switch. Figure 2.15 shows the basic form of such a sensor; it is used for the detection of non-magnetic but conductive materials. They have the advantages of being relatively inexpensive, small in size, with high reliability and can have high sensitivity to small displacements.

2.3.6 Inductive proximity switch

This consists of a coil wound round a core. When the end of the coil is close to a metal object its inductance changes. This change can be monitored by its effect on a resonant circuit and the change used to trigger a switch. It can only be used for the detection of metal objects and is best with ferrous metals.

2.3.7 Optical encoders

An *encoder* is a device that provides a digital output as a result of a linear or angular displacement. Position encoders can be grouped into two categories: *incremental encoders* that detect changes in rotation from some datum position and absolute encoders which give the actual angular position.

Figure 2.16(a) shows the basic form of an *incremental encoder* for the measurement of angular displacement. A beam of light passes through slots in a disc and is detected by a suitable light sensor. When the disc is rotated, a pulsed output is produced by the sensor with the number of pulses being proportional to the angle through which the disc rotates. Thus the angular position of the disc, and hence the shaft rotating it, can be determined by the number of pulses produced since some datum position. In practice three concentric tracks with three sensors are used (Fig. 2.16(b)). The inner track has just one hole and is used to locate the 'home' position of the disc. The other two tracks have a series of equally spaced holes that go completely round the disc but with the holes in the middle track offset from the holes in the outer track by one-half the width of a hole. This offset enables the direction of rotation to be determined. In a clockwise direction the pulses in the outer track lead those in the inner, in the anti-clockwise direction they lag. The resolution is determined by the number of slots on the disc. With 60 slots occurring with 1 revolution then, since 1 revolution is a rotation of 360°, the resolution is 360/60 = 6°.

Figure 2.17 shows the basic form of an *absolute encoder* for the measurement of angular displacement. This gives an output in the form of a binary number of several digits, each such number representing a particular angular position. The rotating disc has three concentric circles of slots and three sensors to detect the light pulses. The slots are arranged in such a way that the sequential output from the sensors is a number in the binary code. Typical encoders tend to have up to 10 or 12 tracks. The number of bits in the binary number will be equal to the number of tracks. Thus with 10 tracks there will be 10 bits and so the number of positions that can be detected is 2^{10}, i.e. 1024, a resolution of $360/1024 = 0.35°$.

Fig. 2.17   A 3-bit absolute encoder

The normal form of binary code is generally not used because changing from one binary number to the next can result in more than one bit changing and if, through some misalignment, one of the bits changes fractionally before the others then an intermediate binary number is momentarily indicated and so can lead to false counting. To overcome this the *Gray code* is generally used. With this code only one bit changes in moving from one number to the next. Figure 2.18 shows the tracks with normal binary code and the Gray code.

Optical encoders, e.g. HEDS-5000 from Hewlett Packard, are supplied for mounting on shafts and contain an LED light source and a code wheel. Interface integrated circuits are also available to decode the encoder and give a binary output suitable for a microprocessor. For an absolute encoder with 7 tracks on its code disc, each track will give one of bits in the binary number and thus we have 2^7 positions specified, i.e. 128.

	Normal binary		Gray code	
0	0000		0000	
1	0001		0001	
2	0010		0011	
3	0011		0010	
4	0100		0110	
5	0101		0111	
6	0110		0101	
7	0111		0100	
8	1000		1100	
9	1001		1101	
10	1010		1111	

Fig. 2.18   Binary and Gray codes

2.3.8 Pneumatic sensors

Pneumatic sensors involve the use of compressed air, displacement or the proximity of an object being transformed into a change in air pressure. Figure 2.19 shows the basic form of such a sensor. Low-pressure air is allowed to escape through a port in the front of the sensor. This escaping air, in the absence of any close-by object, escapes and in doing so also reduces the pressure in the nearby sensor output port. However, if there is a close-by object, the air cannot so readily escape and the result is that the pressure increases in the sensor output port. The output pressure from the sensor thus depends on the proximity of objects.

Air dragged out of port and so drop in system pressure

Escaping
air

Low-pressure
air inlet

Object blocking escaping air increases
pressure in system

Rise
in
pressure

Low-pressure
air inlet

Escaping
air

Fig. 2.19   Pneumatic proximity sensor

Such sensors are used for the measurement of displacements of fractions of millimetres in ranges which typically are about 3 to 12 mm.

2.3.9 Proximity switches

There are a number of forms of switch which can be activated by the presence of an object in order to give a proximity sensor with an output which is either on or off.

The *microswitch* is a small electrical switch which requires physical contact and a small operating force to close the contacts. For example, in the case of determining the presence of an item on a conveyor belt, this might be actuated by the weight of the item on the belt depressing the belt and hence a spring-loaded platform under it, with the movement of this platform then closing the switch. Figure 2.20 shows examples of ways such switches can be actuated.

Button to operate
switch

Switch contacts

(a)

(b)

(c)

Fig. 2.20   (a) Lever-operated,
(b) roller-operated, (c) cam-operated
switches

Fig. 2.21 Reed switch

Fig. 2.22 Using photoelectric sensors to detect objects by (a) the object breaking the beam, (b) it reflecting light

Fig. 2.23 Hall effect

Figure 2.21 shows the basic form of a *reed switch*. It consists of two magnetic switch contacts sealed in a glass tube. When a magnet is brought close to the switch, the magnetic reeds are attracted to each other and close the switch contacts. It is a non-contact proximity switch. Such a switch is very widely used for checking the closure of doors. It is also used with such devices as tachometers which involve the rotation of a toothed wheel past the reed switch. If one of the teeth has a magnet attached to it, then every time it passes the switch it will momentarily close the contacts and hence produce a current/voltage pulse in the associated electrical circuit.

Photosensitive devices can be used to detect the presence of an opaque object by it breaking a beam of light, or infrared radiation, falling on such a device or by detecting the light reflected back by the object (Fig. 2.22).

2.3.10 Hall effect sensors

When a beam of charged particles passes through a magnetic field, forces act on the particles and the beam is deflected from its straight line path. A current flowing in a conductor is like a beam of moving charges and thus can be deflected by a magnetic field. This effect was discovered by E.R. Hall in 1879 and is called the *Hall effect*. Consider electrons moving in a conductive plate with a magnetic field applied at right angles to the plane of the plate (Fig. 2.23). As a consequence of the magnetic field, the moving electrons are deflected to one side of the plate and thus that side becomes negatively charged while the opposite side becomes positively charged since the electrons are directed away from it. This charge separation produces an electric field in the material. The charge separation continues until the forces on the charged particles from the electric field just balance the forces produced by the magnetic field. The result is a transverse potential difference V given by:

$$V = K_H \frac{BI}{t}$$

where B is the magnetic flux density at right angles to the plate, I the current through it, t the plate thickness and K_H a constant called the *Hall coefficient*. Thus if a constant current source is used with a particular sensor, the Hall voltage is a measure of the magnetic flux density.

Hall effect sensors are generally supplied as an integrated circuit with the necessary signal processing circuitry. There are two basic forms of such sensor, linear where the output varies in a reasonably linear manner with the magnetic flux density (Fig. 2.24(a)) and threshold where the output shows a sharp drop at a particular magnetic flux density (Fig. 2.24(b)). The linear output Hall effect sensor 634SS2 gives an output which is fairly linear over the range –40 to +40 mT (–400 to + 400 gauss) at about

10 mV per mT (1 mV per gauss) when there is a supply voltage of 5 V. The threshold Hall effect sensor Allegro UGN3132U gives an output which switches from virtually zero to about 145 mV when the magnetic flux density is about 3 mT (30 gauss). Hall effect sensors have the advantages of being able to operate as switches which can operate up to 100 kHz repetition rate, cost less than electromechanical switches and do not suffer from the problems associated with such switches of contact bounce occurring and hence a sequence of contacts rather than a single clear contact. The Hall effect sensor is immune to environmental contaminants and can be used under severe service conditions.

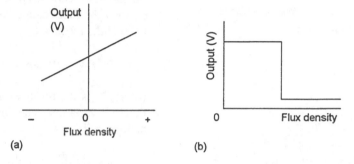

Fig. 2.24   Hall effect sensors:

(a) linear, (b) threshold

Such sensors can be used as position, displacement and proximity sensors if the object being sensed is fitted with a small permanent magnet. As an illustration, such a sensor can be used to determine the level of fuel in an automobile fuel tank. A magnet is attached to a float and as the level of fuel changes so the float distance from the Hall sensor changes (Fig. 2.25). The result is a Hall voltage output which is a measure of the distance of the float from the sensor and hence the level of fuel in the tank.

Another application of Hall effect sensors is in brushless d.c. motors. With such motors it is necessary to determine when the permanent magnet rotor is correctly aligned with the windings on the stator so that the current through the windings can be switched on at the right instant to maintain the rotor rotation. Hall effect sensors are used to detect when the alignment is right.

Fig. 2.25   Fluid level detector

2.4 Velocity and motion

The following are examples of sensors that can be used to monitor linear and angular velocities and detect motion. The application of motion detectors includes security systems used to detect intruders and interactive toys and appliances, e.g. the cash machine screen which becomes active when you get near to it.

2.4.1 Incremental encoder

The incremental encoder described in Section 2.3.7 can be used for a measurement of angular velocity, the number of pulses produced per second being determined.

Pick-up coil

Output

Toothed wheel

Fig. 2.26   Variable reluctance
tachogenerator

2.4.2 Tachogenerator

The tachogenerator is used to measure angular velocity. One form, the *variable reluctance tachogenerator*, consists of a toothed wheel of ferromagnetic material which is attached to the rotating shaft (Fig. 2.26). A pick-up coil is wound on a permanent magnet. As the wheel rotates, so the teeth move past the coil and the air gap between the coil and the ferromagnetic material changes. We have a magnetic circuit with an air gap which periodically changes. Thus the flux linked by a pick-up coil changes. The resulting cyclic change in the flux linked produces an alternating e.m.f. in the coil.

If the wheel contains n teeth and rotates with an angular velocity ω, then the flux change with time for the coil can be considered to be of the form:

$$\Phi = \Phi_0 + \Phi_a \cos n\omega t$$

where Φ_0 is the mean value of the flux and Φ_a the amplitude of the flux variation. The induced e.m.f. e in the N turns of the pick-up coil is thus:

$$e = -N\frac{d\Phi}{dt} = -N\frac{d}{dt}(\Phi_0 + \Phi_a \cos n\omega t) = N\Phi_a n\omega \sin n\omega t$$

and so we can write:

$$e = E_{max} \sin \omega t$$

where the maximum value of the induced e.m.f. E_{max} is $N\Phi_a n\omega$ and so is a measure of the angular velocity.

Instead of using the maximum value of the e.m.f. as a measure of the angular velocity, a pulse-shaping signal conditioner can be used to transform the output into a sequence of pulses which can be counted by a counter, the number counted in a particular time interval being a measure of the angular velocity.

Another form of tachogenerator is essentially an *a.c. generator*. It consists of a coil, termed the rotor, which rotates with the rotating shaft. This coil rotates in the magnetic field produced by a stationary permanent magnet or electromagnet (Fig. 2.27) and so an alternating e.m.f. is induced in it. The amplitude or frequency of this alternating e.m.f. can be used as a measure of the angular velocity of the rotor. The output may be rectified to give a d.c. voltage with a size which is proportional to the angular velocity. Non-linearity for such sensors is typically of the order of ±0.15% of the full range and the sensors typically used for rotations up to about 10 000 rev/minute.

N   S

Rotating
coil

Fig. 2.27   A.C. generator
form of tachogenerator

(a)   (b)

Fig. 2.28   Polarising a pyroelectric
material

2.4.3 Pyroelectric sensors

Pyroelectric materials, e.g. lithium tantalate, are crystalline materials which generate charge in response to heat flow. When such a material is heated to a temperature just below the Curie temperature, this being about 610°C for lithium tantalate, in an electric field and the material cooled while remaining in the field, electric dipoles within the material line up and it becomes polarised (Fig. 2.28, (a) leading to (b)). When the field is then removed the material retains its polarisation; the effect is rather like magnetising a piece of iron by exposing it to a magnetic field. When the pyroelectric material is exposed to infrared radiation, its temperature rises and this reduces the amount of polarisation in the material, the dipoles being shaken up more and losing their alignment (Fig. 2.29).

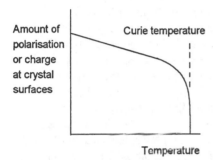

Fig. 2.29   Effect of temperature
on amount of polarisation

A pyroelectric sensor consists of a polarised pyroelectric crystal with thin metal film electrodes on opposite faces. Because the crystal is polarised with charged surfaces, ions are drawn from the surrounding air and electrons from any measurement circuit connected to the sensor to balance the surface charge (Fig. 2.30(a)). If infrared radiation is incident on the crystal and changes its temperature, the polarisation in the crystal is reduced and consequently that is a reduction in the charge at the surfaces of the crystal. There is then an excess of charge on the metal electrodes over that needed to balance the charge on the crystal surfaces (Fig. 2.30(b)). This charge leaks away through the measurement circuit until the charge on the crystal once again is balanced by the charge on the electrodes. The pyroelectric sensor thus behaves as a charge generator which generates charge when there is a change in its temperature as a result of the incidence of infrared radiation. For the linear part of the graph in Figure 2.29, when there is a temperature change the change in charge Δq is proportional to the change in temperature Δt:

Fig. 2.30   Pyroelectric sensor

$$\Delta q = k_p \Delta t$$

where k_p is a sensitivity constant for the crystal. Figure 2.31 shows the equivalent circuit of a pyroelectric sensor, it effectively being a capacitor charged by the excess charge with a resistance R

Fig. 2.31   Equivalent circuit

Fig. 2.32 Dual pyroelectric sensor

to represent either internal leakage resistance or that combined with the input resistance of an external circuit.

To detect the motion of a human or other heat source, the sensing element has to distinguish between general background heat radiation and that given by a moving heat source. A single pyroelectric sensor would not be capable of this and so a dual element is used (Fig. 2.32). One form has the sensing element with a single front electrode but two, separated, back electrodes. The result is two sensors which can be connected so that when both receive the same heat signal their outputs cancel. When a heat source moves so that the heat radiation moves from one of the sensing elements to the other then the resulting current through the resistor alternates from being first in one direction and then reversed to the other direction. Typically a moving human gives an alternating current of the order of 10^{-12} A. The resistance R has thus to be very high to give a significant voltage. For example, 50 GΩ with such a current gives 50 mV. For this reason a JFET transistor is included in the circuit as a voltage follower in order to bring the output impedance down to a few kΩ.

A focusing device is needed to direct the infrared radiation onto the sensor. While parabolic mirrors can be used a more commonly used method is a Fresnel plastic lens. Such a lens also protects the front surface of the sensor and is the form commonly used for sensors to trigger intruder alarms or switch on a light when someone approaches.

2.5 Force

A spring balance is an example of a force sensor in which a force, a weight, is applied to the scale pan and causes a displacement, i.e. the spring stretches. The displacement is then a measure of the force. Forces are commonly measured by the measurement of displacements, the following method illustrating this.

2.5.1 Strain gauge load cell

A very commonly used form of force-measuring transducer is based on the use of electrical resistance strain gauges to monitor the strain produced in some member when stretched, compressed or bent by the application of the force. The arrangement is generally referred to as a *load cell*. Figure 2.33 shows an example of such a cell. This is a cylindrical tube to which strain gauges have been attached. When forces are applied to the cylinder to compress it, then the strain gauges give a resistance change which is a measure of the strain and hence the applied forces. Since temperature also produces a resistance change, the signal conditioning circuit used has to be able to eliminate the effects due to temperature (see Section 3.5.1). Typically such load cells are used for forces up to about 10 MN, the non-linearity error being about ±0.03% of full range, hysteresis error ±0.02% of full

Fig. 2.33 Strain gauge load cell

range and repeatability error ±0.02% of full range. Strain gauge load cells based on the bending of a strain-gauged metal element tend to be used for smaller forces, e.g. with ranges varying from 0 to 5 N up to 0 to 50 kN. Errors are typically a non-linearity error of about ±0.03% of full range, hysteresis error ±0.02% of full range and repeatability error ±0.02% of full range.

2.6 Fluid pressure

(a)

(b)

Fig. 2.34 Diaphragms: (a) flat, (b) corrugated

Fig. 2.35 Diaphragm pressure gauge

Fig. 2.36 Pressure sensor element

Many of the devices used to monitor fluid pressure in industrial processes involve the monitoring of the elastic deformation of diaphragms, capsules, bellows and tubes. The types of pressure measurements that can be required are: absolute pressure where the pressure is measured relative to zero-pressure, i.e. a vacuum, differential pressure where a pressure difference is measured and gauge pressure where the pressure is measured relative to the barometric pressure.

For a diaphragm (Fig. 2.34(a) and (b)), when there is a difference in pressure between the two sides then the centre of the diaphragm becomes displaced. Corrugations in the diaphragm result in a greater sensitivity. This movement can be monitored by some form of displacement sensor, e.g. a strain gauge, as illustrated in Figure 2.35. A specially designed strain gauge is often used, consisting of four strain gauges with two measuring the strain in a circumferential direction while two measure strain in a radial direction. The four strain gauges are then connected to form the arms of a Wheatstone bridge (see Chapter 3). While strain gauges can be stuck on a diaphragm, an alternative is to create a silicon diaphragm with the strain gauges as specially doped areas of the diaphragm.

Another form of silicon diaphragm pressure sensor is used for the Motorola MPX pressure sensors. The strain gauge element is integrated, together with a resistive network, in a single silicon diaphragm chip. When a current is passed through the strain gauge element and pressure applied at right angles to it, a voltage is produced in a transverse direction (Fig. 2.36). This, together with signal conditioning and temperature compensation circuitry, is packaged as the MPX sensor. The output voltage is directly proportional to the pressure. Such sensors are available for use for the measurement of absolute pressure (the MX numbering system ends with A, AP, AS or ASX)), differential pressure (the MX numbering system ends with D or DP) and gauge pressure (the MX numbering system ends with GP, GVP, GS, GVS, GSV or GVSX). For example, the MPX2100 series has a pressure range of 100 kPa and with a supply voltage of 16 V, d.c., gives in the absolute pressure and differential pressure forms a voltage output over the full range of 40 mV. The response time, 10 to 90%, for a step change from 0 to 100 kPa is about 1.0 ms and the output impedance is of the order of 1.4 to 3.0 kΩ. The absolute pressure sensors are used for such applications as altimeters and barometers, the differential pressure sensors for air flow

(a)

(b)

Fig. 2.37 (a) Capsule, (b) bellows

measurements and the gauge pressure sensors for engine pressure and tyre pressure.

Capsules (Fig. 2.37(a)) can be considered to be just two corrugated diaphragms combined and give even greater sensitivity. A stack of capsules is just a bellows (Fig. 2.37(b)) and even more sensitive. Figure 2.38 shows how a bellows can be combined with a LVDT to give a pressure sensor with an electrical output. Diaphragms, capsules and bellows are made from such materials as stainless steel, phosphor bronze, and nickel, with rubber and nylon also being used for some diaphragms. Pressures in the range of about 10^3 to 10^8 Pa can be monitored with such sensors.

Fig. 2.38 LVDT with bellows

A different form of deformation is obtained using a tube with an elliptical cross-section (Fig. 2.39(a)). Increasing the pressure in such a tube causes it to tend to a more circular cross-section. When such a tube is in the form of a C-shaped tube (Fig. 2.39(b)), this being generally known as a *Bourdon tube*, the C opens up to some extent when the pressure in the tube increases. A helical form of such a tube (Fig. 2.39(c)) gives a greater sensitivity. The tubes are made from such materials as stainless steel and phosphor bronze and are used for pressures in the range 10^3 to 10^8 Pa.

Fig. 2.39 Tube pressure sensors

2.6.1 Piezoelectric sensors

Fig. 2.40   Piezoelectricity

Piezoelectric materials when stretched or compressed generate electric charges with one face of the material becoming positively charged and the opposite face negatively charged (Fig. 2.40). As a result a voltage is produced. Piezoelectric materials are ionic crystals which when stretched or compressed results in the charge distribution in the crystal changing so that there is a net displacement of charge with one face of the material becoming positively charged and the other negatively charged. The net charge q on a surface is proportional to the amount x by which the charges have been displaced, and since the displacement is proportional to the applied force F:

$$q = kx = SF$$

where k is a constant and S a constant termed the *charge sensitivity*. The charge sensitivity depends on the material concerned and the orientation of its crystals. Quartz has a charge sensitivity of 2.2 pC/N when the crystal is cut in one particular direction and the forces applied in a specific direction; barium titanate has a much higher charge sensitivity of the order of 130 pC/N and lead zirconate–titanate about 265 pC/N.

Metal electrodes are deposited on opposite faces of the piezoelectric crystal (Fig. 2.41). The capacitance C of the piezoelectric material between the plates is:

Fig. 2.41   Piezoelectric capacitor

$$C = \frac{\varepsilon_0 \varepsilon_r A}{t}$$

where ε_r is the relative permittivity of the material, A is area and t its thickness. Since the charge $q = Cv$, where v is the potential difference produced across a capacitor, then:

$$v = \frac{St}{\varepsilon_0 \varepsilon_r A} F$$

The force F is applied over an area A and so the applied pressure p is F/A and if we write $S_v = (S/\varepsilon_0 \varepsilon_r)$, this being termed the *voltage sensitivity factor*:

$$v = S_v t p$$

The voltage is proportional to the applied pressure. The voltage sensitivity for quartz is about 0.055 V/m Pa. For barium titanate it is about 0.011 V/m Pa.

Piezoelectric sensors are used for the measurement of pressure, force and acceleration. The applications have, however, to be such that the charge produced by the pressure does not have much time to leak off and thus tends to be used mainly for transient rather than steady pressures.

The equivalent electrical circuit for a piezoelectric sensor is a charge generator in parallel with capacitance C_s and in parallel with the resistance R_s arising from leakage through the dielectric (Fig. 2.42(a)). When the sensor is connected via a cable, of capacitance C_c, to an amplifier of input capacitance C_A and resistance R_A we have effectively the circuit shown in Figure 2.42(b) and a total circuit capacitance of $C_s + C_c + C_A$ in parallel with resistance of $R_A R_s/(R_A + R_s)$. When the sensor is subject to pressure it becomes charged, but because of the resistance the capacitor will discharge with time. The time taken for the discharge will depend on the time constant of the circuit.

(a)   (b)

Fig. 2.42 (a) Sensor equivalent circuit,

(b) sensor connected to charge amplifier

2.6.2 Tactile sensor

A tactile sensor is a particular form of pressure sensor. Such a sensor is used on the 'fingertips' of robotic 'hands' to determine when a 'hand' has come into contact with an object. They are also used for 'touch display' screens where a physical contact has to be sensed. One form of tactile sensor uses piezoelectric polyvinylidene fluoride (PVDF) film. Two layers of the film are used and are separated by a soft film which transmits vibrations (Fig. 2.43). The lower PVDF film has an alternating voltage applied to it and this results in mechanical oscillations of the film (the piezoelectric effect described above in reverse). The intermediate film transmits these vibrations to the upper PVDF film. As a consequence of the piezoelectric effect, these vibrations cause an alternating voltage to be produced across the upper film. When pressure is applied to the upper PVDF film its vibrations are affected and the output alternating voltage is changed.

Fig. 2.43   PVDF tactile sensor

2.7 Liquid flow

The traditional methods of measuring the flow rate of liquids involves devices based on the measurement of the pressure drop occurring when the fluid flows through a constriction (Fig. 2.44). For a horizontal tube, where v_1 is the fluid velocity, P_1 the pressure and A_1 the cross-sectional area of the tube prior to the

Fig. 2.44   Fluid flow through a constriction

Fig. 2.45   Orifice plate

Fig. 2.46   Turbine flowmeter

2.8 Liquid level

constriction, v_2 the velocity, P_2 the pressure and A_2 the cross-sectional area at the constriction, ρ the fluid density, then Bernoulli's equation gives

$$\frac{v_1^2}{2g} + \frac{P_1}{\rho g} = \frac{v_2^2}{2g} + \frac{P_2}{\rho g}$$

Since the mass of liquid passing per second through the tube prior to the constriction must equal that passing through the tube at the constriction, we have $A_1 v_1 \rho = A_2 v_2 \rho$. But the quantity Q of liquid passing through the tube per second is $A_1 v_1 = A_2 v_2$. Hence

$$Q = \frac{A_2}{\sqrt{1 - (A_2/A_1)^2}} \sqrt{\frac{2(P_1 - P_2)}{\rho}}$$

Thus the quantity of fluid flowing through the pipe per second is proportional to $\sqrt{}$(pressure difference). Measurements of the pressure difference can thus be used to give a measure of the rate of flow. There are many devices based on this principle, and the following example of the orifice plate is probably one of the most common.

2.7.1 Orifice plate

The orifice plate (Fig. 2.45) is simply a disc, with a central hole, which is placed in the tube through which the fluid is flowing. The pressure difference is measured between a point equal to the diameter of the tube upstream and a point equal to half the diameter downstream. The orifice plate is simple, cheap, with no moving parts, and is widely used. It, however, does not work well with slurries. The accuracy is typically about ±1.5% of full range, it is non-linear, and does produce quite an appreciable pressure loss in the system to which it is connected.

2.7.2 Turbine meter

The turbine flowmeter (Fig. 2.46) consists of a multi-bladed rotor that is supported centrally in the pipe along which the flow occurs. The fluid flow results in rotation of the rotor, the angular velocity being approximately proportional to the flow rate. The rate of revolution of the rotor can be determined using a magnetic pick-up. The pulses are counted and so the number of revolutions of the rotor can be determined. The meter is expensive with an accuracy of typically about ±0.3%.

The level of liquid in a vessel can be measured directly by monitoring the position of the liquid surface or indirectly by measuring some variable related to the height. Direct methods can involve floats; indirect methods include the monitoring of the

Fig. 2.47   Float system

(a)   Atmosphere

(b)   Differential pressure cell

Fig. 2.48   Using a differential
pressure sensor

weight of the vessel by, perhaps, load cells. The weight of the liquid is $Ah\rho g$, where A is the cross-sectional area of the vessel, h the height of liquid, ρ its density and g the acceleration due to gravity. Thus changes in the height of liquid give weight changes. More commonly, indirect methods involve the measurement of the pressure at some point in the liquid, the pressure due to a column of liquid of height h being $h\rho g$, where ρ is the liquid density.

2.8.1 Floats

A direct method of monitoring the level of liquid in a vessel is by monitoring the movement of a float. Figure 2.47 illustrates this with a simple float system. The displacement of the float causes a lever arm to rotate and so move a slider across a potentiometer. The result is an output of a voltage related to the height of liquid. Other forms of this involve the lever causing the core in a LVDT to become displaced, or stretch or compress a strain-gauged element.

2.8.2 Differential pressure

Figure 2.48 shows two forms of level measurement based on the measurement of differential pressure. In Figure 2.48(a), the differ- ential pressure cell determines the pressure difference between the liquid at the base of the vessel and atmospheric pressure, the vessel being open to atmospheric pressure. With a closed or open vessel the system illustrated in (b) can be used. The differential pressure cell monitors the difference in pressure between the base of the vessel and the air or gas above the surface of the liquid.

2.9 Temperature

Fig. 2.49   Bimetallic thermostat

Changes that are commonly used to monitor temperature are the expansion or contraction of solids, liquids or gases, the change in electrical resistance of conductors and semiconductors and thermoelectric e.m.f.s. The following are some of the methods that are commonly used with temperature control systems.

2.9.1 Bimetallic strips

This device consists of two different metal strips bonded together. The metals have different coefficients of expansion and when the temperature changes the composite strip bends into a curved strip, with the higher coefficient metal on the outside of the curve. This deformation may be used as a temperature-controlled switch, as in the simple thermostat which was commonly used with domestic heating systems (Fig. 2.49). The small magnet enables the sensor to exhibit hysteresis, meaning that the switch contacts close at a different temperature from that at which they open.

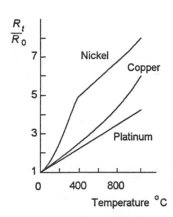

Fig. 2.50   Variation of resistance with temperature for metals

Fig. 2.51   Thermistors

2.9.2 Resistance temperature detectors (RTDs)

The resistance of most metals increases, over a limited temperature range, in a reasonably linear way with temperature (Fig. 2.50). For such a linear relationship:

$$R_t = R_0(1 + at)$$

where R_t is the resistance at a temperature t °C, R_0 the resistance at 0°C and a a constant for the metal termed the temperature coefficient of resistance. Resistance temperature detectors (RTDs) are simple resistive elements in the form of coils of wire of such metals as platinum, nickel or nickel–copper alloys; platinum is the most widely used. Thin film platinum elements are often made by depositing the metal on a suitable substrate, wire-wound elements involving a platinum wire held by a high temperature glass adhesive inside a ceramic tube. Such detectors are highly stable and give reproducible responses over long periods of time. They tend to have response times of the order of 0.5 to 5 s or more.

2.9.3 Thermistors

Thermistors are small pieces of material made from mixtures of metal oxides, such as those of chromium, cobalt, iron, manganese and nickel. These oxides are semiconductors. The material is formed into various forms of element, such as beads, discs and rods (Fig. 2.51). The resistance of conventional metal-oxide thermistors decreases in a very non-linear manner with an increase in temperature, as illustrated in Figure 2.52. Such thermistors having negative temperature coefficients (NTC). Positive temperature coefficient (PTC) thermistors are, however, available. The change in resistance per degree change in temperature is considerably larger than that which occurs with metals. The resistance–temperature relationship for a thermistor can be described by an equation of the form

$$R_t = K e^{\beta/t}$$

where R_t is the resistance at temperature t, with K and β being constants. Thermistors have many advantages when compared with other temperature sensors. They are rugged and can be very small, so enabling temperatures to be monitored at virtually a point. Because of their small size they respond very rapidly to changes in temperature. They give very large changes in resistance per degree change in temperature. Their main disadvantage is their non-linearity.

Fig. 2.52 Variation of resistance with temperature for a typical thermistor

2.9.4 Thermodiodes and transistors

A junction semiconductor diode is widely used as a temperature sensor. When the temperature of doped semiconductors changes, the mobility of their charge carriers changes and this affects the rate at which electrons and holes can diffuse across a p–n junction. Thus when a p–n junction has a potential difference V across it, the current I through the junction is a function of the temperature, being given by:

$$I = I_0(e^{eV/kT} - 1)$$

where T is the temperature on the Kelvin scale, e the charge on an electron, and k and I_0 are constants. By taking logarithms we can write the equation in terms of the voltage as:

$$V = \left(\frac{kT}{e}\right) \ln\left(\frac{I}{I_0} + 1\right)$$

Thus, for a constant current, we have V proportional to the temperature on the Kelvin scale and so a measurement of the potential difference across a diode at constant current can be used as a measure of the temperature. Such a sensor is compact like a thermistor but has the great advantage of giving a response which is a linear function of temperature. Diodes for use as temperature sensors, together with the necessary signal conditioning, are supplied as integrated circuits, e.g. LM3911, and give a very small compact sensor. The output voltage from LM3911 is proportional to the temperature at the rate of 10 mV/°C.

In a similar manner to the thermodiode, for a thermo-transistor the voltage across the junction between the base and the emitter depends on the temperature and can be used as a measure of

Fig. 2.53   LM35

Fig. 2.54   A thermocouple

temperature. A common method is to use two transistors with different collector currents and determine the difference in the base-emitter voltages between them, this difference being directly proportional to the temperature on the Kelvin scale. Such transistors can be combined with other circuit components on a single chip to give a temperature sensor with its associated signal conditioning, e.g. LM35 (Fig. 2.53). This sensor can be used in the range –40°C to 110°C and gives an output of 10 mV/°C.

2.9.5 Thermocouples

If two different metals are joined together, a potential difference occurs across the junction. The potential difference depends on the metals used and the temperature of the junction. A thermocouple is a complete circuit involving two such junctions (Fig. 2.54). If both junctions are at the same temperature there is no net e.m.f. If, however, there is a difference in temperature between the two junctions, there is an e.m.f. The value of this e.m.f. E depends on the two metals concerned and the temperatures t of both junctions. Usually one junction is held at 0°C and then, to a reasonable extent, the following relationship holds:

$$E = at + bt^2$$

where a and b are constants for the metals concerned. Commonly used thermocouples are shown in Table 2.1, with the temperature ranges over which they are generally used and typical sensitivities. These commonly used thermocouples are given reference letters. For example, the iron–constantan thermocouple is called a type J thermocouple. Figure 2.55 shows how the e.m.f. varies with temperature for a number of commonly used pairs of metals.

Table 2.1 Thermocouples

Ref.	Materials	Range °C	μV/°C
B	Platinum 30% rhodium/platinum 6% rhodium	0 to 1800	3
E	Chromel/constantan	–200 to 1000	63
J	Iron/constantan	–200 to 900	53
K	Chromel/alumel	–200 to 1300	41
N	Nirosil/nisil	–200 to 1300	28
R	Platinum/platinum 13% rhodium	0 to 1400	6
S	Platinum/platinum 10% rhodium	0 to 1400	6
T	Copper/constantan	–200 to 400	43

Fig. 2.55 Thermoelectric e.m.f.–temperature graphs

The above is equivalent to:

Fig. 2.56 Law of intermediate temperatures

A thermocouple circuit can have other metals in the circuit and they will have no effect on the thermoelectric e.m.f. provided all their junctions are at the same temperature. A thermocouple can be used with the reference junction at a temperature other than 0°C. The standard tables, however, assume a 0°C junction and hence a correction has to be applied before the tables can be used. The correction is applied using what is known as the *law of intermediate temperatures*, namely

$$E_{t,0} = E_{t,I} + E_{I,0}$$

The e.m.f. $E_{t,0}$ at temperature t when the cold junction is at 0°C equals the e.m.f. $E_{t,I}$ at the intermediate temperature I plus the e.m.f. $E_{I,0}$ at temperature I when the cold junction is at 0°C (Fig. 2.56). To maintain one junction of a thermocouple at 0°C, i.e. have it immersed in a mixture of ice and water, is often not convenient. A compensation circuit can, however, be used to provide an e.m.f. which varies with the temperature of the cold junction in such a way that when it is added to the thermocouple e.m.f. it generates a combined e.m.f. which is the same as would have been generated if the cold junction had been at 0°C (Fig. 2.57). The compensating e.m.f. can be provided by the voltage drop across a resistance thermometer element.

The base-metal thermocouples, E, J, K and T, are relatively cheap but deteriorate with age. They have accuracies which are typically about ±1 to 3%. Noble-metal thermocouples, e.g. R, are more expensive but are more stable with longer life. They have accuracies of the order of ±1% or better.

Fig. 2.57   Cold junction
compensation

Thermocouples are generally mounted in a sheath to give them mechanical and chemical protection. The type of sheath used depends on the temperatures at which the thermocouple is to be used. In some cases the sheath is packed with a mineral which is a good conductor of heat and a good electrical insulator. The response time of an unsheathed thermocouple is very fast. With a sheath this may be increased to as much as a few seconds if a large sheath is used. In some instances a group of thermocouples are connected in series so that there are perhaps ten or more hot junctions sensing the temperature. The e.m.f. produced by each is added together. Such an arrangement is known as a thermopile.

To illustrate the above, consider a type E thermocouple which is to be used for the measurement of temperature with a cold junction at 20°C. What will be the thermoelectric e.m.f. at 200°C? The following is data from standard tables.

Temp. (°C)	0	20	200
e.m.f. (mV)	0	1.192	13.419

Using the law of intermediate temperatures

$$E_{200,0} = E_{200,20} + E_{20,0} = 13.419 - 1.192 = 12.227 \text{ mV}$$

Note that this is not the e.m.f. given by the tables for a temperature of 180°C with a cold junction at 0°C, namely 11.949 mV.

2.10 Light sensors

Photodiodes are semiconductor junction diodes (see Section 7.3.1 for discussion of diodes) which are connected into a circuit in reverse bias, so giving a very high resistance, so that when light falls on the junction the diode resistance drops and the current in the circuit rises appreciably. For example, the current in the absence of light with a reverse bias of 3 V might be 25 μA and

Fig. 2.58   Photo Darlington

when illuminated by 25 000 lumens/m² the current rises to 375 μA. The resistance of the device with no light is $3/(25 \times 10^{-6})$ = 120 kΩ and with light is $3/(375 \times 10^{-6})$ = 8 kΩ. A photodiode can thus be used as a variable resistance device controlled by the light incident on it. Photodiodes have a very fast response to light.

The *phototransistors* (see Section 7.3.3 for a discussion of transistors) have a light-sensitive collector-base p–n junction. When there is no incident light there is a very small collector-to-emitter current. When light is incident, a base current is produced that is directly proportional to the light intensity. This leads to the production of a collector current which is then a measure of the light intensity. Phototransistors are often available as integrated packages with the phototransistor connected in a Darlington arrangement with a conventional transistor (Fig. 2.58). Because this arrangement gives a higher current gain, the device gives a much greater collector current for a given light intensity.

A *photoresistor* has a resistance which depends on the intensity of the light falling on it, decreasing linearly as the intensity increases. The cadmium sulphide photoresistor is most responsive to light having wavelengths shorter than about 515 nm and the cadmium selinide photoresistor for wavelengths less than about 700 nm.

An array of light sensors is often required in a small space in order to determine the variations of light intensity across that space, e.g. in the automatic camera to determine the exposure that will be most appropriate to take account of the varying light intensities across the image. To this end array devices are available with large numbers of photodiodes in an array.

2.11 Selection of sensors

In selecting a sensor for a particular application there are a number of factors that need to be considered:

1   The nature of the measurement required, e.g. the variable to be measured, its nominal value, the range of values, the accuracy required, the required speed of measurement, the reliability required, the environmental conditions under which the measurement is to be made.
2   The nature of the output required from the sensor, this determining the signal conditioning requirements in order to give suitable output signals from the measurement.
3   Then possible sensors can be identified, taking into account such factors as their range, accuracy, linearity, speed of response, reliability, maintainability, life, power supply requirements, ruggedness, availability, cost.

The selection of sensors cannot be taken in isolation from a consideration of the form of output that is required from the system after signal conditioning, and thus there has to be a suitable marriage between sensor and signal conditioner.

To illustrate the above, consider the selection of a sensor for the measurement of the level of a corrosive acid in a vessel. The level can vary from 0 to 2 m in a circular vessel which has a diameter of 1 m. The empty vessel has a weight of 100 kg. The minimum variation in level to be detected is 10 cm. The acid has a density of 1050 kg/m³. The output from the sensor is to be electrical.

Because of the corrosive nature of the acid an indirect method of determining the level seems appropriate. Thus it is possible to use a load cell, or load cells, to monitor the weight of the vessel. Such cells would give an electrical output. The weight of the liquid changes from 0 when empty to, when full, $1050 \times 2 \times \pi(1^2/4) \times 9.8 = 16.2$ kN. Adding this to the weight of the empty vessel gives a weight that varies from about 1 to 17 kN. The resolution required is for a change of level of 10 cm, i.e. a change in weight of $0.10 \times 1050 \times \pi(1^2/4) \times 9.8 = 0.8$ kN. If three load cells are used to support the tank then each will require a range of about 0 to 6 kN with a resolution of 0.27 kN. Manufacturers' catalogues can then be consulted to see if such load cells can be obtained.

2.12 Inputting data by switches

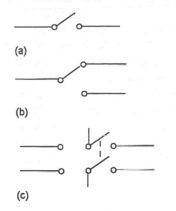

(a)

(b)

(c)

Fig. 2.59   Switches: (a) SPST, (b) SPDT, (c) DPDT

Connection made

Bounces

Final closed contacts

Initially, no connection

Fig. 2.60   Switch bounce on closing a switch

Mechanical switches consist of one or more pairs of contacts which can be mechanically closed or opened and in doing so make or break electrical circuits. Thus 0 or 1 signals can be transmitted by the act of opening or closing a switch.

Mechanical switches are specified in terms of their number of poles and throws. *Poles* are the number of separate circuits that can be completed by the same switching action and *throws* are the number of individual contacts for each pole. Figure 2.59(a) shows a single pole–single throw (SPST) switch, Figure 2.59(b) a single pole–double throw (DPDT) switch and Figure 2.59(c) a double pole–double throw (DPDT) switch.

2.12.1 Debouncing

A problem that occurs with mechanical switches is *switch bounce*. When a mechanical switch is switched to close the contacts, we have one contact being moved towards the other. It hits the other and, because the contacting elements are elastic, bounces. It may bounce a number of times (Fig. 2.60) before finally settling to its closed state after, typically, some 20 ms. Each of the contacts during this bouncing time can register as a separate contact. Thus, to a microprocessor, it might appear that perhaps two or more separate switch actions have occurred. Similarly, when a mechanical switch is opened, bouncing can occur. To overcome this problem either hardware or software can be used.

With software, the microprocessor is programmed to detect if the switch is closed and then wait, say, 20 ms. After checking that

Fig. 2.61   Debouncing a SPDT switch

Fig. 2.62   Debouncing a SPDT switch

Fig. 2.63   Schmitt trigger

Fig. 2.64   Debouncing

bouncing has ceased and the switch is in the same closed position, the next part of the program can take place. The hardware solution to the bounce problem is based on the use of a flip-flop. Figure 2.61 shows a circuit for debouncing a SPDT switch which is based on the use of a SR flip-flop (see Section 14.7). As shown, we have S at 0 and R at 1 with an output of 0. When the switch is moved to its lower position, initially S becomes 1 and R becomes 0. This gives an output of 1. Bouncing in changing S from 1 to 0 to 1 to 0, etc. gives no change in the output. Such a flip-flop can be derived from two NOR or two NAND gates. A SPDT switch can be debounced by the use of a D flip-flop (see Section 14.7). Figure 2.62 shows the circuit. The output from such a flip-flop only changes when the clock signal changes. Thus by choosing a clock period which is greater than the time for which the bounces last, say 20 ms, the bounce signals will be ignored.

An alternative method of debouncing using hardware is to use a *Schmitt trigger*. This device has the 'hysteresis' characteristic shown in Figure 2.63. When the input voltage is beyond an upper switching threshold and giving a low output, then the input voltage needs to fall below the lower threshold before the output can switch to high. Conversely, when the input voltage is below the lower switching threshold and giving a high, then the input needs to rise above the upper threshold before the output can switch to low. Such a device can be used to sharpen slowly changing signals; when the signal passes the switching threshold it becomes a sharply defined edge between two well-defined logic levels. The circuit shown in Figure 2.64 can be used for debouncing, note the circuit symbol for a Schmitt trigger. With the switch open, the capacitor becomes charged and the voltage applied to the Schmitt trigger becomes high and so it gives a low output. When the switch is closed, the capacitor discharges very rapidly and so the first bounce discharges the capacitor; the Schmitt trigger thus switches to give a high output. Successive switch bounces do not have time to recharge the capacitor to the required threshold value and so further bounces do not switch the Schmitt trigger.

2.12.2 Keypads

A keypad is an array of switches, perhaps the keyboard of a computer or the touch input membrane pad for some device such as a microwave oven. A contact type key of the form generally used with a keyboard is shown in Figure 2.65(a), depressing the key plunger forces the contacts together with the spring returning the key to the off position when the key is released. A typical membrane switch (Fig. 2.65(b)) is built up from two wafer-thin plastic films on which conductive layers have been printed. These layers are separated by a spacer layer. When the switch area of the membrane is pressed, the top contact layer closes with the bottom one to make the connection and then opens when the pressure is released.

While each switch in such arrays could be connected to individually give signals when closed, a more economical method is to connect them in an array in that an individual output is not needed for each key but each key gives a unique row–column combination. Figure 2.66 shows the connections for a 16-way keypad.

Problems

Key plunger

Switch contacts

(a)

↓ Applied pressure

Conductive layers

(b)

Fig. 2.65   (a) Contact key,
(b) membrane key

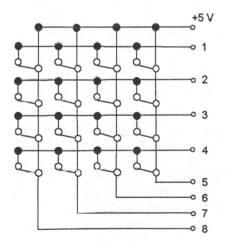

+5 V
1
2
3
4
5
6
7
8

Fig. 2.66   16-way keypad

1   Explain the significance of the following information given in the specification of transducers:
   (a)   A piezoelectric accelerometer.
         Non-linearity: ±0.5% of full range.
   (b)   A capacitive linear displacement transducer.
         Non-linearity and hysteresis: ±0.01% of full range.
   (c)   A resistance strain gauge force measurement transducer.
         Temperature sensitivity: ±1% of full range over normal environmental temperatures.
   (d)   A capacitance fluid pressure transducer.
         Accuracy: ±1% of displayed reading.
   (e)   Thermocouple.
         Sensitivity: nickel chromium–nickel aluminium thermocouple: 0.039 mV/°C when the cold junction is at 0°C.
   (f)   Gyroscope for angular velocity measurement.
         Repeatability: ±0.01% of full range.
   (g)   Inductive displacement transducer.
         Linearity: ±1% of rated load.
   (h)   Load cell.
         Total error due to non-linearity, hysteresis and non-repeatability: ±0.1%.

2   A copper–constantan thermocouple is to be used to measure temperatures between 0 and 200°C. The e.m.f. at 0°C is 0 mV, at 100°C it is 4.277 mV and at 200°C it is 9.286 mV. What will be the non-linearity error at 100°C as a percentage of the full range output if a linear relationship is assumed between e.m.f. and temperature over the full range?

3   A thermocouple element when taken from a liquid at 50°C and plunged into a liquid at 100° C at time $t = 0$ gave the following e.m.f. values. Determine the 95% response time.

Time (s)	0	20	40	60	80	100	120
e.m.f. (mV)	2.5	3.8	4.5	4.8	4.9	5.0	5.0

4   What is the non-linearity error, as a percentage of full range, produced when a 1 kΩ potentiometer has a load of 10 kΩ and is at one-third of its maximum displacement?

5 What will be the change in resistance of an electrical resistance strain gauge with a gauge factor of 2.1 and resistance 50 Ω if it is subject to a strain of 0.001?

6 You are offered a choice of an incremental shaft encoder or an absolute shaft encoder for the measurement of an angular displacement. What is the principal difference between the results that can be obtained by these methods?

7 A shaft encoder is to be used with a 50 mm radius tracking wheel to monitor linear displacement. If the encoder produces 256 pulses per revolution, what will be the number of pulses produced by a linear displacement of 200 mm?

8 A rotary variable differential transformer has a specification which includes the following information:

Ranges: ±30°, linearity error ±0.5% full range
              ±60°, linearity error ±2.0% full range
Sensitivity: 1.1 (mV/V input)/degree
Impedance: Primary 750 Ω, Secondary 2000 Ω

What will be (a) the error in a reading of 40° due to non-linearity when the RDVT is used on the ±60° range, and (b) the output voltage change that occurs per degree if there is an input voltage of 3 V?

9 What are the advantages and disadvantages of the plastic film type of potentiometer when compared with the wire-wound potentiometer?

10 A pressure sensor consisting of a diaphragm with strain gauges bonded to its surface has the following information in its specification:

Ranges: 0 to 1400 kPa, 0 to 35 000 kPa
Non-linearity error: ±0.15% of full range
Hysteresis error: ±0.05% of full range

What is the total error due to non-linearity and hysteresis for a reading of 1000 kPa on the 0 to 1400 kPa range?

11 The water level in an open vessel is to be monitored by a differential pressure cell responding to the difference in pressure between that at the base of the vessel and the atmosphere. Determine the range of differential pressures the cell will have to respond to if the water level can vary between zero height above the cell measurement point and 2 m above it.

12 An iron–constantan thermocouple is to be used to measure temperatures between 0 and 400°C. What will be the non-linearity error as a percentage of the full-scale reading at 100°C if a linear relationship is assumed between e.m.f. and temperature?

E.m.f. at 100°C = 5.268 mV; e.m.f. at 400°C = 21.846 mV

13 A platinum resistance temperature detector has a resistance of 100.00 Ω at 0°C, 138.50 Ω at 100°C and 175.83 Ω at 200°C. What will be the non-linearity error in °C at 100°C if the detector is assumed to have a linear relationship between 0 and 200°C?

14 A strain gauge pressure sensor has the following specification. Will it be suitable for the measurement of pressure of the order of 100 kPa to an accuracy of ±5 kPa in an environment where the temperature is reasonably constant at about 20°C?

Ranges: 2 to 70 MPa, 70 kPa to 1 MPa
Excitation: 10 V d.c. or a.c. (r.m.s.)
Full range output: 40 mV
Non-linearity and hysteresis errors: ±0.5 %
Temperature range: −54 to +120°C
Thermal shift zero: 0.030% full range output/°C
Thermal shift sensitivity: 0.030% full range output/°C

15 A float sensor for the determination of the level of water in a vessel has a cylindrical float of mass 2.0 kg, cross-sectional area 20 cm² and a length of 1.5 m. It floats vertically in the water and presses upwards against a beam attached to its upward end. What will be the minimum and maximum upthrust forces exerted by the float on the beam? Suggest a means by which the deformation of the beam under the action of the upthrust force could be monitored.

16 Suggest a sensor that could be used as part of the control system for a furnace to monitor the rate at which the heating oil flows along a pipe. The output from the measurement system is to be an electrical signal which can be used to adjust the speed of the oil pump. The system must be capable of operating continuously and automatically, without adjustment, for long periods of time.

17 Suggest a sensor that could be used, as part of a control system, to determine the difference in levels between liquids in two containers. The output is to provide an electrical signal for the control system.

18 Suggest a sensor that could be used as part of a system to control the thickness of rolled sheet by monitoring its thickness as it emerges from rollers. The sheet metal is in continuous motion and the measurement needs to be made quickly to enable corrective action to be made quickly. The measurement system has to supply an electrical signal.

3 Signal conditioning

3.1 Signal conditioning

The output signal from the sensor of a measurement system has generally to be processed in some way to make it suitable for the next stage of the operation. The signal may be, for example, too small and have to be amplified, contain interference which has to be removed, be non-linear and require linearisation, be analogue and have to be made digital, be digital and have to be made analogue, be a resistance change and have to be made into a current change, be a voltage change and have to be made into a suitable size current change, etc. All these changes can be referred to as *signal conditioning*. For example, the output from a thermocouple is a small voltage, a few millivolts. A signal conditioning module might then be used to convert this into a suitable size current signal, provide noise rejection, linearisation, and cold junction compensation (i.e. compensating for the cold junction not being at 0°C).

3.1.1 Interfacing with a microprocessor

Input and output devices are connected to a microprocessor system through *ports*. The term *interface* is used for the item that is used to make connections between devices and a port. Thus there could be inputs from sensors, switches, and keyboards and outputs to displays and actuators. The simplest interface could be just a piece of wire. However, the interface often contains signal conditioning and protection, the protection being to prevent damage to the microprocessor system. For example, inputs needing to be protected against excessive voltages or signals of the wrong polarity.

Microprocessors require inputs which are digital, thus a conversion of analogue to digital signal is necessary if the output from a sensor is analogue. However, many sensors generate only a very small signal, perhaps a few millivolts. Such a signal is insufficient to be directly converted from analogue to digital without first being amplified. Signal conditioning might also be needed with digital signals to improve their quality. The interface may thus contain a number of elements.

There is also the output from a microprocessor, perhaps to operate an actuator. A suitable interface is also required here. The actuator might require an analogue signal and so the digital output from the microprocessor needs converting to an analogue signal. There can also be a need for protection to stop any signal becoming inputted back through the output port to damage the microprocessor.

3.1.2 Signal-conditioning processes

The following are some of the processes that can occur in conditioning a signal:

1   *Protection* to prevent damage to the next element, e.g. a microprocessor, as a result of high current or voltage. Thus there can be series current-limiting resistors, fuses to break if the current is too high, polarity protection and voltage limitation circuits (see Section 3.3).

2   Getting the signal into the *right type of signal*. This can mean making the signal into a d.c. voltage or current. Thus, for example, the resistance change of a strain gauge has to be converted into a voltage change. This can be done by the use of a Wheatstone bridge and using the out-of-balance voltage (see Section 3.5). It can mean making the signal digital or analogue (see Section 3.6 for analogue-to-digital and analogue-to-digital converters).

3   Getting the *level* of the signal right. The signal from a thermocouple might be just a few millivolts. If the signal is to be fed into an analogue-to-digital converter for inputting to a microprocessor then it needs to be made much larger, volts rather than millivolts. Operational amplifiers are widely used for amplification (see Section 3.2).

4   Eliminating or reducing *noise*. For example, filters might be used to eliminate mains noise from a signal (see Section 3.4).

5   Signal *manipulation*, e.g. making it a linear function of some variable. The signals from some sensors, e.g. a flowmeter, are non-linear and thus a signal conditioner might be used so that the signal fed on to the next element is linear (see Section 3.2.6).

The following sections outline some of the elements that might be used in signal conditioning.

3.2 The operational amplifier

Fig. 3.1   Pin connections for a 741 operational amplifier

The basis of many signal conditioning modules is the *operational amplifier*. The operational amplifier is a high gain d.c. amplifier, the gain typically being of the order of 100 000 or more, that is supplied as an integrated circuit on a silicon chip. It has two inputs, known as the inverting input (−) and the non-inverting input (+). The output depends on the connections made to these inputs. There are other inputs to the operational amplifier, namely a negative voltage supply, a positive voltage supply and two inputs termed offset null, these being to enable corrections to be made for the non-ideal behaviour of the amplifier (see Section 3.2.8). Figure 3.1 shows the pin connections for a 741 type operational amplifier.

The following indicates the types of circuits that might be used with operational amplifiers when used as signal conditioners. For more details the reader is referred to more specialist texts, e.g. *Feedback Circuits and Op. Amps* by D.H. Horrocks (Chapman and Hall 1990) or *Analysis and Design of Analog Integrated Circuits* by P.R. Gray and R.G. Meyer (Wiley 1993).

3.2.1 Inverting amplifier

Figure 3.2 shows the connections made to the amplifier when used as an *inverting amplifier*. The input is taken to the inverting input through a resistor R_1 with the non-inverting input being connected to ground. A feedback path is provided from the output, via the resistor R_2 to the inverting input. The operational amplifier has a voltage gain of about 100 000 and the change in output voltage is typically limited to about ±10 V. The input voltage must then be between +0.0001 V and −0.0001 V. This is virtually zero and so point X is at virtually earth potential. For this reason it is called a *virtual earth*. The potential difference across R_1 is $(V_{in} − V_X)$. Hence, for an ideal operational amplifier with an infinite gain, and hence $V_X = 0$, the input potential V_{in} can be considered to be across R_1. Thus

$$V_{in} = I_1 R_1$$

The operational amplifier has a very high impedance between its input terminals; for a 741 about 2 MΩ. Thus virtually no current flows through X into it. For an ideal operational amplifier the input impedance is taken to be infinite and so there is no current flow through X. Hence the current I_1 through R_1 must be the current through R_2. The potential difference across R_2 is $(V_X − V_{out})$ and thus, since V_X is zero for the ideal amplifier, the potential difference across R_2 is $−V_{out}$. Thus

$$−V_{out} = I_1 R_2$$

Dividing these two equations:

Fig. 3.2   Inverting amplifier

$$\text{Voltage gain of circuit} = \frac{V_{\text{out}}}{V_{\text{in}}} = -\frac{R_2}{R_1}$$

Thus the voltage gain of the circuit is determined solely by the relative values of R_2 and R_1. The negative sign indicates that the output is inverted, i.e. 180° out of phase, with respect to the input.

To illustrate the above, consider an inverting operational amplifier circuit which has a resistance of 1 MΩ in the inverting input line and a feedback resistance of 10 MΩ. What is the voltage gain of the circuit?

$$\text{Voltage gain of circuit} = \frac{V_{\text{out}}}{V_{\text{in}}} = -\frac{R_2}{R_1} = -\frac{10}{1} = -10$$

Feedback resistor
R_2

Fig. 3.3   Non-inverting amplifier

3.2.2 Non-inverting amplifier

Figure 3.3 shows the operational amplifier connected as a non-inverting amplifier. The output can be considered to be taken from across a potential divider circuit consisting of R_1 in series with R_2. The voltage V_X is then the fraction $R_1/(R_1 + R_2)$ of the output voltage.

$$V_X = \frac{R_1}{R_1 + R_2} V_{\text{out}}$$

Since there is virtually no current through the operational amplifier between the two inputs there can be virtually no potential difference between them. Thus, with the ideal operational amplifier, we must have $V_X = V_{\text{in}}$. Hence

$$\text{Voltage gain of circuit} = \frac{V_{\text{out}}}{V_{\text{in}}} = \frac{R_1 + R_2}{R_1} = 1 + \frac{R_2}{R_1}$$

Fig. 3.4   Voltage follower

A particular form of this amplifier is when the feedback loop is a short circuit, i.e. $R_2 = 0$. Then the voltage gain is 1. The input to the circuit is into a large resistance, the input resistance typically being 2 MΩ. The output resistance, i.e. the resistance between the output terminal and the ground line is, however, much smaller, e.g. 75 Ω. Thus the resistance in the circuit that follows is a relatively small one and is less likely to load that circuit. Such an amplifier is referred to as a *voltage follower*; Figure 3.4 showing the basic circuit.

3.2.3 Summing amplifier

Feedback resistor
R_2

Fig. 3.5   Summing amplifier

Figure 3.5 shows the circuit of a summing amplifier. As with the inverting amplifier (Section 3.2.1), X is a virtual earth. Thus the sum of the currents entering X must equal that leaving it. Hence

$$I = I_A + I_B + I_C$$

But $I_A = V_A/R_A$, $I_B = V_B/R_B$ and $I_C = V_C/R_C$. Also we must have the same current I passing through the feedback resistor. The potential difference across R_2 is $(V_X - V_{out})$. Hence, since V_X can be assumed to be zero, it is $-V_{out}$ and so $I = -V_{out}/R_2$. Thus

$$-\frac{V_{out}}{R_2} = \frac{V_A}{R_A} + \frac{V_B}{R_B} + \frac{V_C}{R_C}$$

The output is thus the scaled sum of the inputs, i.e.

$$V_{out} = -\left(\frac{R_2}{R_A}V_A + \frac{R_2}{R_B}V_B + \frac{R_2}{R_C}V_C\right)$$

If $R_A = R_B = R_C = R_1$ then

$$V_{out} = -\frac{R_1}{R_2}(V_A + V_B + V_C)$$

To illustrate the above, consider the design of a circuit that can be used to produce an output voltage which is the average of the input voltages from three sensors. Assuming that an inverted output is acceptable, a circuit of the form shown in Figure 3.5 can be used. Each of the three inputs must be scaled to 1/3 to give an output of the average. Thus a voltage gain of the circuit of 1/3 for each of the input signals is required. Hence, if the feedback resistance is 4 kΩ the resistors in each input arm will be 12 kΩ.

3.2.4 Integrating amplifier

Consider an inverting operational amplifier circuit with the feedback being via a capacitor, as illustrated in Figure 3.6. Current is the rate of movement of charge q and since for a capacitor the charge $q = Cv$, where v is the voltage across it, then the current through the capacitor $i = dq/dt = C\,dv/dt$. The potential difference across C is $(v_X - v_{out})$ and since v_X is effectively zero, being the virtual earth, it is $-v_{out}$. Thus the current through the capacitor is $-C\,dv_{out}/dt$. But this is also the current through the input resistance R. Hence

$$\frac{v_{in}}{R} = -C\frac{dv_{out}}{dt}$$

Rearranging this gives

$$dv_{out} = -\left(\frac{1}{RC}\right)v_{in}\,dt$$

Integrating both sides gives

$$v_{out}(t_2) - v_{out}(t_1) = -\frac{1}{RC}\int_{t_1}^{t_2} v_{in}\,dt$$

Fig. 3.6   Integrating amplifier

$v_{out}(t_2)$ is the output voltage at time t_2 and $v_{out}(t_1)$ is the output voltage at time t_1. The output is proportional to the integral of the input voltage, i.e. the area under a graph of input voltage with time.

A differentiation circuit can be produced if the capacitor and resistor are interchanged in the circuit for the integrating amplifier.

3.2.5 Differential amplifier

A differential amplifier is one that amplifies the difference between two input voltages. Figure 3.7 shows the circuit. Since there is virtually no current through the high resistance in the operational amplifier between the two input terminals, there is no potential drop and thus both the inputs X will be at the same potential. The voltage V_2 is across resistors R_1 and R_2 in series. Thus the potential V_X at X is

$$\frac{V_X}{V_2} = \frac{R_2}{R_1 + R_2}$$

The current through the feedback resistance must be equal to that from V_1 through R_1. Hence

$$\frac{V_1 - V_X}{R_1} = \frac{V_X - V_{out}}{R_2}$$

This can be rearranged to give

$$\frac{V_{out}}{R_2} = V_X\left(\frac{1}{R_2} + \frac{1}{R_1}\right) - \frac{V_1}{R_1}$$

Hence substituting for V_X using the earlier equation,

$$V_{out} = \frac{R_2}{R_1}(V_2 - V_1)$$

The output is thus a measure of the difference between the two input voltages.

As an illustration of the use of such a circuit with a sensor, Figure 3.8 shows it used with a thermocouple. The difference in voltage between the e.m.f.s of the two junctions of the thermocouple is being amplified. The values of R_1 and R_2 can, for example, be chosen to give a circuit with an output of 10 mV for a temperature difference between the thermocouple junctions of 10°C if such a temperature difference produces an e.m.f. difference between the junctions of 530 µV. For the circuit we have

$$V_{out} = \frac{R_2}{R_1}(V_2 - V_1)$$

Fig. 3.7   Differential amplifier

Fig. 3.8 Differential amplifier
with a thermocouple

$$10 \times 10^{-3} = \frac{R_2}{R_1} \times 530 \times 10^{-6}$$

Hence $R_2/R_1 = 18.9$. Thus if we take for R_1 a resistance of 10 kΩ then R_2 must be 189 kΩ.

A differential amplifier might be used with a Wheatstone bridge (see Section 3.5), perhaps one with strain gauge sensors in its arms, to amplify the out-of-balance potential difference that occurs when the resistance in one or more arms changes. When the bridge is balanced both the output terminals of the bridge are at the same potential; there is thus no output potential difference. The output terminals from the bridge might both be at, say, 5.00 V. Thus the differential amplifier has both its inputs at 5.00 V. When the bridge is no longer balanced we might have one output terminal at 5.01 V and the other at 4.99 V and so the inputs to the differential amplifier are 5.01 and 4.99 V. The amplifier amplifies this difference in the voltages of 0.02 V. The original 5.00 V signal which is common to both inputs is termed the *common mode voltage* V_{CM}. For the amplifier only to amplify the difference between the two signals assumes that the two input channels are perfectly matched and the operational amplifier has the same, high, gain for both of them. In practice this is not perfectly achieved and thus the output is not perfectly proportional to the difference between the two input voltages. Thus we write for the output:

$$V_{out} = G_d \Delta V + G_{CM} V_{CM}$$

where G_d is the gain for the voltage difference ΔV, G_{CM} the gain for the common mode voltage V_{CM}. The smaller the value of G_{CM} the smaller the effect of the common mode voltage on the output. The extent to which an operational amplifier deviates from the ideal situation is specified by the *common mode rejection ratio* (CMRR):

$$\text{CMRR} = \frac{G_d}{G_{CM}}$$

To minimise the effect of the common mode voltage on the output, a high CMRR is required. Common mode rejection ratios are generally specified in decibels (dB). Thus, on the decibel scale a CMRR of, say, 10 000 would be 20 lg 10 000 = 80 dB. A typical operational amplifier might have a CMRR between about 80 and 100 dB.

A common form of *instrumentation amplifier* involves three operational amplifiers (Fig. 3.9), rather than just a single differential amplifier, and is available as a single integrated circuit. Such a circuit is designed to have a high input impedance, typically about 300 MΩ, a high voltage gain and excellent CMRR, typically more than 100 dB. The first stage involves the amplifiers A_1 and A_2, one being connected as an inverting amplifier and the other as a non-inverting amplifier. Amplifier A_3 is a differential amplifier with inputs from A_1 and A_2.

Fig. 3.9 Instrumentation amplifier

Because virtually no current passes through A_3, the current through R_4 will be the same as that through R_5. Hence:

$$\frac{V_{out1} - V_X}{R_4} = \frac{V_X - V_{out}}{R_5}$$

The differential input to A_3 is virtually zero so $V_Y = V_X$. Hence the above equation can be written as:

$$V_{out} = \left(1 + \frac{R_5}{R_4}\right)V_Y - \frac{R_5}{R_4}V_{out1}$$

R_6 and R_7 form a potential divider for the voltage V_{out2} so that:

$$V_Y = \frac{R_6}{R_6 + R_7}V_{out2}$$

Hence we can write:

$$V_{out} = \frac{1 + \dfrac{R_5}{R_4}}{1 + \dfrac{R_7}{R_6}} V_{out2} - \frac{R_5}{R_4} V_{out1}$$

Hence by suitable choice of resistance values we obtain equal multiplying factors for the two inputs to the differential amplifier. This requires:

$$1 + \frac{R_5}{R_4} = \left(1 + \frac{R_7}{R_6}\right)\frac{R_5}{R_4}$$

and hence $R_4/R_5 = R_6/R_7$.

We can apply the principle of superposition, i.e. we can consider the output produced by each source acting alone and then add them to obtain the overall response. Amplifier A_1 has an input of the differential signal V_{in1} on its non-inverting input and amplifies this with a gain of $1 + R_3/R_1$. It also has an input of V_{in2} on its inverting input and this is amplified to give a gain of $-R_3/R_1$. Also the common mode voltage V_{cm} on the non-inverting input is amplified by A_1. Thus the output of A_1 is:

$$V_{out1} = \left(1 + \frac{R_3}{R_1}\right)V_{in1} - \left(\frac{R_3}{R_1}\right)V_{in2} + \left(1 + \frac{R_3}{R_1}\right)V_{cm}$$

Amplifier A_2 likewise gives:

$$V_{out2} = \left(1 + \frac{R_2}{R_1}\right)V_{in2} - \left(\frac{R_2}{R_1}\right)V_{in1} + \left(1 + \frac{R_2}{R_1}\right)V_{cm}$$

The differential input to A_3 is $V_{out1} - V_{out2}$ and so:

$$V_{out2} - V_{out1} = \left(1 + \frac{R_3}{R_1} + \frac{R_2}{R_1}\right)V_{in1} - \left(1 + \frac{R_2}{R_3} + \frac{R_3}{R_1}\right)V_{in2}$$

$$+ \left(\frac{R_3}{R_1} - \frac{R_2}{R_1}\right)V_{cm}$$

With $R_2 = R_3$ the common mode voltage term disappears, thus:

$$V_{out2} - V_{out1} = \left(1 + \frac{2R_2}{R_1}\right)(V_{in1} - V_{in2})$$

The overall gain is thus $(1 + 2R_2/R_1)$ and is generally set by varying R_1.

Figure 3.10 shows the pin connections and some specification details for a low cost, general-purpose instrumentation amplifier (Burr-Brown INA114) using this three op-amps form of design. The gain is set by connecting an external resistor R_G between pins 1 and 8, the gain then being $1 + 50/R_G$ when R_G is in kΩ. The 50 kΩ term arises from the sum of the two internal feedback resistors.

Input impedance, differential
common mode: $10^{10}\,\Omega$ in parallel
with 6 pF

Input common mode range: ±13.5 V
Common mode rejection,
G = 1: 90 dB, G = 1000: 110 dB
Gain range 1 to 10 000
Gain error: 2% max.
Output voltage: ±13.7 V (V_s = ±15 V)

Fig. 3.10   INA114

3.2.6 Logarithmic amplifier

Some sensors have outputs which are non-linear. For example, the output from a thermocouple is not a perfectly linear function of the temperature difference between its junctions. A signal conditioner might then be used to linearise the output from such a sensor. This can be done using an operational amplifier circuit which is designed to have a non-linear relationship between its input and output so that when its input is non-linear the output is linear. This is achieved by a suitable choice of component for the feedback loop.

The logarithmic amplifier shown in Figure 3.11 is an example of such a signal conditioner. The feedback loop contains a diode (or a transistor with a grounded base). The diode has a non-linear characteristic. It might be represented by $V = C \ln I$, where C is a constant. Then, since the current through the feedback loop is the same as the current through the input resistance and the potential difference across the diode is $-V_{out}$, we have

$$V_{out} = -C \ln (V_{in}/R) = K \ln V_{in}$$

where K is some constant. However, if the input V_{in} is provided by a sensor with an input t, where $V_{in} = A\ e^{at}$, with A and a being constants, then

$$V_{out} = K \ln V_{in} = K \ln (A\ e^{at}) = K \ln A + Kat$$

The result is a linear relationship between V_{out} and t.

Fig. 3.11   Logarithmic amplifier

3.2.7 Comparator

A comparator indicates which of two voltages is the larger. An operational amplifier used with no feedback or other components can be used as a comparator. One of the voltages is applied to the inverting input and the other to the non-inverting input (Fig. 3.12(a)). Figure 3.12(b) shows the relationship between the output voltage and the difference between the two input voltages. When the two inputs are equal there is no output. However, when the non-inverting input is greater than the inverting input by more than a small fraction of a volt then the output jumps to a steady positive saturation voltage of typically +10 V. When the inverting input is greater than the non-inverting input then the output jumps to a steady negative saturation voltage of typically –10 V. Such a circuit can be used to determine when a voltage exceeds a certain level, the output then being used to perhaps initiate some action.

As an illustration of such a use, consider the circuit shown in Figure 3.13. This is designed so that when a critical temperature is reached a relay is activated and initiates some response. The circuit has a Wheatstone bridge with a thermistor in one arm. The

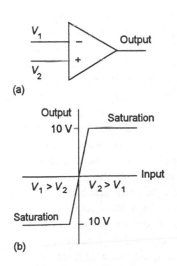

Fig. 3.12   Comparator

resistors in the bridge have their resistances selected so that at the critical temperature the bridge will be balanced. When the temperature is below this value the thermistor resistance R_1 is more than R_2 and the bridge is out-of-balance. As a consequence there is a voltage difference between the inputs to the operational amplifier and it gives an output at its lower saturated level. This keeps the transistor off, i.e. both the base-emitter and base-collector junctions are reverse biased, and so no current passes through the relay coil. When the temperature rises and the resistance of the thermistor falls, the bridge becomes balanced and the operational amplifier then switches to its upper saturation level. Consequently the transistor is switched on, i.e. its junctions become forward biased, and the relay energised.

Fig. 3.13 Temperature switch circuit

As another illustration of the use of a comparator, consider the system used to ensure that in a compact disc player the laser beam is focused on the disc surface. With a CD player, lenses are used to focus a laser beam onto a CD, this having the audio information stored as a sequence of microscopic pits and flats. The light is reflected back from the disc to an array of four photodiodes (Fig. 3.14). The output from these photodiodes is then used to reproduce the sound. The reason for having four photodiodes is that the array can also be used to determine whether the beam of laser light is in focus. When the beam is in focus on the disc then a circular spot of light falls on the photodiode array with equal amounts of light falling on each photodiode. As a result the output from the operational amplifier, which is connected as a comparator, is zero. When the beam is out of focus an elliptical spot of light is produced. This results in different amounts of light falling on each of the photocells. The outputs from the two diagonal sets of cells are compared and, because they are different, the comparator gives an output which indicates that the beam is out of focus and in which direction it is out of focus. The output can then be used to initiate correcting action by adjusting the lenses focusing the beam onto the disc.

Fig. 3.14 Focusing system for a CD player

3.2.8 Amplifier errors

Operational amplifiers are not in the real world the perfect (ideal) element discussed in the previous sections of this chapter. A particularly significant problem is that of the *offset voltage*.

An operational amplifier is a high gain amplifier which amplifies the difference between its two inputs. Thus if the two inputs are shorted we might expect to obtain no output. However, in practice this does not occur and quite a large output voltage might be detected. This effect is produced by imbalances in the internal circuitry in the operational amplifier. The output voltage can be made zero by applying a suitable voltage between the input terminals. This is known as the *offset voltage*. Many operational amplifiers are provided with arrangements for applying such an offset voltage via a potentiometer. With the 741 this is done by connecting a 10 kΩ potentiometer between pins 1 and 5 (see Fig. 3.1) and connecting the sliding contact of the potentiometer to a negative voltage supply (Fig. 3.15). The imbalances within the operational amplifier are corrected by adjusting the position of the slider until with no input to the amplifier there is no output.

For more details of this, and other non-ideal characteristics, the reader is referred to the texts listed in Section 3.1.

Fig. 3.15 Correcting the offset voltage

3.3 Protection

There are many situations where the connection of a sensor to the next unit, e.g. a microprocessor, can lead to the possibility of damage as a result of perhaps a high current or high voltage. A high current can be protected against by the incorporation in the input line of a series resistor to limit the current to an acceptable level and a fuse to break if the current does exceed a safe level. High voltages, and wrong polarity, may be protected against by the use of a Zener diode circuit (Fig. 3.16). Zener diodes behave like ordinary diodes up to some breakdown voltage when they become conducting. Thus to allow a maximum voltage of 5 V but stop voltages above 5.1 V getting through, a Zener diode with a voltage rating of 5.1 V might be chosen. When the voltage rises to 5.1 V the Zener diode breakdown and its resistance drops to a very low value. The result is that the voltage across the diode, and hence that outputted to the next circuit, drops. Because the Zener diode is a diode with a low resistance for current in one direction through it and a high resistance for the opposite direction, it also provides protection against wrong polarity. It is connected with the correct polarity to give a high resistance across the output and so a high voltage drop. When the supply polarity is reversed, the diode has low resistance and so little voltage drop occurs across the output.

In some situations it is desirable to isolate circuits completely and remove all electrical connections between them. This can be done using an *optoisolator*. Thus we might have the output from a microprocessor applied to a light-emitting diode (LED) which

Fig. 3.16 Zener diode protection circuit

(a)

(b)

(c)

(d)

Zero-crossing unit

Fig. 3.17   Optoisolators:
(a) transistor, (b) Darlington, (c) triac,
(d) triac with zero-crossing unit

emits infrared radiation. This radiation is detected by a phototransistor or triac and gives rise to a current which replicates the changes occurring in the voltage applied to the LED. Figure 3.17 shows a number of forms of optoisolator. The term *transfer ratio* is used to specify the ratio of the output current to the input current. Typically, a simple transistor optoisolator (Fig. 3.17(a)) gives an output current which is smaller than the input current and a transfer ratio of perhaps 30% with a maximum value of 7 mA. However, the Darlington form (Fig. 3.17(b)) gives an output current larger than the input current, e.g. the Siemens 6N139 gives a transfer ratio of 800% with a maximum output value of 60 mA. Another form of optoisolator (Fig. 3.17(c)) uses the triac and so can be used with alternating current, a typical triac optoisolator being able to operate with the mains voltage. Yet another form (Fig. 3.17(d)) uses a triac with a zero-crossing unit, e.g. Motorolla MOC3011, to reduce transients and electromagnetic interference.

Optoisolator outputs can be used to directly switch low-power load circuits. Thus a Darlington optoisolator might be used as the interface between a microprocessor and lamps or relays. To switch a high-power circuit, an optocoupler might be used to operate a relay and so use the relay to switch the high-power device.

A protection circuit for a microprocessor input is thus likely to be like that shown in Figure 3.18; to prevent the LED having the wrong polarity or too high an applied voltage, it is also likely to be protected by the Zener diode circuit shown in Figure 3.16 and if there is alternating signal in the input a diode would be put in the input line to rectify it.

Input   Output

Fig. 3.18   Protection circuit

3.4 Filtering

The term *filtering* is used to describe the process of removing a certain band of frequencies from a signal and permitting others to be transmitted. The range of frequencies passed by a filter is known as the *pass band*, the range not passed as the *stop band* and the boundary between stopping and passing as the *cut-off frequency*. Filters are classified according to the frequency ranges they transmit or reject. A *low-pass filter* (Fig. 3.19(a)) has a pass band which allows all frequencies from 0 up to some frequency to be transmitted. A *high-pass filter* (Fig. 3.19(b)) has a pass band which allows all frequencies from some value up to infinity to be transmitted. A *band-pass filter* (Fig. 3.19(c)) allows all the

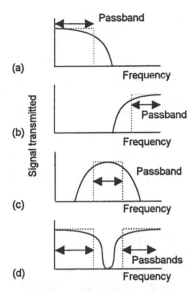

Fig. 3.19 Characteristics of ideal filters: (a) low pass, (b) high pass, (c) band pass, (d) band stop

Fig. 3.20 Low-pass passive filter

3.5 Wheatstone bridge

Fig. 3.21 Low-pass active filter

frequencies within a specified band to be transmitted. A *band-stop filter* (Fig. 3.19(d)) stops all frequencies with a particular band from being transmitted. In all cases the cut-off frequency is defined as being that at which the output voltage is 70.7% of that in the pass band. The term *attenuation* is used for the ratio of input and output powers, this being written as the ratio of the logarithm of the ratio and so gives the attenuation in units of bels. Since this is a rather large unit, decibels (dB) are used and then attenuation in dB = 10 lg (input power/output power). Since the power through an impedance is proportional to the square of the voltage the attenuation in dB = 20 lg (input voltage/output voltage). The output voltage of 70.7% of that in the pass band is thus an attenuation of 3 dB.

The term *passive* is used to describe a filter made up using only resistors, capacitors and inductors, the term *active* being used when the filter also involves an operational amplifier. Passive filters have the disadvantage that the current that is drawn by the item that follows can change the frequency characteristic of the filter. This problem does not occur with an active filter.

Low-pass filters are very commonly used as part of signal conditioning. This is because most of the useful information being transmitted is low frequency. Since noise tends to occur at higher frequencies, a low-pass filter can be used to block it off. Thus a low-pass filter might be selected with a cut-off frequency of 40 Hz, thus blocking off any inference signals from the a.c. mains supply and noise in general. Figure 3.20 shows the basic forms that might be used for a passive low-pass filter and Figure 3.21 a very basic form of an active low-pass filter. For more details of filters the reader is referred to *Filter Handbook* by S. Niewiadomski (Heinemann Newnes 1989).

The *Wheatstone bridge* can be used to convert a resistance change to a voltage change. Figure 3.22 shows the basic form of such a bridge. When the output voltage V_o is zero, then the potential at B must equal that at D. The potential difference across R_1, i.e. V_{AB}, must then equal that across R_3, i.e. V_{AD}. Thus $I_1R_1 = I_2R_2$. It also means that the potential difference across R_2, i.e. V_{BC}, must equal that across R_4, i.e. V_{DC}. Since there is no current through BD then the current through R_2 must be the same as that through R_1 and the current through R_4 the same as that through R_3. Thus $I_1R_2 = I_2R_4$. Dividing these two equations gives

$$\frac{R_1}{R_2} = \frac{R_3}{R_4}$$

The bridge is said to be *balanced*.

Now consider what happens when one of the resistances changes from this balanced condition. The supply voltage V_s is connected

Fig. 3.22   Wheatstone bridge

between points A and C and thus the potential drop across the resistor R_1 is the fraction $R_1/(R_1 + R_2)$ of the supply voltage. Hence

$$V_{AB} = \frac{V_s R_1}{R_1 + R_2}$$

Similarly, the potential difference across R_3 is

$$V_{AD} = \frac{V_s R_3}{R_3 + R_4}$$

Thus the difference in potential between B and D, i.e. the output potential difference V_o, is

$$V_o = V_{AB} - V_{AD} = V_s \left(\frac{R_1}{R_1 + R_2} - \frac{R_3}{R_3 + R_4} \right)$$

This equation gives the balanced condition when $V_o = 0$.

Consider resistance R_1 to be a sensor which has a resistance change. A change in resistance from R_1 to $R_1 + \delta R_1$ gives a change in output from V_o to $V_o + \delta V_o$, where

$$V_o + \delta V_o = V_s \left(\frac{R_1 + \delta R_1}{R_1 + \delta R_1 + R_2} - \frac{R_3}{R_3 + R_4} \right)$$

Hence

$$(V_o + \delta V_o) - V_o = V_s \left(\frac{R_1 + \delta R_1}{R_1 + \delta R_1 + R_2} - \frac{R_1}{R_1 + R_2} \right)$$

If δR_1 is much smaller than R_1 then the above equation approximates to

$$\delta V_o \approx V_s \left(\frac{\delta R_1}{R_1 + R_2} \right)$$

With this approximation, the change in output voltage is thus proportional to the change in the resistance of the sensor. This gives the output voltage when there is no load resistance across the output. If there is such a resistance then the loading effect has to be considered.

To illustrate the above, consider a platinum resistance temperature sensor which has a resistance at 0°C of 100 Ω and forms one arm of a Wheatstone bridge. The bridge is balanced, at this temperature, with each of the other arms also being 100 Ω. If the temperature coefficient of resistance of platinum is 0.0039 /K, what will be the output voltage from the bridge per degree change in temperature if the load across the output can be assumed to be infinite? The supply voltage, with negligible internal resistance, is 6.0 V. The variation of the resistance of the platinum with temperature can be represented by

$$R_t = R_0(1 + at)$$

where R_t is the resistance at $t\,°C$, R_0 the resistance at $0°C$ and a the temperature coefficient of resistance. Thus

$$\text{Change in resistance} = R_t - R_0 = R_0 at$$

$$= 100 \times 0.0039 \times 1 = 0.39\ \Omega/K$$

Since this resistance change is small compared to the $100\ \Omega$, the approximate equation can be used. Hence

$$\delta V_0 \approx V_s\left(\frac{\delta R_1}{R_1 + R_2}\right) = \frac{6.0 \times 0.39}{100 + 100} = 0.012\ \text{V}$$

3.5.1 Temperature compensation

In many measurements involving a resistive sensor the actual sensing element may have to be at the end of long leads. Not only the sensor but the resistance of these leads will be affected by changes in temperature. For example, a platinum resistance temperature sensor consists of a platinum coil at the ends of leads. When the temperature changes, not only will the resistance of the coil change but so also will the resistance of the leads. What is required is just the resistance of the coil and so some means has to be employed to compensate for the resistance of the leads to the coil. One method of doing this is to use three leads to the coil, as shown in Figure 3.23. The coil is connected into the Wheatstone bridge in such a way that lead 1 is in series with the R_3 resistor while lead 3 is in series with the platinum resistance coil R_1. Lead 2 is the connection to the power supply. Any change in lead resistance is likely to affect all three leads equally, since they are of the same material, diameter and length and held close together. The result is that changes in lead resistance occur equally in two arms of the bridge and cancel out if R_1 and R_3 are the same resistance.

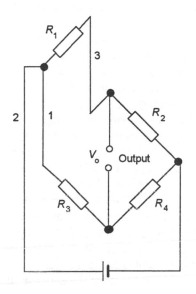

Fig. 3.23   Compensation for leads

The electrical resistance strain gauge is another sensor where compensation has to be made for temperature effects. The strain gauge changes resistance when the strain applied to it changes. Unfortunately, it also changes if the temperature changes. One way of eliminating the temperature effect is to use a *dummy strain gauge*. This is a strain gauge which is identical to the one under strain, the active gauge, and is mounted on the same material but is not subject to the strain. It is positioned close to the active gauge so that it suffers the same temperature changes. Thus a temperature change will cause both gauges to change resistance by the same amount. The active gauge is mounted in one arm of a Wheatstone bridge (Fig. 3.24) and the dummy gauge in another arm so that the effects of temperature-induced resistance changes cancel out.

Fig. 3.24   Compensation with strain gauges

Fig. 3.25 Four active arm strain gauge bridge

Strain gauges are often used with other sensors such as load cells or diaphragm pressure gauges to give measures of the amount of displacement occurring. In such situations, temperature compensation is still required. While dummy gauges could be used, a better solution is to use four strain gauges. Two of them are attached so that when forces are applied they are in tension and the other two in compression. The load cell in Figure 3.25 shows such a mounting. The gauges that are in tension will increase in resistance while those in compression will decrease in resistance. As the gauges are connected as the four arms of a Wheatstone bridge (Fig. 3.25), then since all will be equally affected by any temperature changes the arrangement is temperature compensated. It also gives a much greater output voltage than would occur with just a single active gauge.

To illustrate this, consider a load cell with four strain gauges arranged as shown in Figure 3.25 which is to be used with a four active arm strain gauge bridge. The gauges have a gauge factor of 2.1 and a resistance of 100 Ω. When the load cell is subject to a compressive force, the vertical gauges show compression and, because when you squash an item there is a consequential sideways extension, the horizontal gauges are subject to tensile strain (the ratio of the transverse to longitudinal strains is called Poisson's ratio and is usually about 0.3). Thus, if the compressive gauges suffer a strain of -1.0×10^{-5} and the tensile gauges $+0.3 \times 10^{-5}$, the supply voltage for the bridge is 6 V and the output voltage from the bridge is amplified by a differential operational amplifier circuit, what will be the ratio of the feedback resistance to that of the input resistances in the two inputs of the amplifier if the load is to produce an output of 1 mV?

The change in resistance of a gauge subject to the compressive strain is given by $\Delta R/R = G\varepsilon$:

$$\text{Change in resistance} = G\varepsilon R = -2.1 \times 1.0 \times 10^{-5} \times 100$$

$$= -2.1 \times 10^{-3}\,\Omega$$

For a gauge subject to tension we have

$$\text{Change in resistance} = G\varepsilon R = 2.1 \times 0.3 \times 10^{-5} \times 100$$

$$= 6.3 \times 10^{-4}\,\Omega$$

The out-of-balance potential difference is given by (see Section 3.5)

$$V_o = V_s\left(\frac{R_1}{R_1+R_2} - \frac{R_3}{R_3+R_4}\right)$$

$$= V_s\left(\frac{R_1(R_3+R_4) - R_3(R_1+R_2)}{(R_1+R_2)(R_3+R_4)}\right)$$

$$= V_s\left(\frac{R_1 R_4 - R_2 R_3}{(R_1 + R_2)(R_3 + R_4)}\right)$$

We now have each of the resistors changing. We can, however, neglect the changes in relation to the denominators where the effect of the changes on the sum of the two resistances is insignificant. Thus

$$V_o = V_s\left(\frac{(R_1 + \delta R_1)(R_4 + \delta R_4) - (R_2 + \delta R_2)(R_3 + \delta R_3)}{(R_1 + R_2)(R_3 + R_4)}\right)$$

Neglecting products of δ terms and since we have an initially balanced bridge with $R_1 R_4 = R_2 R_3$, then

$$V_o = \frac{V_s R_1 R_4}{(R_1 + R_2)(R_3 + R_4)} = \left(\frac{\delta R_1}{R_1} - \frac{\delta R_2}{R_2} - \frac{\delta R_3}{R_3} + \frac{\delta R_4}{R_4}\right)$$

Hence

$$V_o = \frac{6 \times 100 \times 100}{200 \times 200}\left(\frac{2 \times 6.3 \times 10^{-4} + 2 \times 2.1 \times 10^{-3}}{100}\right)$$

The output is thus 3.6×10^{-5} V. This becomes the input to the differential amplifier; hence, using the equation developed in Section 3.2.7,

$$V_o = \frac{R_2}{R_1}(V_2 - V_1)$$

$$1.0 \times 10^{-3} = \frac{R_2}{R_1} \times 3.6 \times 10^{-5}$$

Thus $R_2/R_1 = 27.8$.

3.5.2 Thermocouple compensation

A thermocouple gives an e.m.f. which depends on the temperature of its two junctions (see Section 2.9.5). Ideally one junction is kept at 0°C, then the temperature relating to the e.m.f. can be directly read from tables. However, this is not always feasible and the cold junction is often allowed to be at the ambient temperature. To compensate for this a potential difference has to be added to the thermocouple. This must be the same as the e.m.f. that would be generated by the thermocouple with one junction at 0°C and the other at the ambient temperature. Such a potential difference can be produced by using a resistance temperature sensor in a Wheatstone bridge. The bridge is balanced at 0°C and the output voltage from the bridge provides the correction potential difference at other temperatures.

The resistance of a metal resistance temperature sensor can be described by the relationship

$$R_t = R_0(1 + at)$$

where R_t is the resistance at $t\,°C$, R_0 the resistance at $0°C$ and a the temperature coefficient of resistance. Thus

$$\text{change in resistance} = R_t - R_0 = R_0 at$$

The output voltage for the bridge, taking R_1 to be the resistance temperature sensor, is given by

$$\delta V_0 \approx V_s\left(\frac{\delta R_1}{R_1 + R_2}\right) = \frac{V_s R_0 at}{R_0 + R_2}$$

The thermocouple e.m.f. e is likely to vary with temperature t in a reasonably linear manner over the small temperature range being considered – from $0°C$ to the ambient temperature. Thus $e = at$, where a is a constant, i.e. the e.m.f. produced per degree change in temperature. Hence for compensation we must have

$$at = \frac{V_s R_0 at}{R_0 + R_2}$$

and so

$$aR_2 = R_0(V_s a - a)$$

For an iron–constantan thermocouple giving $51\ \mu V/°C$, compensation can be provided by a nickel resistance element with a resistance of $10\ \Omega$ at $0°C$ and a temperature coefficient of resistance of $0.0067\ /K$, a supply voltage for the bridge of $1.0\ V$, and R_2 as $1304\ \Omega$.

3.6 Digital signals

The output from most sensors tends to be in analogue form. Where a microprocessor is used as part of the measurement or control system, the analogue output from the sensor has to be converted into a digital form before it can be used as an input to the microprocessor. Likewise, most actuators operate with analogue inputs and so the digital output from a microprocessor has to be converted into an analogue form before it can be used as an input by the actuator.

The *binary system* is based on just the two symbols or states 0 and 1. These are termed *binary digits* or *bits*. When a number is represented by this system, the digit position in the number indicates the weight attached to each digit, the weight increasing by a factor of 2 as we proceed from right to left:

...	2^3	2^2	2^1	2^0
	bit 3	bit 2	bit 1	bit 0

For example, the decimal number 15 is $2^0 + 2^1 + 2^2 + 2^3 = 1111$ in the binary system. In a binary number the bit 0 is termed the *least significant bit* (LSB) and the highest bit the *most significant bit* (MSB). The combination of bits to represent a number is termed a *word*. Thus 1111 is a four-bit word. The term byte is used for a group of 8 bits. See Chapter 14 for more discussion of binary numbers.

3.6.1 Analogue-to-digital conversion

Analogue-to-digital conversion involves converting analogue signals into binary words. Figure 3.26 shows the basic elements of analogue-to-digital conversion.

Fig. 3.26   Analogue-to-digital conversion

The procedure used is that a clock supplies regular time signal pulses to the analogue-to-digital converter (ADC) and every time it receives a pulse it samples the analogue signal. Figure 3.27 illustrates this analogue-to-digital conversion by showing the types of signals involved at the various stages. Figure 3.27(a) shows the analogue signal and Figure 3.27(b) the clock signal which supplies the time signals at which the sampling occurs. The result of the sampling is a series of narrow pulses (Fig. 3.27(c)). A *sample and hold* unit is then used to hold each sampled value until the next pulse occurs, with the result shown in Figure 3.27(d). The sample and hold unit is necessary because the analogue-to-digital converter requires a finite amount of time, termed the *conversion time*, to convert the analogue signal into a digital one.

The relationship between the sampled and held input and the output for an analogue-to-digital converter is illustrated by the graph shown in Figure 3.28 for a digital output which is restricted to three bits. With three bits there are $2^3 = 8$ possible output levels. Thus, since the output of the ADC to represent the analogue input can be only one of these eight possible levels, there is a range of inputs for which the output does not change. The eight possible output levels are termed *quantisation levels* and the difference in analogue voltage between two adjacent levels is termed the *quantisation interval*. Thus for the ADC given in Figure 3.28, the quantisation interval is 1 V. Because of the step-like nature of the relationship, the digital output is not always proportional to the analogue input and thus there will be error, this being termed the *quantisation error*. When the input is centred over the interval the quantisation error is zero, the maximum error being equal to one-half of the interval or $\pm\frac{1}{2}$ bit.

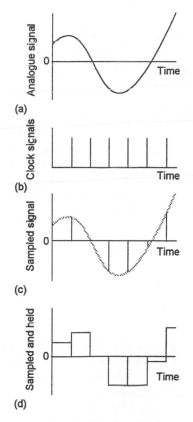

Fig. 3.27   Signals: (a) analogue, (b) clock, (c) sampled, (d) sampled and held

The word length possible determines the *resolution* of the element, i.e. the smallest change in input which will result in a change in the digital output. The smallest change in digital output is one bit in the least significant bit position in the word, i.e. the far right bit. Thus with a word length of n bits the full-scale analogue input V_{FS} is divided into 2^n pieces and so the minimum change in input that can be detected, i.e. the *resolution*, is $V_{FS}/2^n$.

Fig. 3.28   Input–output for an ADC

Thus if we have an analogue-to-digital converter having a word length of 10 bits and the analogue signal input range is 10 V, then the number of levels with a 10-bit word is $2^{10} = 1024$ and thus the resolution is $10/1024 = 9.8$ mV.

Consider a thermocouple giving an output of 0.5 mV/°C. What will be the word length required when its output passes through an analogue-to-digital converter if temperatures from 0 to 200°C are to be measured with a resolution of 0.5°C? The full-scale output from the sensor is $200 \times 0.5 = 100$ mV. With a word length n, this voltage will be divided into $100/2^n$ mV steps. For a resolution of 0.5°C we must be able to detect a signal from the sensor of $0.5 \times 0.5 = 0.25$ mV. Thus we require

$$0.25 = \frac{100}{2^n}$$

Hence $n = 8.6$. Thus a 9-bit word length is required.

3.6.2 Sampling theorem

Analogue-to-digital converters sample analogue signals at regular intervals and convert these values to binary words. How often should an analogue signal be sampled in order to give an output which is representative of the analogue signal?

Figure 3.29 illustrates the problem with different sampling rates being used for the same analogue signal. When the signal is reconstructed from the samples, it is only when the sampling rate is at least twice that of the highest frequency in the analogue signal that the sample gives the original form of signal. This

Fig. 3.29   Effect of sampling frequency: (a) analogue signal, (b) sampled signal, (c) sampled signal

criterion is known as the *Nyquist criterion* or *Shannon's sampling theorem*. When the sampling rate is less than twice the highest frequency the reconstruction can represent some other analogue signal and we obtain a false image of the real signal. This is termed *aliasing*. In Figure 3.29(c) this could be an analogue signal with a much smaller frequency than that of the analogue signal that was sampled.

Whenever a signal is sampled too slowly, there can be a false interpretation of high frequency components as arising from lower frequency aliases. High frequency noise can also create errors in the conversion process. To minimise errors due to both aliasing and high frequency noise, a low-pass filter is used to precede the ADC, the filter having a bandwidth such that it passes only low frequencies for which the sampling rate will not give aliasing errors. Such a filter is termed an *anti-aliasing filter*.

3.6.3 Digital-to-analogue conversion

The input to a digital-to-analogue converter (DAC) is a binary word; the output is an analogue signal that represents the weighted sum of the non-zero bits represented by the word. Thus, for example, an input of 0010 must give an analogue output which is twice that given by an input of 0001. Figure 3.30 illustrates this for an input to a DAC with a resolution of 1 V for unsigned binary words. Each additional bit increases the output voltage by 1 V.

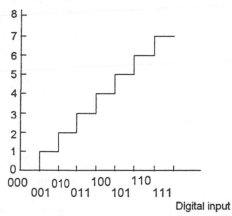

Fig. 3.30   Input–output for a DAC

Consider the situation where a microprocessor gives an output of an 8-bit word. This is fed through an 8-bit digital-to-analogue converter to a control valve. The control valve requires 6.0 V to be fully open. If the fully open state is indicated by 11111111 what will be the output to the valve for a change of 1 bit?

The full-scale output voltage of 6.0 V will be divided into 2^8 intervals. A change of 1 bit is thus a change in the output voltage of $6.0/2^8 = 0.023$ V.

3.6.4 Digital-to-analogue converters

A simple form of digital-to-analogue converter uses a summing amplifier (see Section 3.2.3) to form the weighted sum of all the non-zero bits in the input word (Fig. 3.31). The reference voltage is connected to the resistors by means of electronic switches which respond to binary 1. The values of the input resistances depend on which bit in the word a switch is responding to, the value of the resistor for successive bits from the LSB being halved. Hence the sum of the voltages is a weighted sum of the digits in the word. Such a system is referred to as a *weighted-resistor network*.

A problem with the weighted-resistor network is that accurate resistances have to be used for each of the resistors and it is difficult to obtain such resistors over the wide range needed. As a result this form of DAC tends be limited to 4-bit conversions.

Another, more commonly used, version uses a *R-2R ladder network* (Fig. 3.32). This overcomes the problem of obtaining accurate resistances over a wide range of values, only two values being required. The output voltage is generated by switching sections of the ladder to either the reference voltage or 0 V according to whether there is a 1 or 0 in the digital input.

Fig. 3.31   Weighted-resistor DAC

Fig. 3.32   *R-2R* ladder DAC

Figure 3.33 shows details of the GEC Plessey ZN558D 8-bit latched input digital-to-analogue converter using a R-$2R$ ladder network. After the conversion is complete, the 8-bit result is placed in an internal latch until the next conversion is complete. Data is held in the latch when ENABLE is high, the latch being said to be transparent when ENABLE is low. A *latch* is just a device to retain the output until a new one replaces it. When a DAC has a latch it may be interfaced directly to the data bus of a microprocessor and treated by it as just an address to send data. A DAC without a latch would be connected via a peripheral interface adapter (PIA), such a device providing latching (see Section 18.4). Figure 3.34 shows how the ZN558D might be used with a microprocessor when the output is required to be a voltage which varies between zero and the reference voltage, this being termed *unipolar operation*. With $V_{\text{ref in}} = 2.5$ V, the output range is +5 V when $R_1 = 8$ kΩ and $R_2 = 8$ kΩ and the range is +10V when $R_1 = 16$ kΩ and $R_2 = 5.33$ kΩ.

Fig. 3.33   ZN558D DAC

Fig. 3.34   Unipolar operation

The specifications of DACs include such terms as:

1   The *full-scale output*, i.e. the output when the input word is all 1s. For the ZN558D this is typically 2.550 V.

2   The *resolution*, 8-bit DACS generally being suitable for most microprocessor control systems. The ZN558D is 8-bit.

3   The *settling time*, this being the time taken by the DAC to reach within ½ LSB of its new voltage after a binary change. This is 800 ns for the ZN558D.

4   The *linearity*, this being the maximum deviation from the straight line through zero and the full range of the output. This is a maximum of ±0.5 LSB for the ZN558D.

3.6.5 Analogue-to-digital converters

The input to an analogue-to-digital converter is an analogue signal and the output is a binary word that represents the level of the input signal. There are a number of forms of analogue-to-digital converter, the most common being successive approximations, ramp, dual ramp and flash.

Successive approximations is probably the most commonly used method. Figure 3.35 illustrates the subsystems involved. A voltage is generated by a clock emitting a regular sequence of pulses which are counted, in a binary manner, and the resulting binary word converted into an analogue voltage by a digital-to-analogue converter. This voltage rises in steps and is compared with the analogue input voltage from the sensor. When the clock-generated voltage passes the input analogue voltage the pulses from the clock are stopped from being counted by a gate being closed. The output from the counter at that time is then a digital representation of the analogue voltage. While the comparison could be accomplished by starting the count at 1, the least significant bit, and then proceeding bit by bit upwards, a faster method is by successive approximations. This involves selecting the most significant bit that is less than the analogue value, then adding successive lesser bits for which the total does not exceed the analogue value. For example, we might start the comparison with 1000. If this is too large we try 0100. If this is too small we then try 0110. If this is too large we try 0101. Because each of the bits in the word is tried in sequence, with an n-bit word it only takes n steps to make the comparison. Thus if the clock has a frequency f, the time between pulses is $1/f$. Hence the time taken to generate the word, i.e. the conversion time, is n/f.

Fig. 3.35 Successive
approximations ADC

Figure 3.36 shows the typical form of an 8-bit analogue-to-digital converter (GEC Plessey ZN439) designed for use with microprocessors and using the successive approximations method. Figure 3.37 shows how it can be connected so that it is controlled by a microprocessor and sends its digital output to the microprocessor. All the active circuitry, including the clock, is contained on a single chip. The ADC is first selected by taking the chip select pin low. When the start conversion pin receives a negative-going pulse the conversion starts. At the end of the conversion the status pin goes low. The digital output is sent to an internal buffer where it is held until read as a result of the output enable pin being taken low.

Fig. 3.36   ZN439 ADC

Fig. 3.37 ZN439 connected to a microprocessor

The *ramp* form of analogue-to-digital converter involves an analogue voltage which is increased at a constant rate, a so-called ramp voltage, and applied to a comparator where it is compared with the analogue voltage from the sensor. The time taken for the ramp voltage to increase to the value of the sensor voltage will depend on the size of the sampled analogue voltage. When the ramp voltage starts, a gate is opened which starts a binary counter counting the regular pulses from a clock. When the two voltages are equal, the gate closes and the word indicated by the counter is the digital representation of the sampled analogue voltage. Figure 3.38 indicates the subsystems involved in the ramp form of analogue-to-digital converter.

Fig. 3.38 Ramp ADC

The *dual ramp converter* is more common than the single ramp. Figure 3.39 shows the basic circuit. The analogue voltage is applied to an integrator which drives a comparator. The output from the comparator goes high as soon as the integrator output is more than a few millivolts. When the comparator output is high, an AND gate passes pulses to a binary counter. The counter counts pulses until it overflows. The counter then resets to zero, sends a signal to a switch which disconnects the unknown voltage and connects a reference voltage, and starts counting again. The polarity of the reference voltage is opposite to that of the input voltage. The integrator voltage then decreases at a rate proportional to the reference voltage. When the integrator output reaches zero, the comparator goes low, bringing the AND gate low and so switching the clock off. The count is then a measure of the analogue input voltage. Dual ramp analogue-to-digital converters have excellent noise rejection because the integral action averages out random negative and positive contributions over the sampling period. They are, however, very slow.

The *flash analogue-to-digital converter* is very fast. For an n-bit converter, $2^n - 1$ separate voltage comparators are used in parallel, with each having the analogue input voltage as one input (Fig. 3.40). A reference voltage is applied to a ladder of resistors so that the voltage applied as the other input to each comparator is one bit larger in size than the voltage applied to the previous comparator in the ladder. Thus when the analogue voltage is applied to the ADC, all those comparators for which the analogue voltage is greater than the reference voltage of a comparator will give a high output and those for which it is less will be low. The resulting outputs are fed in parallel to a logic gate system which translates them into a digital word.

Fig. 3.39   Dual ramp ADC

Fig. 3.40 Flash ADC

In considering the specifications of ADCs the following terms will be encountered:

1   *Conversion time*, i.e. the time required to complete a conversion of the input signal. It establishes the upper signal frequency that can be sampled without aliasing; the maximum frequency is $1/(2 \times \text{conversion time})$.

2   *Resolution*, this being the full-scale signal divided by 2^n, where n is the number of bits. It is often just specified by a statement of the number of bits.

3   *Linearity error*, this being the deviation from a straight line drawn through zero and full-scale. It is a maximum of $\pm\frac{1}{2}$ LSB.

Table 3.1 shows some specification details of commonly used analogue-to-digital converters.

Table 3.1 Analogue-to-digital converters

ADC	Type	Resolution (bits)	Conversion time (ns)	Linearity error (LSB)
ZN439	SA	8	5 000	±1/2
ZN448E	SA	8	9 000	±1/2
ADS7806	SA	12	20 000	±1/2
ADS7078C	SA	16	20 000	±1/2
ADC302	F	8	20	±1/2

SA = successive approximations, F = flash.

Fig. 3.41   Sample and hold

3.6.6 Sample and hold amplifiers

It takes a finite time for an ADC to convert an analogue signal to digital and problems can arise if the analogue signal changes during the conversion time. To overcome this, a sample-and-hold system is used to sample the analogue signal and hold it while the conversion takes place.

The basic circuit (Fig. 3.41) consists of an electronic switch to take the sample, with a capacitor for the hold and an operational amplifier voltage follower. The electronic switch is controlled so that the sample is taken at the instant dictated by the control input. When the switch closes, the input voltage is applied across the capacitor and the output voltage becomes the same as the input voltage. If the input voltage changes while the switch is closed the voltage across the capacitor and the output voltage change accordingly. When the switch opens, the capacitor retains its charge and the output voltage remains equal to the input voltage at the instant the switch was opened. The voltage is thus held until such time as the switch closes again. The time required for the capacitor to charge to a new sample of the input analogue voltage is called the *acquisition time* and depends on the value of the capacitance and the circuit resistance when the switch is on. Typical values are of the order of 4 µs.

3.7 Multiplexers

A *multiplexer* is a circuit that is able to have inputs of data from a number of sources and then, by selecting an input channel, give an output from just one of them. In applications where there is a need for measurements to be made at a number of different locations, rather than use a separate ADC and microprocessor for each measurement, a multiplexer can be used to select each input in turn and switch it through a single ADC and microprocessor (Fig. 3.42). The multiplexer is essentially an electronic switching device which enables each of the inputs to be sampled in turn.

As an illustration of the types of analogue multiplexers available, the DG508ACJ has eight input channels with each channel having a 3-bit binary address for selection purposes. The transition time between taking samples is 0.6 µs.

Fig. 3.42   Multiplexer

Fig. 3.43   Two channel multiplexer

3.7.1 Digital multiplexer

Figure 3.43 shows the basic principle of a multiplexer which can be used to select digital data inputs; for simplicity only a two input channel system is shown. The logic level applied to the select input determines which AND gate is enabled so that its data input passes through the OR gate to the output. A number of forms of multiplexers are available in integrated packages. The 151 types enable one line from eight to be selected, the 153 type one line from four inputs which are supplied as data on two lines each, the 157 types one line from two inputs which are supplied as data on four lines.

3.7.2 Time division multiplexing

Often there is a need for a number of peripheral devices to share the same input/output lines from a microprocessor. So that each peripheral can be supplied with different data it is necessary to allocate each a particular time slot during which data is transmitted. This is termed *time division multiplexing*. Figure 3.44 illustrates how this can be used to drive two display devices. In Figure 3.44(a) the system is not time multiplexed, in (b) it is.

(a)

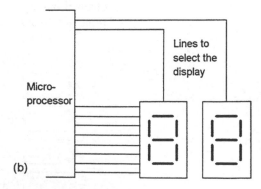

(b)

Fig. 3.44   Time division multiplexing

3.8 Data acquisition

The term *data acquisition*, or DAQ, is used for the process of taking data from sensors and inputting that data into a computer for processing. The sensors are connected, generally via some signal conditioning, to a data acquisition board which is plugged into the back of a computer (Fig. 3.45(a)). The DAQ board is a printed circuit board that, for analogue inputs, basically provides a multiplexer, amplification, analogue-to-digital conversion, registers and control circuitry so that sampled digital signals are applied to the computer system. Figure 3.45(b) shows the basic elements of such a board.

(a)

(b)

Fig. 3.45   DAQ system

Computer software is used to control the acquisition of data via the DAQ board. When the program requires an input from a particular sensor, it activates the board by sending a control word to the control and status register. Such a word indicates the type of operation the board has to carry out. As a consequence the board switches the multiplexer to the appropriate input channel. The input from the sensor connected to that input channel is then passed via an amplifier to the analogue-to-digital converter. After conversion the resulting digital signal is passed to the data register and the word in the control and status register changes to indicate that the signal has arrived. Following that signal, the computer then issues a signal for the data to be read and taken into the computer for processing. This signal is necessary to ensure that the computer does not wait doing nothing while the board carries out its acquisition of data, but uses this to signal to the computer when the acquisition is complete and then the computer can interrupt any program it is implementing, read the data from the DAQ and then continue with its program. A faster system does not involve the computer in the transfer of the data into memory but transfers the acquired data directly from the board to memory without involving the computer, this being termed *direct memory address* (DMA).

The specifications for a DAQ board include the sampling rate for analogue inputs, this might be 100 kS/s (100 thousand samples per second). The Nyquist criteria for sampling indicate that the maximum frequency of analogue signal that can be sampled with such a board is 50 kHz, the sample rate having to be twice the maximum frequency component. In addition to the above basic functions of a DAQ board, it may also supply analogue outputs, timers and counters which can be used to provide triggers for the sensor system.

As an example of a low cost multifunction board for use with an IBM computer, Figure 3.46 shows the basic structure of the National Instruments DAQ board PC-LPM-16. This board has 16 analogue input channels, a sampling rate of 50 kS/s, an 8-bit digital input and an 8-bit digital output and a counter/timer which can give an output. Channels can be scanned in sequence, taking one reading from each channel in turn, or there can be continuous scanning of a single channel.

Fig. 3.46 PC-LPM-16 DAQ board

3.9 Digital signal processing

The term *digital signal processing* or *discrete-time signal processing* is used for the processing applied to a signal by a microprocessor. Digital signals are discrete time signals in that they are not continuous functions of time but exist at only discrete times. Whereas signal conditioning of analogue signals requires components such as amplifiers and filter circuits, digital signal conditioning can be carried out by a program applied to a microprocessor, i.e. processing the signal. To change the characteristics of a filter used with analogue signals it is necessary to change hardware components, whereas to change the characteristics of a digital filter all that is necessary is to change the software, i.e. the program of instructions given to a micro-processor.

With a digital signal processing system there is an input of a word representing the size of a pulse and an output of another word. The output pulse at a particular instant is computed by the system as a result of processing the present input pulse input, together with previous pulse inputs and possibly previous system outputs.

For example, the program used by the microprocessor might read the value of the present input and add to it the previous output value to give the new output. If we consider the present input to be the kth pulse in the input sequence of pulses we can represent this pulse as $x[k]$. The kth output of a sequence of pulses can be represented by $y[k]$. The previous output, i.e. the $(k-1)$th pulse, can be represented by $y[k-1]$. Thus we can describe the program which gives an output obtained by adding to the value of the present input the previous output by:

$$y[k] = x[k] + y[k-1]$$

Such an equation is called a *difference equation*. It gives the relationship between the output and input for a discrete-time system and is comparable with a differential equation which is used to describe the relationship between the output and input for a system having inputs and outputs which vary continuously with time.

For the above difference equation, suppose we have an input of a sampled sine wave signal which gives a sequence of pulses of:

0.5, 1.0, 0.5, −0.5, −1.0, −0.5, 0.5, 1.0, etc.

The $k = 1$ input pulse has a size of 0.5. If we assume that previously the output was zero then $y[k-1] = 0$ and so $y[1] = 0.5 + 0 = 0.5$. The $k = 2$ input pulse has a size of 1.0 and so $y[2] = x[2] + y[2-1] = 1.0 + 0.5 = 1.5$. The $k = 3$ input pulse has a size of 0.5 and so $y[3] = x[3] + y[3-1] = 0.5 + 1.5 = 2.0$. The $k = 4$ input pulse has a size of −0.5 and so $y[4] = x[4] + y[4-1] = -0.5 + 2.0 = 1.5$. The $k = 5$ input pulse has a size of −1.0 and so $y[5] = x[5] + y[5-1] = -1.0 + 1.5 = 0.5$. The output is thus the pulses:

0.5, 1.5, 2.0, 1.5, 0.5, ...

We can continue in this way to obtain the output for all the pulses.

As another example of a difference equation we might have:

$$y[k] = x[k] + ay[k-1] - by[k-2]$$

The output is the value of the current input plus a times the previous output and minus b times the last but one output. If we have $a = 1$ and $b = 0.5$ and consider the input to be the sampled sine wave signal considered above, then the output now becomes:

0.5, 1.5, 1.75, 0.5, −1.37, ...

We can have a difference equation which produces an output which is similar to that which would have been obtained by integrating a continuous-time signal. Integration of a continuous-time signal between two times can be considered to be the area under the continuous-time function between those times. Thus if we consider two discrete-time signals $x[k]$ and $x[k-1]$ occurring with a time interval of T between them (Fig. 3.47), the change in area is $\frac{1}{2}T(x[k] + x[k-1])$. Thus if the output is to be the sum of the previous area and this change in area, the difference equation is:

$$y[k] = y[k-1] + \frac{1}{2}T(x[k] + x[k-1])$$

This is known as *Tustin's approximation* for integration.

Differentiation can be approximated by determining the rate at which the input changes. Thus when the input changes from $x[k-1]$ to $x[k]$ in time T the output is:

$$y[k] = (x[k] - x[k-1])/T$$

Fig. 3.47   Integration

3.10 Pulse-modulation

A problem that is often encountered with dealing with the transmission of low-level d.c. signals from sensors is that the gain of an operational amplifier used to amplify them may drift and so the output drifts. This problem can be overcome if the signal is a sequence of pulses rather than a continuous-time signal.

One way this conversion can be achieved is by chopping the d.c. signal in the way suggested in Figure 3.48. The output from the chopper is a chain of pulses, the heights of which are related to the d.c. level of the input signal. This process is called *pulse amplitude modulation*. After amplification and any other signal conditioning, the modulated signal can be demodulated to give a d.c. output. With pulse amplitude modulation, the height of the pulses is related to the size of the d.c. voltage.

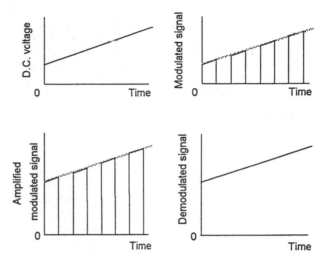

Fig. 3.48   Pulse amplitude modulation

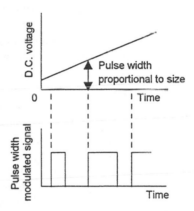

Fig. 3.49   Pulse width modulation

An alternative to this is *pulse width modulation* (PWM) where the width, i.e. duration, of a pulse rather than its amplitude depends on the size of the voltage (Fig. 3.49).

Pulse width modulation is widely used with control systems as a means of controlling the average value of a d.c. voltage. Thus if there is a constant analogue voltage and it is chopped into pulses, by varying the width of the pulses so the average value of the voltage can be changed. Figure 3.50 illustrates this. The term *duty cycle* is used for the fraction of each cycle for which the voltage is high. Thus for a PWM signal where the signal is high for half of each cycle, the duty cycle is ½ or 50%. If it is only on for a quarter of each cycle then the duty cycle is ¼ or 25%.

Fig. 3.50   PWM for voltage control: (a) duty cycle 50%, (b) duty cycle 25%

Problems

1 Design an operational amplifier circuit that can be used to produce an output that ranges from 0 to –5 V when the input goes from 0 to 100 mV.

2 An inverting amplifier has an input resistance of 2 kΩ. Determine the feedback resistance needed to give a voltage gain of 100.

3 Design a summing amplifier circuit that can be used to produce an output that ranges from –1 to –5 V when the input goes from 0 to 100 mV.

4 A differential amplifier is used with a thermocouple sensor in the way shown in Figure 3.8. What values of R_1 and R_2 would give a circuit which has an output of 10 mV for a temperature difference between the thermocouple junctions of 100°C with a copper–constantan thermocouple if the thermocouple is assumed to have a constant sensitivity of 43 μV/°C?

5 The output from the differential pressure sensor used with an orifice plate for the measurement of flow rate is non-linear, the output voltage being proportional to the square of the flow rate. Determine the form of characteristic required for the element in the feedback loop of an operational amplifier signal conditioner circuit in order to linearise this output.

6 A differential amplifier is to have a voltage gain of 100. What will be the feedback resistance required if the input resistances are both 1 kΩ?

7 A differential amplifier has a differential voltage gain of 2000 and a common mode gain of 0.2. What is the common mode rejection ration in dB?

8 Digital signals from a sensor are polluted by noise and mains interference and are typically of the order of 100 V or more. Explain how protection can be afforded for a microprocessor to which these signals are to be inputted.

9 A platinum resistance temperature sensor has a resistance of 120 Ω at 0°C and forms one arm of a Wheatstone bridge. At this temperature the bridge is balanced with each of the other arms being 120 Ω. The temperature coefficient of resistance of the platinum is 0.0039 /K. What will be the output voltage from the bridge for a change in temperature of 20°C? The

loading across the output is effectively open-circuit and the supply voltage to the bridge is from a source of 6.0 V with negligible internal resistance.

10 A diaphragm pressure gauge employs four strain gauges to monitor the displacement of the diaphragm. The four active gauges form the arms of a Wheatstone bridge, in the way shown in Figure 3.24. The gauges have a gauge factor of 2.1 and resistance 120 Ω. A differential pressure applied to the diaphragm results in two of the gauges on one side of the diaphragm being subject to a tensile strain of 1.0×10^{-5} and the two on the other side a compressive strain of 1.0×10^{-5}. The supply voltage for the bridge is 10 V. What will be the voltage output from the bridge?

11 A Wheatstone bridge has a single strain gauge in one arm and the other arms are resistors with each having the same resistance as the unstrained gauge. Show that the output voltage from the bridge is given by $\frac{1}{4}V_sG\varepsilon$, where V_s is the supply voltage to the bridge, G the gauge factor of the strain gauge and ε the strain acting on it.

12 What is the resolution of an analogue-to-digital converter with a word length of 12 bits and an analogue signal input range of 100 V?

13 A sensor gives a maximum analogue output of 5 V. What word length is required for an analogue-to-digital converter if there is to be a resolution of 10 mV?

14 A R-$2R$ DAC ladder of resistors has its output fed through an inverting operational amplifier with a feedback resistance of $2R$. If the reference voltage is 5 V, determine the resolution of the converter.

15 For a binary weighted-resistor DAC how should the values of the input resistances be weighted for a 4-bit DAC?

16 What is the conversion time for a 12-bit ADC with a clock frequency of 1 MHz?

17 In monitoring the inputs from a number of thermocouples the following sequence of modules is used for each thermocouple in its interface with a microprocessor.

Protection, cold junction compensation, amplification, linearisation, sample and hold, analogue-to-digital converter, buffer, multiplexer.

Explain the function of each of the modules.

18 Suggest the modules that might be needed to interface the output of a microprocessor with an actuator.

4 Data presentation systems

4.1 Displays

This chapter is about how data can be displayed, e.g. as digits on a LED display or as a display on a computer screen, and stored, e.g. on a computer hard disc.

Measurement systems consist of three elements: sensor, signal conditioner and display or data presentation element (see Section 1.3). There are a very wide range of elements that can be used for the presentation of data. Traditionally they have been classified into two groups: indicators and recorders. *Indicators* give an instant visual indication of the sensed variable while *recorders* record the output signal over a period of time and give automatically a permanent record. The recorder will be the most appropriate choice if the event is high speed or transient and cannot be followed by an observer, or there are large amounts of data, or it is essential to have a record of the data.

Both indicators and recorders can be subdivided into two groups of devices, *analogue* and *digital*. An example of an analogue indicator is a meter which has a pointer moving across a scale, while a digital meter would be just a display of a series of numbers. An example of an analogue recorder is a chart recorder which has a pen moving across a moving sheet of paper, while a digital recorder has the output printed out on a sheet of paper as a sequence of numbers.

This chapter can also be considered as the completion of the group of chapters concerned with measurement systems, i.e. sensors, signal conditioning and now display, and so the chapter is used to bring the items together in a consideration of examples of complete measurement systems.

4.1.1 Loading

A general point that has to be taken account of when putting together any measurement system is *loading*, i.e. the effect of connecting a load across the output terminals of any element of a measurement system.

Connecting an ammeter into a circuit to make a measurement of the current changes the resistance of the circuit and so changes the current. The act of attempting to make such a measurement has modified the current that was being measured. When a voltmeter is connected across a resistor then we effectively have put two resistances in parallel, and if the resistance of the voltmeter is not considerably higher than that of the resistor the current through the resistor is markedly changed and so the voltage being measured is changed. The act of attempting to make the measurement has modified the voltage that was being measured. Such acts are termed *loading*.

Loading can also occur within a measurement system when the connection of one element to another modifies the character-istics of the preceding element. Consider, for example, a measure-ment system consisting of a sensor, an amplifier and a display element (Fig. 4.1). The sensor has an open-circuit output voltage of V_s and a resistance R_s. The amplifier has an input resistance of R_{in}. This is thus the load across the sensor. Hence the input voltage from the sensor is divided so that the potential difference across this load, and so the input voltage V_{in} to the amplifier, is

$$V_{in} = \frac{V_s R_{in}}{R_s + R_{in}}$$

If the amplifier has a voltage gain of G then the open-circuit voltage output from it will be GV_{in}. If the amplifier has an output resistance of R_{out} then the output voltage from the amplifier is divided so that the potential difference V_d across the display element, resistance R_d, is:

$$V_d = \frac{GV_{in}R_d}{R_{out} + R_d} = \frac{GV_s R_{in}R_d}{(R_{out} + R_d)(R_s + R_{in})}$$

$$= \frac{GV_s}{\left(\dfrac{R_{out}}{R_d} + 1\right)\left(\dfrac{R_s}{R_{in}} + 1\right)}$$

Thus if loading effects are to be negligible we require $R_{out} \gg R_d$ and $R_s \gg R_{in}$.

Fig. 4.1 Measurement system loading

Sensor   Amplifier   Display

4.2 Data presentation elements

This section is a brief overview of commonly used examples of data presentation elements.

4.2.1 Analogue and digital meters

The *moving coil meter* is an analogue indicator with a pointer moving across a scale. The basic instrument movement is a d.c. microammeter with shunts, multipliers and rectifiers being used to convert it to other ranges of direct current and measurement of alternating current, direct voltage, and alternating voltage. With alternating current and voltages, the instrument is restricted to between about 50 Hz and 10 kHz. The accuracy of such a meter depends on a number of factors, among which are temperature, the presence nearby of magnetic fields or ferrous materials, the way the meter is mounted, bearing friction, inaccuracies in scale marking during manufacture, etc. In addition there are errors involved in reading the meter, e.g. parallax errors when the position of the pointer against the scale is read from an angle other than directly at right angles to the scale and errors arising from estimating the position of the pointer between scale markings. The overall accuracy is generally of the order of ±0.1 to ±5%. The time taken for a moving coil meter to reach a steady deflection is typically in the region of a few seconds. The low resistance of the meter can present loading problems.

The *digital voltmeter* gives its reading in the form of a sequence of digits. Such a form of display eliminates parallax and interpolation errors and can give accuracies as high as ±0.005%. The digital voltmeter is essentially just a sample and hold unit feeding an analogue-to-digital converter with its output counted by a counter (Fig. 4.2). It has a high resistance, of the order of 10 MΩ, and so loading effects are less likely than with the moving coil meter with its lower resistance. Thus, if a digital voltmeter specification includes the statement 'sample rate approximately 5 readings per second' then this means that every 0.2 s the input voltage is sampled. It is the time taken for the instrument to process the signal and give a reading. Thus, if the input voltage is changing at a rate which results in significant changes during 0.2 s then the voltmeter reading can be in error. A low cost digital voltmeter has typically a sample rate of 3 per second and an input impedance of 100 MΩ.

Fig. 4.2   Principle of digital voltmeter

For details of the 'mechanics' of meters the reader is referred to such texts as *Electrical and Electronic Measurement and Testing* by W. Bolton (Longman 1992) or *Electronic Instruments and Measurement Techniques* by F.F. Mazda (Cambridge University Press 1987).

4.2.2 Analogue chart recorders

There are three basic types of analogue chart recorders: the direct reading recorder, the galvanometric recorder and the potentiometric or closed-loop recorder. The data can be recorded on paper by fibre-tipped ink pens, by the impact of a pointer pressing a carbon ribbon against the paper, by the use of thermally sensitive paper which changes colour when a heated pointer moves across it, by a beam of ultraviolet light falling on paper sensitive to it and by a tungsten wire stylus moving across the surface of specially coated paper, a thin layer of aluminium over coloured dye, and the electrical discharge removing the aluminium and exposing the dye.

The *direct reading recorder* (Fig. 4.3) has a pen or stylus directly moved by the displacement action of the measurement system. For temperature measurement this might be the displacement of a bimetallic strip, for a pressure gauge the displacement of a Bourdon tube. A circular chart is used and rotates at a constant rate, typically one revolution in 12 hours, 24 hours or 7 days. The pen moves along curved radial lines and thus paper with curved lines has to be used for the plotting. This makes interpolation difficult. Simultaneous recording with up to four separate variables is possible with four separate pens. The instrument is fairly robust with an accuracy of the order of ±0.5% of the full-scale deflection.

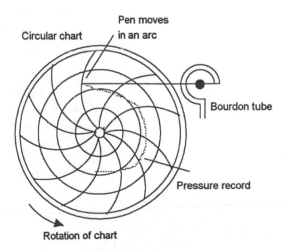

Fig. 4.3   Direct reading recorder

Curvilinear chart paper

Fig. 4.4   Galvanometric recorder

The *galvanometric type* of chart recorder (Fig. 4.4) works on the same principle as the moving coil meter movement. The coil is suspended between two fixed points by a suspension wire. When a current passes through the coil a torque acts on it, causing it to rotate and twist the suspension. The coil rotates to an angle at which the torque is balanced by the opposing torque resulting from the twisting of the suspension. The rotation of the coil results in a pen being moved across a chart.

If R is the length of the pointer and θ the angular deflection of the coil, then the displacement y of the pen is $y = R \sin \theta$. Since θ is proportional to the current i through the coil, then y is proportional to sin i. This is a non-linear relationship. However, if the angular deflections are restricted to less than $\pm 10°$, then the relationship is reasonably linear, the non-linearity error being less than 0.5%. A greater problem is, however, the fact that the pen moves in an arc rather than a straight line and thus curvilinear paper (Fig. 4.4) has to be used for the plotting. With such forms of chart there are difficulties in interpolation for points between the lines.

Figure 4.5 illustrates the general principles of the *potentiometric recorder*. Such a recorder is sometimes referred to as a *closed-loop recorder* or a *closed-loop servo recorder*. The position of the pen is monitored by means of a slider which moves along a linear potentiometer. The position of the slider determines the potential applied to an operational amplifier. The operational amplifier subtracts the slider signal, which is obtained from the input signal, from the sensor/signal conditioner. The output from the amplifier is thus a signal that is related to the difference between the pen and sensor signals. This signal is used to operate a servo motor which in turn controls the movement of the pen across the chart. The pen thus ends up moving to a position where there is no difference between the pen and sensor signals. The pen is thus made to track the sensor signal.

Potentiometric recorders have high input resistances, higher accuracies than galvanometric recorders (about $\pm 0.1\%$ of full-scale reading) but considerably slower response times. Response times are typically of the order of 1 to 2 s and can only be used for d.c. signals or very low frequencies, up to about 2 Hz. They can thus only be used for slowly changing signals. Because of friction, a minimum current is required to operate the motor and thus there is an error due to the recorder not responding to a small input signal. This error is known as the *dead band*. Typically it is about $\pm 0.3\%$ of the range of the instrument. Thus, for a range of 5 mV the dead band amounts to about ± 0.015 mV.

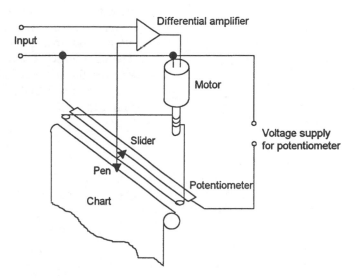

Fig. 4.5   Potentiometric recorder

4.2.3 Cathode-ray oscilloscope

The cathode-ray oscilloscope is a voltage-measuring instrument which is capable of displaying extremely high frequency signals. A general-purpose instrument can respond to signals up to about 10 MHz while more specialist instruments can respond up to about 1 GHz. Double-beam oscilloscopes enable two separate traces to be observed simultaneously on the screen while storage oscilloscopes enable the trace to remain on the screen after the input signal has ceased, only being removed by a deliberate action of erasure. Digital storage oscilloscopes digitise the input signal and store the digital signal in a memory. It can then be analysed and manipulated and the analogue display on the oscilloscope screen obtained from reconstructing the analogue signal. Permanent records of traces can be made with special-purpose cameras that attach directly to the oscilloscope.

A general-purpose oscilloscope is likely to have vertical deflection, i.e. Y-deflection, sensitivities which vary between 5 mV per scale division to 20 V per scale division. In order that a.c. components can be viewed in the presence of high d.c. voltages, a blocking capacitor can be switched into the input line. When the amplifier is in its a.c. mode its bandwidth typically extends from about 2 Hz to 10 MHz and when in the d.c. mode from d.c. to 10 MHz. The Y-input impedance is typically about 1 MΩ shunted with about 20 pF capacitance. When an external circuit is connected to the Y-input, problems due to loading and interference can distort the input signal. While interference can be reduced by the use of coaxial cable, the capacitance of the coaxial cable and any probe attached to it can be enough, particularly at low frequencies, to introduce a relatively low impedance across the input impedance of the oscilloscope and so introduce significant loading. A number of probes exist for connection to the input cable and which are designed to increase the input impedance and avoid this loading problem. A passive voltage probe that is often used is a 10 to 1 attenuator (Fig. 4.6). This has a 9 MΩ resistor and variable capacitor in the probe tip. However, this not only reduces the capacitive loading but also the voltage sensitivity and so an active voltage probe using an FET is often used.

Fig. 4.6   Passive voltage probe

For details of the 'mechanics' of cathode-ray oscilloscopes the reader is referred to such texts as *Newnes Instrumentation and Measurement* by W. Bolton (Newnes 1991, 1996, 2000), *Principles of Electronic Instrumentation and Measurement* by H.M. Berlin and F.C. Getz (Merrill 1988) or *Electronic Instruments and Measurement Techniques* by F.F. Mazda (Cambridge University Press 1987).

4.2.4 Visual display unit

Output data is increasingly being presented using a television type of display, termed a visual display unit (VDU). This builds up a picture on a cathode-ray tube screen by moving the spot formed by the electron beam in a series of horizontal scan lines, one after another down the screen. The image is built up by varying the intensity of the spot on the screen as each line is scanned. This raster form of display is termed *non-interlaced* (Fig. 4.7(a)). To reduce the effects of flicker two scans down the screen are used to trace a complete picture. On the first scan all the odd-numbered lines are traced out and on the second the even-numbered lines are traced. This technique is called *interlaced scanning* (Fig. 4.7(b)).

The screen of the visual display unit is coated with a large number of phosphor dots, these dots forming the *pixels*. The term pixel is used for the smallest addressable dot on any display device. A text character or a diagram is produced on the screen by selectively lighting these dots. Figure 4.8 shows how, for a 7 by 5 matrix, characters are built up by the electron beam moving in its zigzag path down the screen. The input data to the VDU is usually in digital *ASCII (American Standard Code for*

Fig. 4.7   (a) Non-interlaced,
(b) interlaced displays

7 by 5 matrix
of pixels

1
2
3   Lines of
4   the scan
5   down the
6   screen
7

Fig. 4.8   Character build-up
by selective lighting

Information Interchange) format. This is a 7-bit code and so can be used to represent $2^7 = 128$ characters. It enables all the standard keyboard characters to be covered, as well as some control functions such as RETURN which is used to indicate the return from the end of a line to the start of the next line. Table 4.1 gives an abridged list of the code.

4.2.5 Printers

Printers provide a record of data on paper. There are a number of versions of such printers: the dot matrix printer, the ink/bubble jet printer and the laser printer.

The *dot matrix printer* has a print head (Fig. 4.9) which consists of either 9 or 24 pins in a vertical line. Each pin is controlled by an electromagnet which when turned on propels the pin onto the inking ribbon. This transfers a small blob of ink onto the paper behind the ribbon. A character is formed by moving the print head in horizontal lines back-and-forth across the paper and firing the appropriate pins.

Fig. 4.9   Dot matrix print head mechanism

Table 4.1 ASCII code

Character	ASCII	Character	ASCII	Character	ASCII
A	100 0001	N	100 1110	0	011 0000
B	100 0010	O	100 1111	1	011 0001
C	100 0011	P	101 0000	2	011 0010
D	100 0100	Q	101 0001	3	011 0011
E	100 0101	R	101 0010	4	011 0100
F	100 0110	S	101 0011	5	011 0101
G	100 0111	T	101 0100	6	011 0110
H	100 1000	U	101 0101	7	011 0111
I	100 1001	V	101 0110	8	011 1000
J	100 1010	W	101 0111	9	011 1001
K	100 1011	X	101 1000		
L	100 1100	Y	101 1001		
M	100 1101	Z	101 1010		

Fig. 4.10   Producing a stream of drops

Heater not on

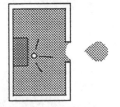

Heater on and vapour bubble being produced

Vapour bubble bursts and ejects ink drop

Fig. 4.11   Principle of the bubble jet

The *ink jet printer* uses a conductive ink which is forced through a small nozzle to produce a jet of very small drops of ink of constant diameter at a constant frequency. With one form a constant stream of ink passes along a tube and is pulsed to form fine drops by a piezoelectric crystal which vibrates at a frequency of about 100 kHz (Fig. 4.10). Another form uses a small heater in the print head with vaporised ink in a capillary tube, so producing gas bubbles which push out drops of ink (Fig. 4.11). In one printer version each drop of ink is given a charge as a result of passing through a charging electrode and the charged drops are deflected by passing between plates between which an electric field is maintained; in another version a vertical stack of nozzles is used and each jet is just switched on or off on demand. Inkjet printers can give colour prints by the use of three different colour ink jet systems. The fineness of the drops is such that prints can be produced with more than 600 dots per inch.

The *laser printer* has a photosensitive drum which is coated with a selenium-based light-sensitive material (Fig. 4.12). In the dark the selenium has a high resistance and consequently becomes charged as it passes close to the charging wire; this is a wire at a high voltage and off which charge leaks. A light beam is made to scan along the length of the drum by a small rotating eight-sided mirror. When light strikes the selenium its resistance drops and it can no longer remain charged. By controlling the brightness of the beam of light, so points on the drum can be discharged or left charged. As the drum passes the toner reservoir, the charged areas attract particles of toner which thus stick to the areas that have not been exposed to light and do not stick on the areas that have been exposed to light. The paper is given a charge as it passes another charging wire, the so-called corona wire, so that as it passes close to the drum it attracts the toner off the drum. A hot fusing roller is then used to melt the toner particles so that, after passing between rollers, they firmly adhere to the paper. General-use laser printers are currently able to produce 600 dots per inch.

Fig. 4.12   Basic elements of a laser printer

4.3 Magnetic recording

Fig. 4.13   Basis of magnetic recording head

Fig. 4.14   Basis of magnetic replay head

Fig. 4.15   Bit cells

Magnetic recording is used for the storage of data on the floppy discs and hard discs of computers. The basic principles are that a recording head, which responds to the input signal, produces corresponding magnetic patterns on a thin layer of magnetic material and a read head gives an output by converting the magnetic patterns on the magnetic material to electrical signals. In addition to these heads the systems requires a transport system which moves the magnetic material in a controlled way under the heads.

Figure 4.13 shows the basic elements of the recording head, it consists of a core of ferromagnetic material which has a non-magnetic gap. When electrical signals are fed to the coil which is wound round the core, magnetic flux is produced in the core. The proximity of the magnetic coated plastic to the non-magnetic gap means that the magnetic flux readily follows a path through the core and that part of the magnetic coating in the region of the gap. When there is magnetic flux passing through a region of the magnetic coating it becomes permanently magnetised. Hence a magnetic record is produced of the electrical input signal. Reversing the direction of the current reverses the flux direction.

The replay head (Fig. 4.14) has a similar construction to that of the recording head. When a piece of magnetised coating bridges the non-magnetised gap then magnetic flux is induced in the core. Flux changes in the core induce e.m.t.s in the coil wound round the core. Thus the output from the coil is an electrical signal which is related to the magnetic record on the coating

4.3.1 Magnetic recording codes

Digital recording involves the recording of signals as a coded combination of bits. A bit cell is the element of the magnetic coating where the magnetism is either completely saturated in one direction or completely saturated in the reverse direction. Saturation is when the magnetising field has been increased to such an extent that the magnetic material has reached its maximum amount of magnetic flux and further increases in magnetising current produce no further change.

The bit cells on the magnetic surface might then appear in the form shown in Figure 4.15. An obvious method of putting data on the magnetic material might seem to be to use the magnetic flux in one direction to represent a 0 and in the reverse direction a 1. However, it is necessary to read each cell and thus accurate timing points are needed in order to clearly indicate when sampling should take place. Problems can arise if some external clock is used to give the timing signals, a small mismatch between the timing signals and the rate at which the magnetic surface is moving under the read head can result in perhaps a cell being missed or even read twice. Synchronisation is essential. Such synchronisation is achieved by using the bit cells themselves

Fig. 4.16 Non-return-to-zero recording

Fig. 4.17 Phase encoding

Fig. 4.18 Frequency modulation

Fig. 4.19 Modified FM

to generate the signals for taking samples. One method is to use transitions on the magnetic surface from saturation in one direction to saturation in the other, i.e. where the demarcation between two bits is clearly evident, to give feedback to the timing signal generation in order to adjust it so that it is in synchronisation with the bit cells.

If the flux reversals do not occur sufficiently frequently, this method of synchronisation can still result in errors occurring. One way of overcoming this problem is to use a form of encoding. The following are some of the methods commonly used:

1 *Non-return-to-zero* (NRZ)
   With this system the flux is recorded on the tape such that no change in flux represents 0 and a change in flux 1 (Fig. 4.16). It is, however, not self-clocking.

2 *Phase encoding* (PE)
   Phase encoding has the advantage of being self-clocking with no external clock signals being required. Each cell is split in two with one half having positive saturation flux and the other a negative saturation flux. A digit 0 is then recorded as a half-bit positive saturation followed by a half-bit negative saturation; a 1 digit is represented by a half-bit negative saturation followed by a half-bit positive saturation. The mid-cell transition of positive to negative thus indicates a 0 and a negative to positive transition a 1 (Fig. 4.17).

3 *Frequency modulation* (FM)
   This is self-clocking and similar to phase encoding but there is always a flux direction reversal at the beginning of each cell (Fig. 4.18). For a 0 bit there is then no additional flux reversal during the cell but for a 1 there is an additional flux reversal during the cell.

4 *Modified frequency modulation* (MFM)
   This is a modification of the frequency modulation code, the difference being that the flux reversal at the beginning of each bit code is only present if the current and previous bit was 0 (Fig. 4.19). This means that only one flux reversal is required for each bit. This and the run length limited code are the codes generally used for magnetic discs.

5 *Run length limited* (RLL)
   This is a group of self-clocking codes which specify a minimum and maximum distance, i.e. run, between flux reversals. The maximum run is short enough to ensure that the flux reversals are sufficiently frequent for the code to be self-clocking. A commonly used form of this code is $RLL_{2,7}$, the 2,7 indicating the minimum distance between flux reversals is to be 2 bits and the maximum is to be 7. The

sequence of codes is described as a sequence of S-codes and R-codes. An S-code, a space code, has no reversals of flux while an R-code, a reversal code, has a reversal during the bit. Two S/R-codes are used to represent each bit. The bits are grouped into sequences of 2, 3 and 4 bits and a code assigned to each group, the codes being:

Bit sequence	Code sequence
10	SRSS
11	RSSS
000	SSSRSS
010	RSSRSS
011	SSRSSS
0010	SSRSSRSS
0011	SSSSRSSS

Figure 4.20 shows the coding for the sequence 0110010, it being broken into groups 011 and 0010 and so represented by SSRSSSSRSSRSS. There are at least two S-codes between R-codes and there can be no more than seven S-codes between R-codes.

SS RS SS SS RS SR SS
0 1 1 0 0 1 0

Fig. 4.20 RLL code

The optimum code is the one that allows the bits to be packed as close as possible and which can be read without error. The read heads can locate reversals quite easily but they must not be too close together. The RLL code has the advantage of being more compact than the other codes, PE and FM taking up the most space. MFM and NRZ take up the same amount of space. NRZ has the disadvantage of, unlike the other codes, not being self-clocking.

4.3.2 Magnetic discs

Digital recording is very frequently to a floppy or hard disc. The digital data is stored on the disc surface along concentric circles called tracks, a single disc having many such tracks. A single read/write head is used for each disc surface and the heads are moved, by means of a mechanical actuator, backwards and forwards to access different tracks. The disc is spun by the drive and the read/write heads read or write data into a track.

Fig. 4.21   Floppy disc

Fig. 4.22   Hard disc

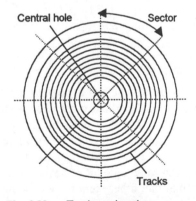

Fig. 4.23   Tracks and sectors

4.4 Displays

The 3½ inch floppy disc (Fig. 4.21) used in the personal computer has 135 tracks per inch and can store 1.4 Mbytes of data. No part of the disc is exposed when it is outside the computer, there being a sliding protective metal cover which only opens to reveal the magnetic surface when the disc is in the computer.

Hard discs (Fig. 4.22) are sealed units with data stored on the disc surface along concentric circles. A hard disc assembly has more than one such disc and the data is stored on magnetic coatings on both sides of the discs. The discs are rotated at high speeds and the tracks accessed by moving the read/write heads. Large amounts of data can be stored on such assemblies of discs; storages of the order of many Gbytes are now common.

The disc surface is divided into sectors (Fig. 4.23) and so a unit of information on a disc has an address consisting of a track number and a sector number. A floppy disc has normally between 8 and 18 sectors and about 100 tracks. A hard disc might have about 2000 tracks per surface and 32 sectors. To seek data the head must be moved to over the required track, the time this takes being termed the *seek time*, and then wait there until the required segment moves under it, this time being termed the *latency*. In order that an address can be identified it is necessary for information to have been recorded on the disc to identify segments and tracks. The writing of this information is called *formatting* and has to be carried out before data can be stored on a disc. The technique usually used is to store this location information on the tracks so that when data is stored the sequence of information on a track becomes:

> index marker,
> sector 0 header, sector 0 data, sector 0 trailer,
> sector 1 header, sector 1 data, sector 1 trailer,
> sector 2 header, sector 2 data, sector 2 trailer,
> etc.

The index marker contains the track number with the sector header identifying the sector. The sector trailer contains information, e.g. a cyclic redundancy check (see Section 20.4) which can be used to check that a sector was read correctly.

Many display systems use light indicators to indicate on-off status or give alphanumeric displays. The term *alphanumeric* is a contraction of the terms alphabetic and numeric and describes displays of the letters of the alphabet and numbers 0 to 9 with decimal points. One form of such a display involves seven 'light' segments to generate the alphabetic and numeric characters. Figure 4.24 shows the segments and Table 4.2 shows how a 4-bit binary code input can be used to generate inputs to switch on the various segments.

Fig. 4.24   Seven-segment display

Table 4.2   Seven-segment display

Binary input				Segments activated							Number displayed
				a	b	c	d	e	f	g	
0	0	0	0	1	1	1	1	1	1	0	0
0	0	0	1	0	1	1	0	0	0	0	1
0	0	1	0	1	1	0	1	1	0	1	2
0	0	1	0	1	1	1	1	0	0	1	3
0	1	0	0	0	1	1	0	0	1	1	4
0	1	0	1	1	0	1	1	0	1	1	5
0	1	1	0	0	0	1	1	1	1	1	6
0	1	1	1	1	1	1	0	0	0	0	7
1	0	0	0	1	1	1	1	1	1	1	8
1	0	0	1	1	1	1	0	0	1	1	9

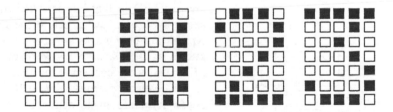

Fig. 4.25   7 by 5 dot matrix display

Another format involves a 7 by 5 or 9 by 7 dot matrix (Fig. 4.25). The characters are then generated by the excitation of appropriate dots.

4.4.1 Light indicators

The light indicators for such displays might be neon lamps, incandescent lamps, light-emitting diodes (LEDs) or liquid crystal displays (LCDs). *Neon lamps* need high voltages and low currents and can be powered directly from the mains voltage but can only be used to give a red light. *Incandescent lamps* can be

Fig. 4.26   Light-emitting diode

used with a wide range of voltages but need a comparatively high current. They emit white light so use lenses to generate any required colour. Their main advantage is their brightness.

Light-emitting diodes require low voltages and low currents and are cheap. These diodes when forward biased emit light over a certain band of wavelengths. Figure 4.26 shows the basic form of a LED, the light emitted from the diode being enhanced in one direction by means of reflectors. Commonly used LED materials are gallium arsenide, gallium phosphide and alloys of gallium arsenide with gallium phosphide. The most commonly used LEDs can give red, yellow or green colours. With microprocessor-based systems, LEDs are the most common form of indicator used.

4.4.2 LED displays

A current-limiting resistor is generally required with a LED in order to limit the current to below the maximum rated current of about 10 to 30 mA. Typically a LED might give a voltage drop across it of 2.1 V when the current is limited to 20 mA. Thus when, say, a 5 V output is applied, 2.9 V has to be dropped across a series resistor. This means a resistance of 2.9/0.020 = 145 Ω is required and so a standard resistor of 150 Ω is likely to be used. Some LEDs are supplied with built-in resistors so they can be directly connected to microprocessor systems.

LEDs are available as single light displays, seven- and sixteen-segment alphanumeric displays, in dot matrix format and bar graph form.

Figure 4.27 shows how seven LEDs, to give the seven segments of a display of the form shown in Figure 4.24, might be connected to a driver so that when a line is driven low, a voltage is applied and the LED in that line is switched on. The voltage has to be above a 'turn-on' value before the LED emits significant light; typical turn-on voltages are about 1.5 V. Such an arrangement is known as the *common anode* form of connection since all the LED anodes are connected together. An alternative arrangement is the *common cathode* (Fig. 4.28). The elements in common anode form are made active by the input going low, in the common cathode type by going high. Common anode is the usual choice since the direction of current flow and the size of current involved are usually most appropriate.

Example of such types of display are the seven-segment 7.6 mm and 10.9 mm high intensity displays of Hewlett Packard which are available as either common anode or common cathode form. In addition to the seven segments to form the characters there is either a left-hand or right-hand decimal point. By illuminating different segments of the display the full range of numbers and a small range of alphabetical characters can be formed.

Often the output from the driver is not in the normal binary form but in *binary coded decimal* (BCD) (see Section 14.2). With BCD, each decimal digit is coded separately in binary. For

Fig. 4.27   Common anode connection for LEDs

Fig. 4.28   Common cathode connection for LEDs

example, the decimal number 15 has the 1 coded as 0001 and the 5 as 0101 to give the BCD code of 0001 0101. The driver output has then to be decoded into the required format for the LED display. The 7447 is a commonly used decoder for driving displays (Fig. 4.29).

See Section 18.3.4 for a discussion of the interfacing of LED displays to microprocessors.

Fig. 4.29   Decoder with seven-segment display

4.4.3 A 5 by 7 dot matrix LED display

Figure 4.30 shows the basic form used for a 5 by 7 dot matrix display. The array consists of five column connectors, each connecting the anodes of seven LEDs. Each row connects to the cathodes of five LEDs. To turn on a particular LED, power is applied to its column and its row is grounded. Such a display enables all the ASCII characters to be produced

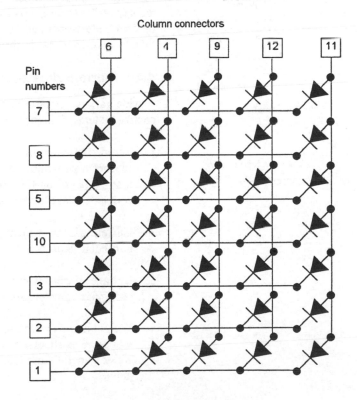

Fig. 4.30   Dot matrix display

(a)

(b)

Fig. 4.31 Liquid crystal: (a) no electric field, (b) electric field

4.4.4 Liquid crystal displays

Liquid crystal displays do not produce any light of their own but rely on reflected light or transmitted light. The liquid crystal material is a compound with long rod-shaped molecules which is sandwiched between two sheets of polymer containing microscopic grooves. The upper and lower sheets are grooved in directions at 90° to each other. The molecules of the liquid crystal material align with the grooves in the polymer and adopt a smooth 90° twist between them (Fig. 4.31).

When plane polarised light is incident on the liquid crystal material its plane of polarisation is rotated as it passes through the material. Thus if it is sandwiched between two sheets of polariser with their transmission directions at right angles, the rotation allows the light to be transmitted and so the material appears light.

However, if an electric field is applied across the material, the molecules become aligned with the field and the light passing through the top polariser is not rotated and cannot pass through the lower polariser but becomes absorbed. The material then appears dark.

The arrangement is put between two sheets of glass on which are transparent electrodes in the shape of the seven-segment display and thus the application of voltages to the various display elements results in them appearing black against the lighter display where there is no electric field. This is the form of display used in battery-operated devices such as watches and calculators. Five by seven dot matrix forms are also available.

4.4.5 Alarm indicators

A wide variety of alarm systems are used with measurement and control systems. Commonly met ones are:

1   Temperature alarms which respond when the temperature reaches a particular value or falls to some other value. These may be based on the use of a resistance element or thermocouple to sense the temperature.

2   Current alarms which respond when the current reaches a particular value or falls below some other value.

3   Voltage alarms which respond when the voltage reaches a particular value or falls below some other value.

4   Weight alarms which respond when the weight in a container reaches a particular value or falls below some other value. These generally use load cells with electrical resistance strain gauges.

Alarm indicators take an analogue input from some sensor, possibly via a signal conditioner, and turn it into an on-off signal for some indicator. Figure 4.32 shows the basic form of alarm system. The input is compared with the alarm set point. The comparator takes two inputs and gives an output when, for example, input A is greater than input B. Thus when the set point is exceeded a logic 0 or 1 signal passes to the logic unit which then gives an output which triggers the switching unit and switches on, or off, an indicator. The indicator can take a variety of forms, e.g. a bell, a horn, a klaxon, a coloured light, a flashing light, a back lighted display (the light comes on behind a message on a screen).

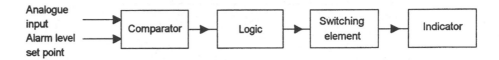

Fig. 4.32   An alarm system

4.5 Data acquisition systems

Automated data acquisition systems can take the form of a dedicated instrument termed a data logger or a personal computer using plug-in DAQ boards.

4.5.1 Data loggers

Figure 4.33 shows the basic elements of a *data logger*. Such a unit can monitor the inputs from a large number of sensors. Inputs from individual sensors, after suitable signal conditioning, are fed into the multiplexer. The multiplexer is used to select one signal which is then fed, after amplification, to the analogue-to-digital converter. The digital signal is then processed by a microprocessor. The microprocessor is able to carry out simple arithmetic operations, perhaps taking the average of a number of measurements. The output from the system might be displayed on a digital meter that indicates the output and channel number, used to give a permanent record with a printer, stored on a floppy disc or transferred to perhaps a computer for analysis.

Because data loggers are often used with thermocouples, there are often special inputs for thermocouples, these providing cold junction compensation and linearisation. The multiplexer can be switched to each sensor in turn and so the output consists of a sequence of samples. Scanning of the inputs can be selected by programming the microprocessor to switch the multiplexer to just sample a single channel, carry out a single scan of all channels, a continuous scan of all channels, or perhaps carry out a periodic scan of all channels, say every 1, 5, 15, 30 or 60 minutes.

Fig. 4.33   Data logger system

Typically a data logger may handle 20 to 100 inputs, though some may handle considerably more, perhaps 1000. It might have a sample and conversion time of 10 μs and be used to make perhaps 1000 readings per second. The accuracy is typically about 0.01% of full-scale input and linearity is about ± 0.005% of full-scale input. Cross-talk is typically 0.01% of full-scale input on any one input. The term *cross-talk* is used to describe the interference that can occur when one sensor is being sampled as a result of signals from other sensors.

4.5.2 Computer with plug-in boards

Figure 4.34 shows the basic elements of a data acquisition system using plug-in boards with a computer. The signal conditioning prior to the inputs to the board depends on the sensors concerned, e.g. it might be for thermocouples – amplification, cold junction compensation and linearisation; for strain gauges – Wheatstone bridge, voltage supply for bridge and linearisation; for RTDs – current supply, circuitry and linearisation.

In selecting the DAQ board to be used the following criteria have to be considered:

Fig. 4.34   Data acquisition system

1   What type of computer software system is being used, e.g. Windows, MacOS?
2   What type of connector is the board to be plugged into, e.g. PCMCIA for laptops, NuBus for MacOS, PCI?
3   How many analogue inputs will be required and what are their ranges?
4   How many digital inputs will be required?
5   What resolution will be required?
6   What is the minimum sampling rate required?
7   Are any timing or counting signals required?

All DAQ boards use *drivers*, software generally supplied by the board manufacturer with a board, to communicate with the computer and tell it what has been inserted and how the computer can communicate with the board. Before a board can be used three parameters have to be set. These are the addresses of the input and output channels, the interrupt level and the channel to be used for direct memory access. With 'plug-and-play' boards for use with Windows software, these parameters are set by the software; otherwise microswitches have to be set on the card in accordance with the instructions in the manual supplied with the board.

Application software can be used to assist in the designing of measurement systems and the analysis of the data. As an illustration of the type of application software available, LabVIEW is a graphical programming software package that has been developed by National Instruments for data acquisition and instrument control. LabVIEW programs are called *virtual instruments* because in appearance and operation they imitate actual instruments. A virtual instrument has three parts, a front panel which is the interactive user interface and simulates the front panel of an instrument by containing control knobs, push buttons and graphical displays, a block diagram which is the source code for the program with the programming being done graphically by drawing lines between connection points on selected icons on the computer screen, and representation as an icon and connector which can provide a graphical representation of the virtual instrument if it is wanted for use in other block diagrams.

Figure 4.35 shows the icon selected for a virtual instrument where one analogue sample is obtained from a specified input channel, the icon having been selected from the Analog Input palette. The 'device' is the device number assigned to the DAQ board, the 'channel' is the source of the data, a 'sample' is one analogue-to-digital conversion, and 'high limit' and 'low limit' are the voltage limits expected for the signal (the default is +10 V and −10 V and changing these values automatically changes the gain of the amplifier on the DAQ board).

Fig. 4.35   Analogue input icon

If we want a waveform from each channel in a designated channel string then the icon shown in Figure 4.36 can be selected. For each input channel a set of samples is acquired over a period of time, at a specified sampling rate, and gives a waveform output showing how the analogue quantity varies with time.

Fig. 4.36   Analogue input icon for sampling a number of channels

By connecting other icons to, say, the above icon, a block diagram can be built up which might take the inputs from a number of analogue channels, sample them in sequence and display the results as a sequence of graphs. The type of front panel display we might have for a simple DAQ acquisition of samples and display is shown in Figure 4.37. By using the up and down arrows the parameters can be changed and the resulting display viewed.

The above is just a simple illustration of what is possible. For more details the reader is referred to the *LabVIEW Manual* or *LabVIEW for Everyone* by L.K. Wells and J. Travis (Prentice-Hall 1997) or *LabVIEW Graphical Programming* by G.W. Johnson (McGraw-Hill 1994).

4.6 Measurement systems

The following examples illustrate some of the points involved in the design of measurement systems for particular applications.

Fig. 4.37   Front panel

Fig. 4.38   Load cell

4.6.1 Load cell for use as a link to detect load lifted

A link-type load cell, of the form shown in Figure 4.38, has four strain gauges attached to its surface and can be inserted in between the cable lifting a load and the load to give a measure of the load being lifted. Two of the strain gauges are in the longitudinal axis direction and two in a transverse direction. When the link is subject to tensile forces, the axial gauges will be in tension and the transverse gauges in compression. Suppose we have the design criteria for the load cell of a sensitivity such that there is an output of about 30 mV when the stress applied to the link is 500 MPa. We will assume the strain gauges may be assumed to have gauge factors of 2.0 and resistances of 100 Ω.

When a load F is applied to the link then, since the elastic modulus E is stress/strain and stress is force per unit area, the longitudinal axis strain ε_l is F/AE and the transverse strain ε_t is $-vF/AE$, where A is the cross-sectional area and v is Poisson's ratio for the link material. The responses of the strain gauges (see Section 2.3.1) to these strains are

$$\frac{\delta R_1}{R_1} = \frac{\delta R_4}{R_4} = G\varepsilon_l = \frac{GF}{AE}$$

$$\frac{\delta R_3}{R_3} = \frac{\delta R_2}{R_2} = G\varepsilon_t = -\frac{vGF}{AE}$$

The output voltage from the Wheatstone bridge (see Section 3.5.1l) is given by

$$V_o = \frac{V_s R_1 R_4}{(R_1 + R_2)(R_3 + R_4)} \left(\frac{\delta R_1}{R_1} - \frac{\delta R_2}{R_2} - \frac{\delta R_3}{R_3} + \frac{\delta R_4}{R_4} \right)$$

With $R_1 = R_2 = R_3 = R_4 = R$, and with $\delta R_1 = \delta R_4$ and $\delta R_2 = \delta R_3$, then

$$V_o = \frac{V_s}{2R}(\delta R_1 - \delta R_2) = \frac{V_s GF}{2AE}(1 + v)$$

Suppose we consider steel for the link. Then tables give E as about 210 GPa and v about 0.30. Thus with a stress (= F/A) of 500 MPa we have, for strain gauges with a gauge factor of 2.0,

$$V_o = 3.09 \times 10^{-3} V_s$$

For a bridge voltage with a supply voltage of 10 V this would be an output voltage of 30.9 mV. No amplification is required if this is the only load value required; if, however, this is a maximum value and we want to determine loads below this level then we might use a differential amplifier. The output can be displayed on a high-resistance voltmeter, high resistance to avoid loading problems. A digital voltmeter might thus be suitable.

4.6.2 Temperature alarm system

A measurement system is required which will set off an alarm when the temperature of a liquid rises above 40°C. The liquid is normally at 30°C. The output from the system must be a 1 V signal to operate the alarm.

Since the output is to be electrical and a reasonable speed of response is likely to be required, an obvious possibility is an electrical resistance element. To generate a voltage output the resistance element could be used with a Wheatstone bridge. The output voltage will probably be less than 1 V for a change from 30 to 40°C but a differential amplifier could be used to enable the required voltage to be obtained. A comparator can then be used to compare the value with the set value for the alarm.

Suppose a nickel element is used. Nickel has a temperature coefficient of resistance of 0.0067 /K. Thus if the resistance element is taken as being 100 Ω at 0°C then its resistance at 30°C will be

$$R_{30} = R_0(1 + at) = 100(1 + 0.0067 \times 30) = 120.1 \ \Omega$$

and at 40°C

$$R_{40} = 100(1 + 0.0067 \times 40) = 126.8 \ \Omega$$

Thus there is a change in resistance of 6.7 Ω. If this element forms one arm of a Wheatstone bridge which is balanced at 30°C, then the output voltage V_o is given by (see Section 3.5)

$$\delta V_o = \frac{V_s \delta R_1}{R_1 + R_2}$$

With the bridge balanced at 30°C and, say, all the arms the same value and a supply voltage of 4 V, then

$$\delta V_o = \frac{4 \times 6.7}{126.8 + 120.1} = 0.109$$

To amplify this to 1 V we can use a differential amplifier (see Section 3.2.5)

$$V_o = \frac{R_2}{R_1}(V_2 - V_1)$$

$$1 = \frac{R_2}{R_1} \times 0.109$$

Hence $R_2/R_1 = 9.17$ and so if we use an input resistance of 1 kΩ the feedback resistance must be 9.17 kΩ.

4.6.3 Angular position of a pulley wheel

A potentiometer is to be used to monitor the angular position of a pulley wheel. Consider the items that might be needed to enable there to be an output to a recorder of 10 mV per degree if the potentiometer has a full-scale angular rotation of 320°.

When the supply voltage V_s is connected across the potentiometer we will need to safeguard it and the wiring against possible high currents and so a resistance R_s can be put in series with the potentiometer R_p. The total voltage drop across the potentiometer is thus $V_s R_p/(R_s + R_p)$. For an angle θ with a potentiometer having a full-scale angular deflection of θ_F we will obtain an output from the potentiometer of

$$V_\theta = \frac{\theta}{\theta_F} \frac{V_s R_p}{R_s + R_p}$$

Suppose we consider a potentiometer with a resistance of 4 kΩ and let R_s be 2 kΩ. Then for 1 mV per degree we have

$$0.01 = \frac{1}{320} \frac{4V_s}{4+2}$$

Hence we would need a supply voltage of 4.8 V. To prevent loading of the potentiometer by the resistance of the recorder, a voltage follower circuit can be used. Thus the circuit might be of the form shown in Figure 4.39.

Fig. 4.39   Pulley wheel monitor

4.6.4 Temperature measurement to give a binary output

Consider the requirement for a temperature measurement system for temperatures in the range 0 to 100°C and which will give an 8-bit binary output with a change in 1 bit corresponding to a temperature change of 1°C. The output is intended for inputting to a microprocessor as part of a temperature control system.

A linear temperature sensor is required and so the thermo-transistor LM35 can be used (see Section 2.9.4). LM35 gives an output of 10 mV/°C when it has a supply voltage of 5 V. If we apply the output from LM35 to an 8-bit analogue-to-digital converter then a digital output can be obtained. We need the resolution of the ADC to be 10 mV so that each step of 10 mV will generate a change in output of 1 bit. Suppose we use a successive approximations ADC, e.g. ADC0801, then this requires an input of a reference voltage which when subdivided into $2^8 = 256$ bits gives 10 mV per bit. Thus a reference voltage of 2.56 V is required. For this to be obtained the reference voltage input to the ADC0801 has to be $V_{ref}/2$ and so an accurate input voltage of 1.28 V is required. Such a voltage can be obtained by using a potentiometer circuit across the 5 V supply with a voltage follower to avoid loading problems. Because the voltage has to remain steady at 1.28 V, even if the 5 V supply voltage fluctuates, a voltage regulator is likely to be used, e.g. a 2.45 V voltage regulator ZN458/B. Thus the circuit might be as in Figure 4.40.

Fig. 4.40   Temperature sensor

4.7 Testing and calibration

Testing a measurement system installation falls into three stages:

1 *Pre-installation testing*
   This is the testing of each instrument for correct calibration and operation prior to it being installed.

2 *Piping and cabling testing*
   In the case of pneumatic lines this involves, prior to the connection of the instruments, blowing through with clear, dry air prior to connection and pressure testing to ensure they are leak free. With process piping, all the piping should be flushed through and tested prior to the connection of instruments. With instrument cables, all should be checked for continuity and insulation resistance prior to the connection of any instruments.

3 *Precommissioning*
   This involves testing that the installation is complete, all instrument components are in full operational order when interconnected and all control room panels or displays function.

4.7.1 Calibration

Calibration consists of comparing the output of a measurement system and its subsystems against standards of known accuracy. The standards may be other instruments which are kept specially for calibration duties or some means of defining standard values. In many companies some instruments and items such as standard resistors and cells are kept in a company standards department and used solely for calibration purposes. The relationship between the calibration of an instrument in everyday use and national standards is likely to be:

1 National standards are used to calibrate standards for calibration centres.
2 Calibration centre standards are used to calibrate standards for instrument manufacturers.
3 Standardised instruments from instrument manufacturers are used to provide in-company standards.
4 In-company standards are used to calibrate process instruments.

There is a simple traceability chain from the instrument used in a process back to national standards. For a more detailed discussion of calibration the reader is referred to *Measurement and Calibration for Quality Assurance* by A.S. Morris (Prentice-Hall 1991).

The following are some examples of calibration procedures that might be used in-company:

1 *Voltmeters*
These can be checked against standard voltmeters or standard cells giving standard e.m.f.s.

2 *Ammeters*
These can be checked against standard ammeters.

3 *Gauge factor of strain gauges*
This can be checked by taking a sample of gauges from a batch and applying measured strains to them when mounted on some test piece. The resistance changes can be measured and hence the gauge factor computed.

4 *Wheatstone bridge circuits*
The output from a Wheatstone bridge can be checked when a standard resistance is introduced into one of the arms.

5 *Load cells*
For low-capacity load cells, dead-weight loads using standard weights can be used.

6 *Pressure sensors*
Pressure sensors can be calibrated by using a dead-weight tester (Fig. 4.41). The calibration pressures are generated by adding standard weights W to the piston tray. After the weights are placed on the tray, a screw-driven plunger is forced into the hydraulic oil in the chamber to lift the piston-weight assembly. The calibration pressure is then W/A, where A is the cross-sectional area of the piston. Alternatively the dead-weight tester can be used to calibrate a pressure gauge and this gauge can be used for the calibration of other gauges.

Fig. 4.41 Dead-weight calibration for pressure gauges

7 *Temperature sensors*
These can be calibrated by immersion in a melt of a pure metal or water. The temperature of the substance is then slowly reduced and a temperature–time record obtained. When the substance changes state from liquid to solid, the temperature remains constant. Its value can be looked up from tables and hence an accurate reference temperature for calibration obtained. Alternatively, the temperature at which a liquid boils can be used. However, the boiling point depends on the atmospheric pressure and corrections have to be applied if it differs from the standard atmospheric pressure. Alternatively, in-company the readings given by the measurement system can be compared with those of a standard thermometer.

Problems

1 Explain the significance of the following terms taken from the specifications of display systems:
   (a) Closed-loop servo recorder: dead band ±0.2% of span.
   (b) The hard disc has two discs with four read/write heads, one for each surface of the discs. Each surface has 614 tracks and each track 32 sectors.
   (c) Data logger: number of inputs 100, cross-talk on any one input 0.01% of full-scale input.
   (d) Double-beam oscilloscope: vertical deflection with two identical channels, bandwidth d.c. to 15 MHz, deflection factor of 10 mV/div to 20 V/div in 11 calibrated steps, time base of 0.5 μs/div to 0.5 s/div in 19 calibrated steps.

2 Explain the problems of loading when a measurement system is being assembled from a sensor, signal conditioner and display.

3 Suggest a display unit that could be used to give:
   (a) A permanent record of the output from a thermocouple.
   (b) A display which enables the oil pressure in a system to be observed.
   (c) A record to be kept of the digital output from a microprocessor.
   (d) The transient voltages resulting from monitoring of the loads on an aircraft during simulated wind turbulence.

4 A cylindrical load cell, of the form shown in Figure 2.33, has four strain gauges attached to its surface. Two of the gauges are in the circumferential direction and two in the longitudinal axis direction. When the cylinder is subject to a compressive load, the axial gauges will be in compression while the circumferential ones will be in tension. If the material of the cylinder has a cross-sectional area A and an elastic modulus E, then a force F acting on the cylinder will give a strain acting on the axial gauges of $-F/AE$ and on the circumferential gauges of $+vF/AE$, where v is Poisson's ratio for the material. Design a complete measurement system, using load cells, which could be used to monitor the mass of water in a tank. The tank itself has a mass of 20 kg and the water when at the required level 40 kg. The mass is to be monitored to an accuracy of ±0.5 kg. The strain gauges have a gauge factor of 2.1 and are all of the same resistance of 120.0 Ω. For all other items, specify what your design requires. If you use mild steel for the load cell material, then the tensile modulus may be taken as 210 GPa and Poisson's ratio 0.30.

5 Design a complete measurement system involving the use of a thermocouple to determine the temperature of the water in a boiler and give a visual indication on a meter. The temperature will be in the range 0 to 100°C and is required to an accuracy of ±1% of full-scale reading. Specify the materials to be used for the thermocouple and all other items necessary. In advocating your design you must consider the

problems of cold junction and non-linearity. You will probably need to consult thermocouple tables. The following data is taken from such tables, the cold junction being at 0°C, and may be used as a guide.

Materials	e.m.f. in mV at				
	20°C	40°C	60°C	80°C	100°C
Copper–constantan	0.789	1.611	2.467	3.357	4.277
Chromel–constantan	1.192	2.419	3.683	4.983	6.317
Iron–constantan	1.019	2.058	3.115	4.186	5.268
Chromel–alumel	0.798	1.611	2.436	3.266	4.095
Platinum–10% Rh, Pt	0.113	0.235	0.365	0.502	0.645

6 Design a measurement system which could be used to monitor the temperatures, of the order of 100°C, in positions scattered over a number of points in a plant and present the results on a control panel.

7 A suggested design for the measurement of liquid level in a vessel involves a float which in its vertical motion bends a cantilever. The degree of bending of the cantilever is then taken as a measure of the liquid level. When a force F is applied to the free end of a cantilever of length L, the strain on its surface a distance x from the clamped end is given by

$$\text{Strain} = \frac{6(L-x)}{wt^2E}$$

where w is the width of the cantilever, t its thickness and E the elastic modulus of the material. Strain gauges are to be used to monitor the bending of the cantilever with two strain gauges being attached longitudinally to the upper surface and two longitudinally to the lower surface. The gauges are then to be incorporated into a four-gauge Wheatstone bridge and the output voltage, after possible amplification, then taken as a measure of the liquid level. Determine the specifications required for the components of this system if there is to be an output of 10 mV per 10 cm change in level.

8 Design a static pressure measurement system based on a sensor involving a 40 mm diameter diaphragm across which there is to be a maximum pressure difference of 500 MPa. For a diaphragm where the central deflection y is much smaller than the thickness t of the diaphragm,

$$y \approx \frac{3r^2P(1-v^2)}{16Et^3}$$

where r is the radius of the diaphragm, P the pressure difference, E the modulus of elasticity and v Poisson's ratio.

Explain how the deflection y will be converted into a signal that can be displayed on a meter.

9   Suggest the elements that might be considered for the measurement systems to be used to:

(a) Monitor the pressure in an air pressure line and present the result on a dial, no great accuracy being required.

(b) Continuously monitor and record the temperature of a room with an accuracy of $\pm 1°C$.

(c) Monitor the weight of lorries passing over a weighing platform.

(d) Monitor the angular speed of rotation of a shaft.

5 Pneumatic and hydraulic actuation systems

5.1 Actuation systems

Actuation systems are the elements of control systems which are responsible for transforming the output of a microprocessor or control system into a controlling action on a machine or device. Thus, for example, we might have an electrical output from the controller which has to be transformed into a linear motion to move a load. Another example might be where an electrical output from the controller has to be transformed into an action which controls the amount of liquid passing along a pipe.

In this chapter pneumatic and hydraulic actuation systems are discussed. In Chapter 6 mechanical actuator systems are discussed and in Chapter 7 electrical actuation systems. For a more detailed consideration of pneumatic and hydraulic systems the reader is referred to more specialist books such as *Pneumatic and Hydraulic Systems* by W. Bolton (Butterworth-Heinemann 1997), *Power Pneumatics* by M.J. Pinches and B.J. Callear (Prentice-Hall 1996), *Pneumatic Control for Industrial Automation* by P. Rohner and G. Smith (Wiley 1987, 1990) or *Industrial Hydraulic Control* by P. Rohner (Wiley 1984, 1986, 1988, 1995).

5.2 Pneumatic and hydraulic systems

Pneumatic signals are often used to control final control elements, even when the control system is otherwise electrical. This is because such signals can be used to actuate large valves and other high power control devices and so move significant loads. The main drawback with pneumatic systems is, however, the compressibility of air. Hydraulic signals can be used for even higher power control devices but are more expensive than pneumatic systems and there are hazards associated with oil leaks which do not occur with air leaks.

Fig. 5.1   Hydraulic power supply

Fig. 5.2   Accumulator

5.2.1 Power supplies

With a hydraulic system, pressurised oil is provided by a pump driven by an electric motor. The pump pumps oil from a sump through a non-return valve and an accumulator to the system, from which it returns to the sump. Figure 5.1 illustrates the arrangement. A pressure relief valve is included, this being to release the pressure if it rises above a safe level, the non-return valve is to prevent the oil being back driven to the pump and the accumulator is to smooth out any short-term fluctuations in the output oil pressure. Essentially the accumulator is just a container in which the oil is held under pressure against an external force, Figure 5.2 showing the most commonly used form which is gas pressurised and involves gas within a bladder in the chamber containing the hydraulic fluid, an older type involved a spring-loaded piston. If the oil pressure rises then the bladder contracts, increases the volume the oil can occupy and so reduces the pressure. If the oil pressure falls, the bladder expands to reduce the volume occupied by the oil and so increases its pressure.

With a pneumatic power supply (Fig. 5.3) an electric motor drives an air compressor. The air inlet to the compressor is likely to be filtered and via a silencer to reduce the noise level. A pressure relief valve provides protection against the pressure in the system rising above a safe level. Since the air compressor increases the temperature of the air there is likely to be a cooling system and to remove contamination and water from the air a filter with water trap. An air receiver increases the volume of air in the system and smoothes out any short-term pressure fluctuations.

Fig. 5.3   Pneumatic power supply

5.3 Directional control valves

Fig. 5.4   Spool valve

Pneumatic and hydraulic systems use directional control valves to direct the flow of fluid through a system. They are not intended to vary the rate of flow of fluid but are either completely open or completely closed, i.e. on/off devices. Such on/off valves are widely used to develop sequenced control systems (see later in this chapter). They might be activated to switch the fluid flow direction by means of mechanical, electrical or fluid pressure signals.

A common type of directional control valve is the *spool valve*. A spool moves horizontally within the valve body to control the flow. Figure 5.4 shows a particular form. In (a) the air supply is connected to port 1 and port 3 is closed. Thus the device connected to port 2 can be pressurised. When the spool is moved to the left (Fig. 5.4(b)) the air supply is cut off and port 2 is connected to port 3. Port 3 is a vent to the atmosphere and so the air pressure in the system attached to port 2 is vented. Thus the movement of the spool has allowed the air to firstly flow into the system and then be reversed and flow out of the system. *Rotary spool valves* have a rotating spool which, when it rotates, opens and closes ports in a similar way.

Another common form of directional control valve is the *poppet valve*. Figure 5.5 shows one form. This valve is normally in the closed condition, there being no connection between port 1 to which the pressure supply is connected and port 2 to which the system is connected. In poppet valves, balls, discs or cones are used in conjunction with valve seats to control the flow. In the figure a ball is shown. When the push-button is depressed, the ball is pushed out of its seat and flow occurs as a result of port 1 being connected to port 2. When the button is released, the spring forces the ball back up against its seat and so closes off the flow.

1 Pressure supply port

Fig. 5.5   Poppet valve

(a)   (b)

(c)

Fig. 5.6   (a) Flow path,
(b) flow shut-off, (c) initial connections

Fig. 5.7   Valve actuation symbols

2(A)

1(P)

Fig. 5.8   2/2 valve

5.3.1 Valve symbols

The symbol used for control valves consists of a square for each of its switching positions. Thus for the poppet valve shown in Figure 5.5, there are two positions: one with the button not pressed and one with it pressed. Thus a two-position valve will have two squares, a three-position valve three squares. Arrow-headed lines (Fig. 5.6(a)) are used to indicate the directions of flow in each of the positions, with blocked-off lines closed flow lines (Fig. 5.6(b)). The initial position of the valve has the connections (Fig. 5.6(c)) to the ports shown; in Fig. 5.6(c) the valve has four ports.

Ports are labelled by a number or a letter according to their function. The ports are labelled 1 (or P) for pressure supply, 3 (or T) for hydraulic return port, 3 or 5 (or R or S) for pneumatic exhaust ports, and 2 or 5 (or B or A) for output ports.

Figure 5.7 shows examples of some of the symbols which are used to indicate the various ways the valves can be actuated. More than one of these symbols might be used with the valve symbol.

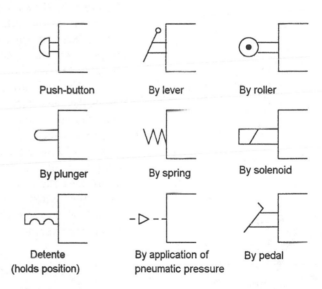

Push-button   By lever   By roller

By plunger   By spring   By solenoid

Detente   By application of   By pedal
(holds position)   pneumatic pressure

As an illustration of how these various symbols can be combined to describe how a valve operates, Figure 5.8 shows the symbol for the 2 port 2 position poppet valve of Figure 5.6. Note that a 2 port 2 position valve would be described as a 2/2 valve, the first number indicating the number of ports and the second number the number of positions.

As a further illustration, Figure 5.9 shows a solenoid operated spool valve and Figure 5.10 its symbol. The valve is actuated by a current passing through a solenoid and returned to its original position by a spring.

Fig. 5.9   Single-solenoid valve

Fig. 5.10   3/2 valve

Fig. 5.11   4/2 valve

Figure 5.11 shows the symbol for a 4/2 valve. The connections are shown for the initial state, i.e. 1(P) is connected to 2(A) and 3(R) closed. When the solenoid is activated it gives the state indicated by the symbols used in the square to which it is attached, i.e. we now have 1(P) closed and 2(A) connected to 3(R). When the current through the solenoid ceases, the spring pushes the valve back to its initial position. The spring movement gives the state indicated by the symbols used in the square to which it is attached.

Figure 5.12 shows a simple example of an application of valves in a pneumatic lift system. Two push-button 2/2 valves are used. When the button on the up valve is pressed, the load is lifted. When the button on the down valve is pressed, the load is lowered. Note that with pneumatic systems an open arrow is used to indicate a vent to the atmosphere.

Fig. 5.12   Lift system

Fig. 5.13   Pilot-operated system

Fig. 5.14   Directional valve

5.4 Pressure control valves

Fig. 5.15   Pressure limiting valve

5.3.2 Pilot-operated valves

The force required to move the ball or shuttle in a valve can often be too large for manual or solenoid operation. To overcome this problem a *pilot-operated system* is used where one valve is used to control a second valve. Figure 5.13 illustrates this. The pilot valve is small capacity and can be operated manually or by a solenoid. It is used to allow the main valve to be operated by the system pressure. The pilot pressure line is indicated by dashes. The pilot and main valves can be operated by two separate valves but they are often combined in a single housing.

5.3.3 Directional valves

Figure 5.14 shows a simple *directional valve* and its symbol. Free flow can only occur in one direction through the valve, that which results in the ball being pressed against the spring. Flow in the other direction is blocked by the spring forcing the ball against its seat.

There are three main types of pressure control valves:

1 *Pressure regulating valves*
   These are used to control the operating pressure in a circuit and maintain it at a constant value.

2 *Pressure limiting valves*
   These are used as safety devices to limit the pressure in a circuit to below some safe value. The valve opens and vents to the atmosphere, or back to the sump, if the pressure rises above the set safe value.

3 *Pressure sequence valves*
   These valves are used to sense the pressure of an external line and give a signal when it reaches some preset value.

5.4.1 Pressure limiting valve

Figure 5.15 shows a *pressure limiting/relief valve* which has one orifice which is normally closed. When the inlet pressure overcomes the force exerted by the spring, the valve opens and vents to the atmosphere, or back to the sump. This can be used as a pressure relief valve to safeguard a system against excessive pressures.

5.4.2 Pressure sequence valve

With the pressure limiting valve of Figure 5.15, the limiting pressure is set by the pressure at the inlet to the valve. We can adapt such a valve to give a sequence valve. This can be used to allow flow to occur to some part of the system when the pressure has risen to the required level. For example, in an automatic machine we might require some operation to start when the clamping pressure applied to a workpiece is at some particular value. Figure 5.16 shows the symbol for a sequence valve, the valve switching on when the inlet pressure reaches a particular value and allowing the pressure to be applied to the system that follows.

Figure 5.17 shows a system where such a sequential valve is used. When the 4/3 valve first operates, the pressure is applied to cylinder 1 and its ram moves to the right. While this is happening the pressure is too low to operate the sequence valve and so no pressure is applied to cylinder 2. When the ram of cylinder 1 reaches the end stop, then the pressure in the system rises and, at an appropriate level, triggers the sequence valve to open and so apply pressure to cylinder 2 to start its ram in motion.

Fig. 5.16 Pressure sequence valve

5.5 Cylinders

The *hydraulic* or *pneumatic cylinder* is an example of a linear actuator. The principles and form are the same for both hydraulic and pneumatic versions, differences being purely a matter of size as a consequence of the higher pressures used with hydraulics. The cylinder consists of a cylindrical tube along which a piston/ram can slide.

Fig. 5.17 Sequential system

The term *single acting* is used when the control pressure is applied to just one side of the piston, a spring often being used to provide the opposition to the movement of the piston. For the single-acting cylinder shown in Figure 5.18, when a current passes through the solenoid, the valve switches position and pressure is applied to move the piston along the cylinder. When the current through the solenoid ceases, the valve reverts to its initial position and the air is vented from the cylinder. As a consequence the spring returns the piston back along the cylinder.

The term *double acting* is used when the control pressures are applied to each side of the piston. A difference in pressure between the two sides then results in motion of the piston, the piston being able to move in either direction along the cylinder as a result of high pressure signals. For the double-acting cylinder shown in Figure 5.19, current through one solenoid causes the piston to move in one direction with current through the other solenoid reversing the direction of motion.

Fig. 5.18   Control of a single-acting cylinder

Fig. 5.19   Control of a double-acting cylinder

The choice of cylinder is determined by the force required to move the load and the speed required. Hydraulic cylinders are capable of much larger forces than pneumatic cylinders. However, pneumatic cylinders are capable of greater speeds. The force produced by a cylinder is equal to the cross-sectional area of the cylinder multiplied by the working pressure, i.e. the pressure difference between the two sides of the piston, in the cylinder. A cylinder for use with a working pneumatic pressure of 500 kPa and having a diameter of 50 mm will thus give a force of 982 N. A hydraulic cylinder with the same diameter and a working pressure of 15 000 kPa will give a force of 29.5 kN.

If the flow rate of hydraulic liquid into a cylinder is a volume of Q per second, then the volume swept out by the piston in a time of 1 s must be Q. But for a piston of cross-sectional area A this is a movement through a distance of v in 1 s, where we have $Q = Av$. Thus the speed v of a hydraulic cylinder is equal to the flow rate of liquid Q through the cylinder divided by the cross-sectional area A of the cylinder. Thus for a hydraulic cylinder of diameter 50 mm and a hydraulic fluid flow of 7.5×10^{-3} m³/s the speed is 3.8 m/s. The speed of a pneumatic cylinder cannot be calculated in this way since its speed depends on the rate at which air can be vented ahead of the advancing piston. A valve to adjust this can be used to regulate the speed.

To illustrate the above consider the problem of a hydraulic cylinder to be used to move a work piece in a manufacturing operation through a distance of 250 mm in 15 s. If a force of 50 kN is required to move the work piece, what is the required working pressure and hydraulic liquid flow rate if a cylinder with a piston diameter of 150 mm is available? The cross-sectional area of the piston is $\frac{1}{4}\pi \times 0.150^2 = 0.0177$ m². The force produced by the cylinder is equal to the product of the cross-sectional area of the cylinder and the working pressure. Thus the working pressure is $50 \times 10^3/0.0177 = 2.8$ MPa. The speed of a hydraulic cylinder is equal to the flow rate of liquid through the cylinder divided by the cross-sectional area of the cylinder. Thus the required flow rate is $(0.250/15) \times 0.0177 = 2.95 \times 10^{-4}$ m³/s.

5.5.1 Cylinder sequencing

Many control systems employ pneumatic or hydraulic cylinders as the actuating elements and require a sequence of extensions and retractions of the cylinders to occur. For example, we might have two cylinders A and B and require that when the start button is pressed, the piston of cylinder A extends and then, when it is fully extended, the piston of cylinder B extends. When this has happened and both are extended we might need the piston of cylinder A to retract, and when it is fully retracted we might then have the piston of B retract. In discussions of sequential control with cylinders it is common practice to give each cylinder a reference letter A, B, C, D, etc., and to indicate the state of each cylinder by using a + sign if it is extended or a − sign if retracted. Thus the above required sequence of operations is A+, B+, A−, B−. Figure 5.20 shows a circuit that could be used to generate this sequence.

The sequence of operations is:

1   Initially both the cylinders have retracted pistons. Start push-button on valve 1 is pressed. This applies pressure to valve 2, as initially limit switch b− is activated, hence valve 3 is switched to apply pressure to cylinder A for extension.

2   Cylinder A extends, releasing limit switch a−. When cylinder A is fully extended, limit switch a+ operates. This switches valve 5 and causes pressure to be applied to valve 6 to switch it and so apply pressure to cylinder B to cause its piston to extend.

3   Cylinder B extends, releasing limit switch b−. When cylinder B is fully extended, limit switch b+ operates. This switches valve 4 and causes pressure to be applied to valve 3 and so applies pressure to cylinder A to start its piston retracting.

4   Cylinder A retracts, releasing limit switch a+. When cylinder A is fully retracted, limit switch a− operates. This switches valve 7 and causes pressure to be applied to valve 5 and so applies pressure to cylinder B to start its piston retracting.

Fig. 5.20   Two-actuator sequential operation

5   Cylinder B retracts, releasing limit switch b+. When cylinder B is fully retracted, limit switch b– operates to complete the cycle.

The cycle can be started again by pushing the start button. If we wanted the system to run continuously then the last movement in the sequence would have to trigger the first movement.

An alternative way of realising the above sequence involves the air supply being switched on and off to valves in groups and is termed *cascade control*. This avoids a problem that can occur with circuits, formed in the way shown in Figure 5.20, of air becoming trapped in the pressure line to control a valve and so preventing the valve from switching. With cascade control, the sequence of operations is divided into groups with no cylinder letter appearing more than once in each group. Thus for the sequence A+, B+, B–, A– we can have the groups A+, B+ and A–, B–. A valve is then used to switch the air supply between the two groups, i.e. air to the group A+B+ and then the air switched to the group with A–B–. A start/stop valve is included in the line that selects the first group, and if the sequence is to be continuously repeated, the last operation has to supply a signal to start the sequence over again. The first function in each group is initiated by that group supply being switched on; further actions within the group are controlled by switch-operated valves, and the last valve operation initiates the next group to be selected. Figure 5.21 shows the pneumatic circuit.

5.6 Process control valves

Process control valves are used to control the rate of fluid flow and are used where, perhaps, the rate of flow of a liquid into a tank has to be controlled. The basis of such valves is an actuator being used to move a plug into the flow pipe and so alter the cross-section of the pipe through which the fluid can flow.

A common form of pneumatic actuator used with process control valves is the *diaphragm actuator*. Essentially it consists of a diaphragm with the input pressure signal from the controller on one side and atmospheric pressure on the other, this difference in pressure being termed the *gauge pressure*. The diaphragm is made of rubber which is sandwiched in its centre between two circular steel discs. The effect of changes in the input pressure is thus to move the central part of the diaphragm, as illustrated in Figure 5.22. This movement is communicated to the final control element by a shaft which is attached to the diaphragm.

The force F acting on the shaft is the force that is acting on the diaphragm and is thus the gauge pressure P multiplied by the diaphragm area A. A restoring force is provided by a spring. Thus if the shaft moves through a distance x, and assuming the compression of the spring is proportional to the force, i.e. $F = kx$ with k being a constant, then $kx = PA$ and thus the displacement of the shaft is proportional to the gauge pressure.

Fig. 5.21 Cascade control used to give A+, B+, B−, A−

Fig. 5.22 Pneumatic diaphragm actuator

Fig. 5.23 Diaphragm-operated
process control valve

To illustrate the above, consider the problem of a diaphragm actuator to be used to open a control valve if a force of 500 N must be applied to the valve. What diaphragm area is required for a control gauge pressure of 100 kPa? The force F applied to the diaphragm of area A by a pressure P is given by $P = F/A$. Hence $A = 500/(100 \times 10^3) = 0.005$ m^2.

5.6.1 Valve bodies and plugs

Figure 5.23 shows a cross-section of a valve for the control of rate of flow of a fluid. The pressure change in the actuator causes the diaphragm to move and so consequently the valve stem. The result of this is a movement of the inner-valve plug within the valve body. The plug restricts the fluid flow and so its position determines the flow rate.

There are many forms of valve body and plug. Figure 5.24 shows some forms of valve bodies. The term *single seated* is used for a valve where there is just one path for the fluid through the valve and so just one plug is needed to control the flow. The term *double seated* is used for a valve where the fluid on entering the valve splits into two streams, as in Figure 5.23, with each stream passing through an orifice controlled by a plug. There are thus two plugs with such a valve.

A single-seated valve has the advantage that it can be closed more tightly than a double-seated one but the disadvantage that the force on the plug due to the flow is much higher and so the diaphragm in the actuator has to exert considerably higher forces on the stem. This can result in problems in accurately positioning the plug. Double-seated valves thus have an advantage here. The form of the body also determines whether an increasing air pressure will result in the valve opening or closing.

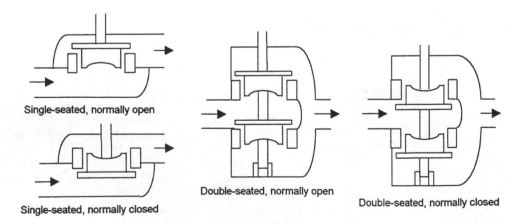

Fig. 5.24 Valve bodies

The shape of the plug determines the relationship between the stem movement and the effect on the flow rate. Figure 5.25 shows three commonly used types and Figure 5.26 how the percentage by which the volumetric rate of flow is related to the percentage displacement of the valve stem.

With the *quick-opening* type a large change in flow rate occurs for a small movement of the valve stem. Such a plug is used where on/off control of flow rate is required.

With the *linear-contoured* type, the change in flow rate is proportional to the change in displacement of the valve stem, i.e.

Change in flow rate = k (change in stem displacement)

where k is a constant. If Q is the flow rate at a valve stem displacement S and Q_{max} is the maximum flow rate at the maximum stem displacement S_{max}, then we have

$$\frac{Q}{Q_{max}} = \frac{S}{S_{max}}$$

or percentage change in the flow rate equals the percentage change in the stem displacement.

To illustrate the above consider the problem of an actuator which has a stem movement at full travel of 30 mm. It is mounted on a linear plug valve which has a minimum flow rate of 0 and a maximum flow rate of 40 m³/s. What will be the flow rate when the stem movement is (a) 10 mm, (b) 20 mm? Since the percentage flow rate is the same as the percentage stem displacement, then: (a) a percentage stem displacement of 33% gives a percentage flow rate of 33%, i.e. 13 m³/s; (b) a percentage stem displacement of 67% gives a percentage flow rate of 67%, i.e. 27 m³/s.

With the *equal percentage* type of plug, equal percentage changes in flow rate occur for equal changes in the valve stem position, i.e.

$$\frac{\Delta Q}{Q} = k \Delta S$$

where ΔQ is the change in flow rate at a flow rate of Q and ΔS the change in valve position resulting from this change. If we write this expression for small changes and then integrate it we obtain

$$\int_{Q_{min}}^{Q} \frac{1}{Q} \, dQ = k \int_{S_{min}}^{S} dS$$

Hence

$$\ln Q - \ln Q_{min} = k(S - S_{min})$$

If we consider the flow rate Q_{max} which is given by S_{max} then

Quick-opening

Linear-contoured

Equal percentage

Fig. 5.25   Plug shapes

Fig. 5.26   Flow characteristics with different plugs

$$\ln Q_{max} - \ln Q_{min} = k(S_{max} - S_{min})$$

Eliminating k from these two equations gives

$$\frac{\ln Q - \ln Q_{min}}{\ln Q_{max} - \ln Q_{min}} = \frac{S - S_{min}}{S_{max} - S_{min}}$$

$$\ln \frac{Q}{Q_{min}} = \frac{S - S_{min}}{S_{max} - S_{min}} \ln \frac{Q_{max}}{Q_{min}}$$

and so

$$\frac{Q}{Q_{min}} = \left(\frac{Q_{max}}{Q_{min}}\right)^{(S-S_{min})/(S_{max}-S_{min})}$$

The term *rangeability R* is used for the ratio Q_{max}/Q_{min}.

To illustrate the above, consider the problem of an actuator which has a stem movement at full travel of 30 mm. It is mounted with a control valve having an equal percentage plug and which has a minimum flow rate of 2 m³/s and a maximum flow rate of 24 m³/s. What will be the flow rate when the stem movement is (a) 10 mm, (b) 20 mm? Using the equation

$$\frac{Q}{Q_{min}} = \left(\frac{Q_{max}}{Q_{min}}\right)^{(S-S_{min})/(S_{max}-S_{min})}$$

we have for (a) $Q = 2 \times (24/2)^{10/30} = 4.6$ m³/s and for (b) $Q = 2 \times (24/2)^{20/30} = 10.5$ m³/s

The relationship between the flow rate and the stem displacement is the inherent characteristic of a valve. It is only realised in practice if the pressure losses in the rest of the pipework, etc., are negligible compared with the pressure drop across the valve itself. If there are large pressure drops in the pipework so that, for example, less than half the pressure drop occurs across the valve then a linear characteristic might become almost a quick-opening characteristic. The linear characteristic is thus widely used when a linear response is required and most of the system pressure is dropped across the valve. The effect of large pressure drops in the pipework with an equal percentage valve is to make it more like a linear characteristic. For this reason, if a linear response is required when only a small proportion of the system pressure is dropped across the valve, then an equal percentage value might be used.

5.6.2 Control valve sizing

The term *control valve sizing* is used for the procedure of determining the correct size of valve body. The equation relating the rate of flow of liquid Q through a wide open valve to its size is

$$Q = A_V \sqrt{\frac{\Delta P}{\rho}}$$

where A_V is the valve flow coefficient, ΔP the pressure drop across the valve and ρ the density of the fluid. This equation is sometimes written, with the quantities in SI units, as

$$Q = 2.37 \times 10^{-5} C_V \sqrt{\frac{\Delta P}{\rho}}$$

where C_V is the valve flow coefficient. Alternatively it may be found written as

$$Q = 0.75 \times 10^{-6} C_V \sqrt{\frac{\Delta P}{G}}$$

where G is the specific gravity or relative density. These last two forms of the equation derive from its original specification in terms of US gallons. Table 5.1 shows some typical values of A_V, C_V and valve size.

To illustrate the above, consider the problem of determining the valve size for a valve that is required to control the flow of water when the maximum flow required is 0.012 m³/s and the permissible pressure drop across the valve at this flow rate is 300 kPa. Using the equation

Table 5.1 Flow coefficients and valve sizes

Flow	Valve size (mm)							
coefficients	480	640	800	960	1260	1600	1920	2560
C_V	8	14	22	30	50	75	110	200
$A_V \times 10^{-5}$	19	33	52	71	119	178	261	474

$$Q = A_V \sqrt{\frac{\Delta P}{\rho}}$$

then, since the density of water is 1000 kg/m³,

$$A_V = Q \sqrt{\frac{\rho}{\Delta P}} = 0.012 \sqrt{\frac{1000}{300 \times 10^3}} = 69.3 \times 10^{-5}$$

Thus, using Table 5.1, the valve size is 960 mm.

5.6.3 Example of fluid control system

Figure 5.27 shows the essential features of a system for the control of a variable such as the level of a liquid in a container by controlling the rate at which liquid enters it. The output from the liquid level sensor, after signal conditioning, is transmitted to the current to pressure converter as a current of 4 to 20 mA. It is then

Fig. 5.27   Fluid control system

converted into a gauge pressure of 20 to 100 kPa which then actuates a pneumatic control valve and so controls the rate at which liquid is allowed to flow into the container.

Figure 5.28 shows the basic form of a current to pressure converter. The input current passes through coils mounted on a core which is attracted towards a magnet, the extent of the attraction depending on the size of the current. The movement of the core causes movement of the lever about its pivot and so the movement of a flapper above the nozzle. The position of the flapper in relation to the nozzle determines the rate at which air can escape from the system and hence the air pressure in the system. Springs on the flapper are used to adjust the sensitivity of the converter so that currents of 4 to 20 mA produce gauge pressures of 20 to 100 kPa. These are the standard values that are generally used in such systems.

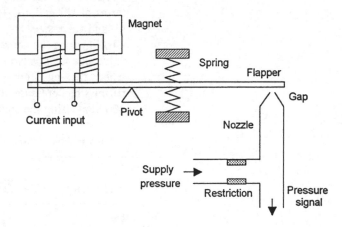

Fig. 5.28 Current to pressure converter

5.7 Rotary actuators

A linear cylinder can, with suitable mechanical linkages, be used to produce rotary movement through angles less than 360°, Figure 5.29 illustrating such an arrangement. Another alternative is a *semi-rotary actuator* involving a vane (Fig. 5.30). A pressure difference between the two ports causes the vane to rotate and so give a shaft rotation which is a measure of the pressure difference. Depending on the pressures, so the vane can be rotated clockwise or anti-clockwise.

For rotation through angles greater than 360° a pneumatic motor can be used; one form of such is the *vane motor* (Fig. 5.31). An eccentric rotor has slots in which vanes are forced outwards against the walls of the cylinder by the rotation. The vanes divide the chamber into separate compartments which increase in size from the inlet port round to the exhaust port. The air entering such a compartment exerts a force on a vane and causes the rotor to rotate. The motor can be made to reverse its direction of rotation by using a different inlet port.

Fig. 5.29 A linear cylinder used to produce rotation

Fig. 5.30   Vane-type semi-rotary actuator

Fig. 5.31   Vane motor

Problems

1 Describe the basic details of (a) a poppet valve, (b) a shuttle valve.

2 Explain the principle of a pilot-operated valve.

3 Explain how a sequential valve can be used to initiate an operation only when another operation has been completed.

4 Draw the symbols for (a) a pressure relief valve, (b) a 2/2 valve which has actuators of a push-button and a spring, (c) a 4/2 valve, (d) a directional valve.

5 State the sequence of operations that will occur for the cylinders A and B in Figure 5.32 when the start button is pressed. a–, a+, b– and b+ are limit switches to detect when the cylinders are fully retracted and fully extended.

6 Design a pneumatic valve circuit to give the sequence A+, followed by B+ and then simultaneously followed by A– and B–.

7 A force of 400 N is required to open a process control valve. What area of diaphragm will be needed with a diaphragm actuator to open the valve with a control gauge pressure of 70 kPa?

8 A pneumatic system is operated at a pressure of 1000 kPa. What diameter cylinder will be required to move a load requiring a force of 12 kN?

9 A hydraulic cylinder is to be used to move a work piece in a manufacturing operation through a distance of 50 mm in 10 s. A force of 10 kN is required to move the work piece. Determine the required working pressure and hydraulic liquid flow rate if a cylinder with a piston diameter of 100 mm is available.

10 An actuator has a stem movement which at full travel is 40 mm. It is mounted with a linear plug process control valve which has a minimum flow rate of 0 and a maximum flow rate of 0.20 m³/s. What will be the flow rate when the stem movement is (a) 10 mm, (b) 20 mm?

Fig. 5.32   Problem 5

11   An actuator has a stem movement which at full travel is 40 mm. It is mounted on a process control valve with an equal percentage plug and which has a minimum flow rate of 0.2 m³/s and a maximum flow rate of 4.0 m³/s. What will be the flow rate when the stem movement is (a) 10 mm, (b) 20 mm?

12   What is the process control valve size required for a valve that is required to control the flow of water when the maximum flow required is 0.002 m³/s and the permissible pressure drop across the valve at this flow rate is 100 kPa? The density of water is 1000 kg/m³.

6 Mechanical actuation systems

6.1 Mechanical systems

This chapter is a consideration of *mechanisms*; mechanisms are devices which can be considered to be motion converters in that they transform motion from one form to some other required form. They might, for example, transform linear motion into rotational motion, or motion in one direction into a motion in a direction at right angles, or perhaps a linear reciprocating motion into rotary motion, as in the internal combustion engine where the reciprocating motion of the pistons is converted into rotation of the crank and hence the drive shaft.

Mechanical elements can include the use of linkages, cams, gears, rack-and-pinion, chains, belt drives, etc. For example, the rack-and-pinion can be used to convert rotational motion to linear motion. Parallel shaft gears might be used to reduce a shaft speed. Bevel gears might be used for the transmission of rotary motion through 90°. A toothed belt or chain drive might be used to transform rotary motion about one axis to motion about another. Cams and linkages can be used to obtain motions which are prescribed to vary in a particular manner. This chapter is a consideration of the basic characteristics of a range of such mechanisms.

Many of the actions which previously were obtained by the use of mechanisms are, however, often nowadays being obtained by the use of microprocessor systems. For example, cams on a rotating shaft were previously used for domestic washing machines in order to give a timed sequence of actions such as opening a valve to let water into the drum, switching the water off, switching a heater on, etc. Modern washing machines use a microprocesssor-based system with the microprocessor programmed to switch on outputs in the required sequence.

Mechanisms still, however, have a role in mechatronics systems. For example, the mechatronics system in use in an automatic camera for adjusting the aperture for correct exposures involves a mechanism for adjusting the size of the diaphragm. While electronics might now be used often for many functions that previously were fulfilled by mechanisms, mechanisms might still be used to provide such functions as:

1 Force amplification, e.g. that given by levers.
2 Change of speed, e.g. that given by gears.
3 Transfer of rotation about one axis to rotation about another, e.g. a timing belt.
4 Particular types of motion, e.g. that given by a quick-return mechanism.

The term *kinematics* is used for the study of motion without regard to forces. When we consider just the motions without any consideration of the forces or energy involved then we are carrying out a kinematic analysis of the mechanism. This chapter is an introduction to such a consideration. For more detail the reader is referred to general texts for mechanical engineers, such as *Mechanical Science* by W. Bolton (Blackwell Scientific Publications 1993, 1998), or more specialist texts on the principles of machines, such as *Design of Machinery* by R.L. Norton (McGraw-Hill 1992).

6.2 Types of motion

A rigid body can have a very complex motion which might seem difficult to describe. However, the motion of any rigid body can be considered to be a combination of translational and rotational motions. By considering the three dimensions of space, a translation motion can be considered to be a movement which can be resolved into components along one or more of the three axes (Fig. 6.1(a)). A rotation can be considered as a rotation which has components rotating about one or more of the axes (Fig. 6.1(b)).

A complex motion may be a combination of translational and rotational motions. For example, think of the motion required for you to pick up a pencil from a table. This might involve your hand moving at a particular angle towards the table, rotation of the hand, and then all the movement associated with opening your fingers and moving them to the required positions to grasp the pencil. This is a sequence of quite complex motions. However, we can break down all these motions into combinations of translational and rotational motions. Such an analysis is particularly relevant if we are not moving a human hand to pick up the pencil but instructing a robot to carry out the task. Then it really is necessary to break down the motion into combinations of translational and rotational motions so that we can design mechanisms to carry out each of these components of the motion. For example, among the sequence of control signals sent to a mechanism might be such groupings of signals as those to instruct joint 1 to rotate by 20° and link 2 to be extended by 4 mm for translational motion.

Fig. 6.1 Types of motion

6.2.1 Freedom and constraints

An important aspect in the design of mechanical elements is the orientation and arrangement of the elements and parts. A body that is free in space can move in three, independent, mutually perpendicular directions and rotate in three ways about those directions (Fig. 6.1). It is said to have six degrees of freedom. The number of *degrees of freedom* are the number of components of motion that are required in order to generate the motion. If a joint is constrained to move along a line then its translational degrees of freedom are reduced to one. Figure 6.2(a) shows a joint with just this one translational degree of freedom. If a joint is constrained to move on a plane then it has two translational degrees of freedom. Figure 6.2(b) shows a joint which has one translational degree of freedom and one rotational degree of freedom.

The problem in design is often to reduce the number of degrees of freedom and this then requires an appropriate number and orientation of constraints. Without any constraints a body would have six degrees of freedom. A constraint is needed for each degree of freedom that is to be prevented from occurring. Provided we have no redundant constraints then the number of degrees of freedom would be 6 minus the number of constraints. However, redundant constraints often occur and so for constraints on a single rigid body we have the basic rule:

$$6 - \text{number of constraints} = \text{number of degrees of freedom}$$
$$- \text{number of redundancies}$$

Thus if a body is required to be fixed, i.e. have zero degrees of freedom, then if no redundant constraints are introduced the number of constraints required is 6.

A concept that is used in design is that of the *principle of least constraint*. This states that in fixing a body or guiding it to a particular type of motion, the minimum number of constraints should be used, i.e. there should be no redundancies. This is often referred to as *kinematic design*.

For example, to have a shaft which only rotates about one axis with no translational motions, we have to reduce the number of degrees of freedom to 1. Thus the minimum number of constraints to do this is 5. Any more constraints than this will give redundancies. The mounting that might be used to mount the shaft has a ball bearing at one end and a roller bearing at the other (Fig. 6.3). The pair of bearings together prevent translational motion at right angles to the shaft, the y-axis, and rotations about the z-axis and the y-axis. The ball bearing prevents translational motion along the x-axis and along the z-axis. Thus there is a total of five constraints. This leaves just one degree of freedom, the required rotation about the x-axis. If there had been a roller bearing at each end of the shaft then both

(a)

(b)

Fig. 6.2   Joints with: (a) one, (b) two degrees of freedom

Fig. 6.3   Shaft with no redundancies

the bearings could have prevented translational motion along the x-axis and the z-axis and thus there would have been redundancy. Such redundancy might cause damage. If ball bearings are used at both ends of the shaft, then in order to prevent redundancy one of the bearings would have its outer race not fixed in its housing so that it could slide to some extent in an axial direction.

6.2.2 Loading

Mechanisms are structures and as such transmit and support loads. Analysis is thus necessary to determine the loads to be carried by individual elements. Then consideration can be given to the dimensions of the element so that it might, for example, have sufficient strength and perhaps stiffness under such loading.

6.3 Kinematic chains

When we consider the movements of a mechanism without any reference to the forces involved, we can treat the mechanism as being composed of a series of individual links. Each part of a mechanism which has motion relative to some other part is termed a *link*. A link need not necessarily be a rigid body but it must be a resistant body which is capable of transmitting the required force with negligible deformation. For this reason it is usually taken as being represented by a rigid body which has two or more points of attachment to other links, these being termed *nodes*. Each link is capable of moving relative to its neighbouring links. Figure 6.4 shows examples of links with two, three and four nodes. A *joint* is a connection between two or more links at their nodes and which allows some motion between the connected links. Levers, cranks, connecting rods and pistons, sliders, pulleys, belts and shafts are all examples of links.

A sequence of joints and links is known as a *kinematic chain*. For a kinematic chain to transmit motion, one link must be fixed. Movement of one link will then produce predictable relative movements of the others. It is possible to obtain from one kinematic chain a number of different mechanisms by having a different link as the fixed one.

As an illustration of a kinematic chain, consider a motor car engine where the reciprocating motion of a piston is transformed into rotational motion of a crankshaft on bearings mounted in a fixed frame (Fig. 6.5(a)). We can represent this as being four connected links (Fig. 6.5(b)). Link 1 is the crankshaft, link 2 the connecting rod, link 3 the fixed frame and link 4 the slider, i.e. piston, which moves relative to the fixed frame (see Section 6.3.2 for further discussion).

The designs of many mechanisms are based on two basic forms of kinematic chains, the four-bar chain and the slider–crank chain. The following illustrates some of the forms such chains can take.

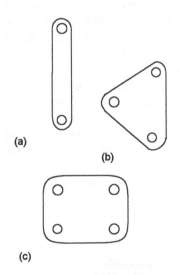

(a)

(b)

(c)

Fig. 6.4   Links: (a) with two nodes, (b) with three nodes, (c) with four nodes

Fig. 6.5 Simple engine mechanism

6.3.1 The four-bar chain

The four-bar chain consists of four links connected to give four joints about which turning can occur. Figure 6.6 shows a number of forms of the four-bar chain produced by altering the relative lengths of the links. If the sum of the length of the shortest link plus the length of the longest link is less than or equal to the sum of the lengths of the other two links then at least one link will be capable of making a full revolution with respect to the fixed link. If this condition is not met then no link is capable of a complete revolution. This is known as the Grashof condition. In Figure 6.6(a), link 3 is fixed and the relative lengths of the links are such that links 1 and 4 can oscillate but not rotate. The result is a *double-lever mechanism*. By shortening link 4 relative to link 1, then link 4 can rotate (Fig. 6.6(b)) with link 1 oscillating and the result is termed a *lever–crank mechanism*. With links 1 and 4 the same length and both able to rotate (Fig. 6.6(c)), then the result is a *double-crank mechanism*. By altering which link is fixed, other forms of mechanism can be produced.

Figure 6.7 illustrates how such a mechanism can be used to advance the film in a cine camera. As link 1 rotates so the end of link 2 locks into a sprocket of the film, pulls it forward before releasing and moving up and back to lock into the next sprocket.

Fig. 6.6 Examples of four-bar chains

Fig. 6.7 Cine film advance mechanism

Some linkages may have *toggle positions*. These are positions where the linkage will not react to any input from one of its links. Figure 6.8 illustrates such a toggle, being the linkage used to control the movement of the tailgate of a truck so that when link 2 reaches the horizontal position no further load on link 2 will cause any further movement. There is another toggle position for the linkage and that is when links 3 and 4 are both vertical and the tailgate is vertical.

6.3.2 The slider–crank mechanism

This form of mechanism consists of a crank, a connecting rod and a slider and is the type of mechanism described in Figure 6.5 which showed the simple engine mechanism. With that configuration, link 3 is fixed, i.e. there is no relative movement between the centre of rotation of the crank and the housing in which the piston slides. Link 1 is the crank that rotates, link 2 the connecting rod and link 4 the slider which moves relative to the fixed link. When the piston moves backwards and forwards, i.e. link 4 moves backwards and forwards, then the crank, link 1, is forced to rotate. Hence the mechanism transforms an input of backwards and forwards motion into rotational motion.

Tailgate in toggled position

Fig. 6.8 Toggle linkage

Figure 6.9 shows another form of this type of mechanism, a *quick-return mechanism*. It consists of a rotating crank, link AB, which rotates round a fixed centre, an oscillating lever CD, which is caused to oscillate about C by the sliding of the block at B along CD as AB rotates, and a link DE which causes E to move backwards and forwards. E might be the ram of a machine and have a cutting tool attached to it. The ram will be at the extremes of its movement when the positions of the crank are AB_1 and AB_2. Thus as the crank moves anti-clockwise from B_1 to B_2 the ram makes a complete stroke, the cutting stroke. When the crank continues its movement from B_2 anti-clockwise to B_1 then the ram again makes a complete stroke in the opposite direction, the return stroke. With the crank rotating at constant speed, then, because the angle of crank rotation required for the cutting stroke is greater than the angle for the return stroke, the cutting stroke takes more time than the return stroke. Hence the term, quick-return for the mechanism.

Fig. 6.9 Quick-return mechanism

6.4 Cams

Fig. 6.10   Cam and cam follower

A *cam* is a body which rotates or oscillates and in doing so imparts a reciprocating or oscillatory motion to a second body, called the *follower*, with which it is in contact (Fig. 6.10). As the cam rotates so the follower is made to rise, dwell and fall, the lengths of times spent at each of these positions depending on the shape of the cam. The rise section of the cam is the part that drives the follower upwards, its profile determining how quickly the cam follower will be lifted. The fall section of the cam is the part that lowers the follower, its profile determining how quickly the cam follower will fall. The dwell section of the cam is the part that allows the follower to remain at the same level for a significant period of time. The dwell section of the cam is where it is circular with a radius that does not change.

The cam shape required to produce a particular motion of the follower will depend on the shape of the cam and the type of follower used. Figure 6.11 shows the types of follower displacement diagrams that can be produced with different shaped cams and either point or knife followers. The radial distance from the axis of rotation of the cam to the point of contact of the cam with the follower gives the displacement of the follower with reference to the axis of rotation of the cam. The figures show how these radial distances, and hence follower displacements, vary with the angle of rotation of the cams.

Fig. 6.11   Cams: (a) eccentric, (b) heart-shaped, (c) pear-shaped

Fig. 6.12   Cam followers: (a) point, (b) knife, (c) roller, (d) sliding and oscillating, (e) flat, (f) mushroom

The eccentric cam (Fig. 6.11(a)) is a circular cam with an offset centre of rotation. It produces an oscillation of the follower which is simple harmonic motion and is often used with pumps. The heart-shaped cam (Fig. 6.11(b)) gives a follower displacement which increases at a constant rate with time before decreasing at a constant rate with time, hence a uniform speed for the follower. The pear-shaped cam (Fig. 6.11(c)) gives a follower motion which is stationary for about half a revolution of the cam and rises and falls symmetrically in each of the remaining quarter revolutions. Such a pear-shaped cam is used for engine valve control. The dwell holds the valve open while the petrol/air mixture passes into the cylinder. The longer the dwell, i.e. the greater the length of the cam surface with a constant radius, the more time is allowed for the cylinder to be completely charged with flammable vapour.

Figure 6.12 shows a number of examples of different types of cam followers. Roller followers are essentially ball or roller bearings. They have the advantage of lower friction than a sliding contact but can be more expensive. Flat-faced followers are often used because they are cheaper and can be made smaller than roller followers. Such followers are widely used with engine valve cams. While cams can be run dry, they are often used with lubrication and may be immersed in an oil bath.

6.5 Gear trains

Fig. 6.13   Rolling cylinders

Gear trains are mechanisms which are very widely used to transfer and transform rotational motion. They are used when a change in speed or torque of a rotating device is needed. For example, the car gearbox enables the driver to match the speed and torque requirements of the terrain with the engine power available.

Rotary motion can be transferred from one shaft to another by a pair of rolling cylinders (Fig. 6.13); however, there is a possibility of slip. The transfer of the motion between the two cylinders depends on the frictional forces between the two surfaces in contact. Slip can be prevented by the addition of meshing teeth to the two cylinders and the result is then a pair of meshed gear wheels.

Gears can be used for the transmission of rotary motion between parallel shafts (Fig. 6.14(a)) and for shafts which have axes inclined to one another (Fig. 6.14(b)). The term *bevel gears* is used when the lines of the shafts intersect, as illustrated in Figure 6.14(b). When two gears are in mesh, the larger gear wheel is often called the *spur* or *crown wheel* and the smaller one the *pinion*. Gears for use with parallel shafts may have axial teeth with the teeth cut along axial lines parallel to the axis of the shaft (Fig. 6.15(a)). Such gears are then termed *spur gears*. Alternatively they may have helical teeth with the teeth being cut on a helix (Fig. 6.15(b)) and are then termed *helical gears*. Helical gears have the advantage that there is a gradual engagement of any individual tooth and consequently there is a smoother drive and generally prolonged life of the gears. However, the inclination of the teeth to the axis of the shaft results in an axial force component on the shaft bearing. This can be overcome by using double helical teeth (Fig. 6.15(c)).

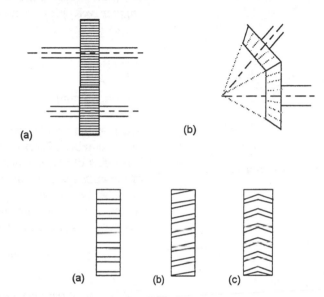

(a)　　　　　　　　　　　　(b)

Fig. 6.14 Gear axes: (a) parallel, (b) inclined to one another

Fig. 6.15 Teeth: (a) axial, (b) helical, (c) double helical

(a)　　　　(b)　　　　(c)

Another form of gear is the *rack-and-pinion* (Fig. 6.16), this being essentially two intermeshed gears with one having a base circle of infinite radius. Such gears can be used to transform either linear motion to rotational motion or rotational motion to linear motion.

Consider two meshed gear wheels A and B (Fig. 6.17). If there are 40 teeth on wheel A and 80 teeth on wheel B, then wheel A must rotate through two revolutions in the same time as wheel B rotates through one. Thus the angular velocity ω_A of wheel A must be twice that ω_B of wheel B, i.e.

Fig. 6.16 Rack-and-pinion

$$\frac{\omega_A}{\omega_B} = \frac{\text{number of teeth on B}}{\text{number of teeth on A}} = \frac{80}{40} = 2$$

Since the number of teeth on a wheel is proportional to its diameter, we can write:

$$\frac{\omega_A}{\omega_B} = \frac{\text{number of teeth on B}}{\text{number of teeth on A}} = \frac{d_B}{d_A}$$

Fig. 6.17 Two meshed gears

Thus for the data we have been considering, wheel B must have twice the diameter of wheel A. The term *gear ratio* is used for the ratio of the angular speeds of a pair of intermeshed gear wheels. Thus the gear ratio for this example is 2.

6.5.1 Gear trains

The term *gear train* is used to describe a series of intermeshed gear wheels. The term *simple gear train* is used for a system where each shaft carries only one gear wheel, as in Figure 6.18. For such a gear train, the overall gear ratio is the ratio of the angular velocities at the input and output shafts and is thus ω_A/ω_C.

$$G = \frac{\omega_A}{\omega_C}$$

Consider a simple gear train consisting of wheels A, B and C, as in Figure 6.18, with A having 9 teeth and C having 27 teeth. Then, as the angular velocity of a wheel is inversely proportional to the number of teeth on the wheel, the gear ratio is 27/9 = 3. The effect of wheel B is purely to change the direction of rotation of the output wheel compared with what it would have been with just the two wheels A and C intermeshed. The intermediate wheel, B, is termed the *idler wheel*.

We can rewrite this equation for the overall gear ratio G as

$$G = \frac{\omega_A}{\omega_C} = \frac{\omega_A}{\omega_B} \times \frac{\omega_B}{\omega_C}$$

But ω_A/ω_B is the gear ratio for the first pair of gears and ω_B/ω_C the gear ratio for the second pair of gears. Thus the overall gear ratio for a simple gear train is the product of the gear ratios for each successive pair of gears.

The term *compound gear train* is used to describe a gear train when two wheels are mounted on a common shaft. Figure 6.19(a) and (b) shows two examples of such a compound gear train. The gear train in Figure 6.19(b) enables the input and output shafts to be in line. An alternative way of achieving this is the epicyclic gear train discussed in the next section.

When two gear wheels are mounted on the same shaft they have the same angular velocity. Thus, for both of the compound gear trains in Figure 6.19, $\omega_B = \omega_C$. The overall gear ratio G is thus

$$G = \frac{\omega_A}{\omega_D} = \frac{\omega_A}{\omega_B} \times \frac{\omega_B}{\omega_C} \times \frac{\omega_C}{\omega_D} = \frac{\omega_A}{\omega_B} \times \frac{\omega_C}{\omega_D}$$

For the arrangement shown in Figure 6.19(b), for the input and output shafts to be in line we must also have for the radii of the gears:

$$r_A + r_B = r_D + r_C$$

Consider a compound gear train of the form shown in Figure 6.19(a), with A, the first driver, having 15 teeth, B 30 teeth, C 18 teeth and D, the final driven wheel, 36 teeth. Since the angular

Driver   Idler   Driven

Fig. 6.18   Simple gear train

(a)

(b)

Fig. 6.19   Compound gear trains

velocity of a wheel is inversely proportional to the number of teeth on the wheel, the overall gear ratio is

$$G = \frac{30}{15} \times \frac{36}{18} = 4$$

Thus, if the input to wheel A is an angular velocity of 160 rev/min, then the output angular velocity of wheel D is 160/4 = 40 rev/min.

A simple gear train of spur, helical or bevel gears is usually limited to an overall gear ratio of about 10. This is because of the need to keep the gear train down to a manageable size if the number of teeth on the pinion is to be kept above a minimum number which is usually about 10 to 20. Higher gear ratios can, however, be obtained with compound gear trains. This is because the gear ratio is the product of the individual gear ratios of parallel gear sets.

6.6 Ratchet and pawl

Fig. 6.20   Ratchet and pawl

Ratchets can be used to lock a mechanism when it is holding a load. Figure 6.20 shows a ratchet and pawl. The mechanism consists of a wheel, called a *ratchet*, with saw-shaped teeth which engage with an arm called a *pawl*. The arm is pivoted and can move back and forth to engage the wheel. The shape of the teeth is such that rotation can occur in only one direction. Rotation of the ratchet wheel in a clockwise direction is prevented by the pawl and can only take place when the pawl is lifted. The pawl is normally spring loaded to ensure that it automatically engages with the ratchet teeth.

Thus a winch used to wind up a cable on a drum may have a ratchet and pawl to prevent the cable unwinding from the drum when the handle is released.

6.7 Belt and chain drives

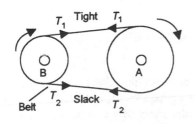

Fig. 6.21   Belt drive

Belt drives are essentially just a pair of rolling cylinders, as described in Figure 6.13 and Section 6.5, with the motion of one cylinder being transferred to the other by a belt (Fig. 6.21). Belt drives use the friction that develops between the pulleys attached to the shafts and the belt around the arc of contact in order to transmit a torque. Since the transfer relies on frictional forces then slip can occur. The transmitted torque is due to the differences in tension that occur in the belt during operation. This difference results in a tight side and a slack side for the belt. If the tension on the tight side is T_1, and that on the slack side T_2, then with pulley A in Figure 6.21 as the driver:

Torque on A = $(T_1 - T_2)r_A$

where r_A is the radius of pulley A. For the driven pulley B we have:

Torque on B = $(T_1 - T_2)r_B$

where r_B is the radius of pulley B. Since the power transmitted is the product of the torque and the angular velocity, and since the angular velocity is v/r_A for pulley A and v/r_B for pulley B, where v is the belt speed, then for either pulley we have:

Power = $(T_1 - T_2)v$

As a method of transmitting power between two shafts, belt drives have the advantage that the length of the belt can easily be adjusted to suit a wide range of shaft-to-shaft distances and the system is automatically protected against overload because slipping occurs if the loading exceeds the maximum tension that can be sustained by frictional forces. If the distances between shafts is large, a belt drive is more suitable than gears, but over small distances gears are to be preferred. Different size pulleys can be used to give a gearing effect. However, the gear ratio is limited to about 3 because of the need to maintain an adequate arc of contact between the belt and the pulleys.

The belt drive shown in Figure 6.21 gives the driven wheel rotating in the same direction as the driver wheel. Figure 6.22 shows two types of reversing drives. With both forms of drive, both sides of the belt come into contact with the wheels and so V-belts or timing belts cannot be used.

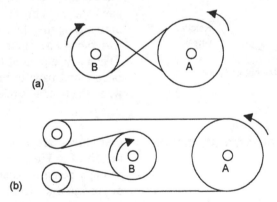

(a)

(b)

Fig. 6.22 Reversed belt drives: a) crossed belt, (b) open belt

6.7.1 Types of belts

The four main types of belts (Fig. 6.23) are:

1   *Flat*
    The belt has a rectangular cross-section. Such a drive has an efficiency of about 98% and produces little noise. They can transmit power over long distances between pulley centres. Crowned pulleys are used to keep the belts from running off the pulleys.

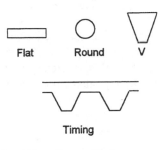

Flat   Round   V

Timing

Fig. 6.23   Types of belt

2   *Round*
The belt has a circular cross-section and is used with grooved pulleys.

3   *V*
V-belts are used with grooved pulleys and are less efficient than flat belts but a number of them can be used on a single wheel and so give a multiple drive.

4   *Timing*
Timing belts require toothed wheels, having teeth which fit into the grooves on the wheels. The timing belt, unlike the other belts, does not stretch or slip and consequently transmits power at a constant angular velocity ratio. The teeth make it possible for the belt to be run at slow or fast speeds.

6.7.2 Chains

Slip can be prevented by the use of chains which lock into teeth on the rotating cylinders to give the equivalent of a pair of intermeshing gear wheels. A chain drive has the same relationship for gear ratio as a simple gear train. The drive mechanism used with a bicycle is an example of a chain drive. Chains enable a number of shafts to be driven by a single wheel and so give a multiple drive. They are not as quiet as timing belts but can be used for larger torques.

6.8 Bearings

Whenever there is relative motion of one surface in contact with another, either by rotating or sliding, the resulting frictional forces generate heat which wastes energy and results in wear. The function of a *bearing* is to guide with minimum friction and maximum accuracy the movement of one part relative to another.

Of particular importance is the need to give suitable support to rotating shafts, i.e. support radial loads. The term *thrust bearing* is used for bearings that are designed to withstand forces along the axis of a shaft when the relative motion is primarily rotation. The following sections outline the characteristics of commonly used forms of bearings.

6.8.1 Plain journal bearings

Journal bearings are used to support rotating shafts which are loaded in a radial direction; the term *journal* is used for a shaft. The bearing basically consists of an insert of some suitable material which is fitted between the shaft and the support (Fig. 6.24). Rotation of the shaft results in its surface sliding over that of the bearing surface. The insert may be a white metal, aluminium alloy, copper alloy, bronze or a polymer such as nylon

Fig. 6.24   Plain journal bearing

Fig. 6.25   Hydrodynamic journal bearing

Fig. 6.26   Basic elements of a ball bearing

or PTFE. The insert provides lower friction and less wear than if the shaft just rotated in a hole in the support. The bearing may be a dry rubbing bearing or lubricated. Plastics such as nylon and PTFE are generally used without lubrication, the coefficient of friction with such materials being exceptionally low. A widely used bearing material is sintered bronze, this is bronze with a porous structure which allows it to be impregnated with oil and so the bearing has a 'built in' lubricant.

The lubricant may be:

1   *Hydrodynamic*
    The *hydrodynamic journal bearing* consists of the shaft rotating continuously in oil in such a way that it rides on oil and is not supported by metal (Fig. 6.25). The load is carried by the pressure generated in the oil as a result of the shaft rotating.

2   *Hydrostatic*
    A problem with hydrodynamic lubrication is that the shaft only rides on oil when it is rotating and when at rest there is metal-to-metal contact. To avoid excessive wear at start-up and when there is only a low load, oil is pumped into the load-bearing area at a high-enough pressure to lift the shaft off the metal when at rest.

3   *Solid-film*
    This is a coating of a solid material such as graphite or molybdenum disulphide.

4   *Boundary layer*
    This is a thin layer of lubricant which adheres to the surface of the bearing.

6.8.2 Ball and roller bearings

With this type of bearing, the main load is transferred from the rotating shaft to its support by rolling contact rather than sliding contact. A rolling element bearing consists of four main elements: an inner race, an outer race, the rolling element of either balls or rollers, and a cage to keep the rolling elements apart (Fig. 6.26). The inner and outer races contain hardened tracks in which the rolling elements roll.

There are a number of forms of ball bearings:

1   *Deep-groove* (Fig. 6.27(a))
    This is good at withstanding radial loads but is only moderately good for axial loads. It is a versatile bearing which can be used with a wide range of load and speed.

Fig. 6.27   Types of ball bearings

2   *Filling-slot* (Fig. 6.27(b))
    This is able to withstand higher radial loads than the deep-groove equivalent but cannot be used when there are axial loads.

3   *Angular contact* (Fig. 6.27(c))
    This is good for both radial and axial loads and is better for axial loads than the deep-groove equivalent.

4   *Double-row* (Fig. 6.27(d))
    Double-row ball bearings are made in a number of types and are able to withstand higher radial loads than their single-row equivalents. The figure shows a double-row deep-groove ball bearing, there being double-row versions of each of the above single-row types.

5   *Self-aligning* (Fig. 6.27(e))
    Single-row bearings can withstand a small amount of shaft misalignment but where there can be severe misalignment a self-aligning bearing is used. This is able to withstand only moderate radial loads and is fairly poor for axial loads.

6   *Thrust, grooved race* (Fig. 6.27(f))
    These are designed to withstand axial loads but are not suitable for radial loads.

There are also a number of forms of roller bearing, the following being common examples:

1   *Straight roller* (Fig. 6.28(a))
    This is better for radial loads than the equivalent ball bearing but is not generally suitable for axial loads. They will carry a greater load than ball bearings of the same size because of

Fig. 6.28   Roller bearings

their greater contact area. However, they are not tolerant of misalignment.

2 *Taper roller* (Fig. 6.28(b))
This is good for radial loads and good in one direction for axial loads.

3 *Needle roller* (Fig. 6.28(c))
This has a roller with a high length/diameter ratio and tends to be used in situations where there is insufficient space for the equivalent ball or roller bearing.

6.8.3 Selection of bearings

In general, dry sliding bearings tend to be only used for small diameter shafts with low load and low speed situations, ball and roller bearings, i.e. bearings involving rolling, with a much wider range of diameter shafts and higher load and higher speed, and hydrodynamic bearings for the high loads with large diameter shafts. Figure 6.29 shows a chart indicating the selection of bearings based on their load–shaft speed characteristics for a number of different diameter shafts (the data used is based on the paper by M.J. Neale in *Proc. I. Mech. E.*, 182(3A), 547 (1967)). Thus suppose we want a bearing for a 25 mm diameter shaft rotating at 10 rev/s and carrying a radial load of 10 000 N. This is beyond the limit for a dry sliding bearing and is a point on the graph below the line for rolling bearings for such a diameter and speed, hence rolling bearings can be used.

Fig. 6.29 Journal bearing selection

Failure of ball and roller bearings generally occurs as a result of fatigue. With fatigue failures there is always a scatter of values at which failure of an individual item occurs. The life of a bearing is thus defined as the number of millions of shaft revolutions that 90% of the bearings are expected to exceed before failing. This life L_{10} depends on the applied load F. For ball bearings the relationship is:

$$L_{10} = \left(\frac{C}{F}\right)^3$$

where C is a constant for a particular form of bearing. For roller bearings:

$$L_{10} = \left(\frac{C}{F}\right)^{10/3}$$

Manufacturers often tabulate data for bearings in terms of the number of hours of life at a particular speed given in units of rev/min. The life in hours = $10^6/(3600 \times n/60) \times L_0$ in millions of revs = $(16\ 667/n) \times L_0$ in millions of revs; n is the number of revolutions per minute. For example, a particular ball bearing may be rated as 3000 h at 500 rev/min for a radial loading of 10 kN. This gives L_0 as 90 million revs and hence C as 44.8 kN. Thus with a load of, say, 20 kN at 400 rev/min then the life we can expect is 11.2 million revolutions or 468 h. If this is not long enough we need to select a ball bearing with a higher rating.

6.9 Mechanical aspects of motor selection

A motor drive system is mechanically required to rotate a shaft and its attached load. Factors that have to be considered are moments of inertia and torque.

6.9.1 Moments of inertia

The torque required to give a load with moment of inertia I_L an angular acceleration a is $I_L a$. The torque required to accelerate the motor shaft is $T_M = I_M a_M$ and that required to accelerate the load is $T_L = I_L a_L$. The motor shaft will, in the absence of gearing, have the same angular acceleration and same angular velocity. The power needed to accelerate the system as a whole is $T_M \omega + T_L \omega$, where ω is the angular velocities. Thus:

$$\text{power} = (I_M + I_L) a \omega$$

This power is produced by the motor torque T_M and thus the power must equal $T_M \omega$. Hence:

$$T = (I_M + I_L) a$$

The torque to obtain a given angular acceleration will be minimised when $I_M = I_L$. Thus, for optimum performance, the moment of inertia of the load should be similar to that of the motor.

Consider a gear system with the motor shaft rotating at a different angular speed to the shaft rotating the load. The gear ratio $G = \omega_L/\omega_M = a_L/a_M$, where ω_L is the angular velocity of the load, ω_M the angular velocity of the motor, a_L the angular acceleration of the load and a_M the angular acceleration of the motor. The load shaft will have an angular acceleration of $a_L = Ga_M$. The torque required to accelerate the motor shaft is $T_M = I_M a_M$ and that required to accelerate the load is $T_L = I_L a_L$. The power needed to accelerate the system as a whole is $T_M \omega_M + T_L \omega_L$, where ω is the angular velocities. But $G = \omega_L/\omega_M$ and so the power is:

$$\text{power} = (I_M + G^2 I_L)a_M \omega_M$$

This power is produced by the motor torque T_M and thus the power must equal $T_M \omega_M$. Hence:

$$T_M = (I_M + G^2 I_L)a_M$$

Thus the effect of using the gearing is to give the load an effective moment of inertia of $G^2 I_L$. The torque to give a particular angular acceleration will be minimised when $I_M = G^2 I_L$.

6.9.2 Torque

Figure 6.30 shows the operating curves for a typical motor. For continuous running the stall torque value should not be exceeded. This is the maximum torque value at which overheating will not occur. For intermittent use, greater torques are possible. As the angular speed is increased so the ability of the motor to deliver torque diminishes. Thus if higher speeds and torques are required than given by a particular motor, a more powerful motor needs to be selected.

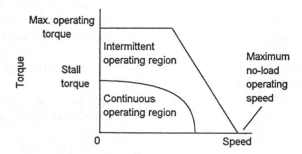

Fig. 6.30 Torque–speed graph

Problems

Fig. 6.31   Problem 5

Fig. 6.32   Problem 6

1. Explain the terms (a) mechanism, (b) kinematic chain.
2. Explain what is meant by the four-bar chain.
3. By examining the following mechanisms, state the number of degrees of freedom each has.
   (a) A car hood hinge mechanism.
   (b) An estate car tailgate mechanism.
   (c) A windscreen wiper mechanism.
   (d) Your knee.
   (e) Your ankle.
4. Analyse the motions of the following mechanisms and state whether they involve pure rotation, pure translation or are a mixture of rotation and translation components.
   (a) The keys on a computer keyboard.
   (b) The pen in an XY plotter.
   (c) The hour hand of a clock.
   (d) The pointer on a moving coil ammeter.
   (e) An automatic screwdriver.
5. For the mechanism shown in Figure 6.31, the arm AB rotates at a constant rate. B and F are sliders moving along CD and AF. Describe the behaviour of this mechanism.
6. Describe how the displacement of the cam follower shown in Figure 6.32 will vary with the angle of rotation of the cam.
7. A circular cam of diameter 100 mm has an eccentric axis of rotation which is offset 30 mm from the centre. When used with a knife follower with its line of action passing through the centre of rotation, what will be the difference between the maximum and minimum displacements of the follower?
8. Design a cam follower system to give constant follower speeds over follower displacements varying from 40 to 100 mm.
9. Design a mechanical system which can be used to:
   (a) Operate a sequence of microswitches in a timed sequence.
   (b) Move a tool at a steady rate in one direction and then quickly move it back to the beginning of the path.
   (c) Transform a rotation into a linear back-and-forth movement with simple harmonic motion.
   (d) Transform a rotation through some angle into a linear displacement.
   (e) Transform a rotation of a shaft into rotation of another, parallel shaft some distance away.
   (f) Transform a rotation of one shaft into rotation of another, close shaft which is at right angles to it.
10. A compound gear train consists of the final driven wheel with 15 teeth which meshes with a second wheel with 90 teeth. On the same shaft as the second wheel is a wheel with 15 teeth. This meshes with a fourth wheel, the first driver, with 60 teeth. What is the overall gear ratio?

11 Which types of bearings are likely to be the most suitable for the following situations?

(a) A 50 mm diameter shaft which is carrying a load of 10 000 N and rotating at 100 rev/s.

(b) A 10 mm diameter shaft which is carrying a load of 1000 N and rotating at 5 rev/min.

7 Electrical actuation systems

7.1 Electrical systems

In any discussion of electrical systems used as actuators for control, the discussion has to include:

1 *Switching devices* such as mechanical switches, e.g. relays, or solid-state switches, e.g. diodes, thyristors, and transistors, where the control signal switches on or off some electrical device, perhaps a heater or a motor.

2 *Solenoid type devices* where a current through a solenoid is used to actuate a soft iron core, as, for example, the solenoid operated hydraulic/pneumatic valve where a control current through a solenoid is used to actuate a hydraulic/pneumatic flow.

3 *Drive systems*, such as d.c. and a.c. motors, where a current through a motor is used to produce rotation.

This chapter is an overview of such devices and their characteristics.

7.2 Mechanical switches

Mechanical switches are elements which are often used as sensors to give inputs to systems (see Section 2.12), e.g. keyboards. In this chapter we are concerned with their use as actuators to perhaps switch on electric motors or heating elements, or switch on the current to actuate solenoid valves controlling hydraulic or pneumatic cylinders. The electrical *relay* is an example of a mechanical switch used in control systems as an actuator.

Fig. 7.1   (a) A relay and (b) a driver circuit

7.2.1 Relays

Relays are electrically operated switches in which changing a current in one electrical circuit switches a current on or off in another circuit. For the relay shown in Figure 7.1(a), when there is a current through the solenoid of the relay, a magnetic field is produced which attracts the iron armature, moves the push-rod, and so closes the normally open (NO) switch contacts and opens the normally closed (NC) switch contacts.

Relays are often used in control systems; the output from the controller is a relatively small current and a much larger current is needed to switch on or off the final correction element, e.g. the current required by an electric heater in a temperature control system or a motor. In such a situation they are likely to be used in conjunction with transistors and Figure 7.1(b) shows the type of circuit that might be used. Because relays are inductances, they can generate a back voltage when the energising current is switched off or when their input switches from a high to low signal. As a result damage can occur in the connecting circuit. To overcome this problem, a diode is connected across the relay. When the back e.m.f. occurs, the diode conducts and shorts it out.

As an illustration of the ways relays can be used in control systems, Figure 7.2 shows how two relays might be used to control the operation of pneumatic valves which in turn control the movement of pistons in three cylinders A, B and C. The sequence of operation is:

1   When the start switch is closed, current is applied to the A and B solenoids and results in both A and B extending, i.e. A+ and B+.
2   The limit switches a+ and b+ are then closed; the a+ closure results in a current flowing through relay coil 1 which then closes its contacts and so supplies current to the C solenoid and results in it extending, i.e. C+.
3   Its extension causes limit switch c+ to close and so current to switch the A and B control valves and hence retraction of cylinders A and B, i.e. A− and B−.
4   Closing limit switch a− passes a current through relay coil 2; its contacts close and allows a current to valve C and cylinder C to retract, i.e. C−.

The sequence thus given by this system is A+ and B+ concurrently, then C+, followed by A− and B− concurrently and finally C−.

Time-delay relays are control relays that have a delayed switching action. The time delay is usually adjustable and can be initiated when a current flows through the relay coil or when it ceases to flow through the coil.

Fig. 7.2   Relay controlled system

7.3 Solid-state switches

There are a number of solid-state devices which can be used to electronically switch circuits. These include:

1   Diodes
2   Thyristors and triacs.
3   Bipolar transistors.
4   Power MOSFETs.

7.3.1 Diodes

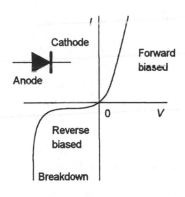

Fig. 7.3   Diode characteristic

The *diode* has the characteristic shown in Figure 7.3 and so allows a significant current in one direction only. A diode can thus be regarded as a 'directional element', only passing a current when forward biased, i.e. with the anode being positive with respect to the cathode. If the diode is sufficiently reverse biased, i.e. a very high voltage, it will break down. If an alternating voltage is applied across a diode, it can be regarded as only switching on when the direction of the voltage is such as to forward bias it and being off in the reverse biased direction. The result is that the current through the diode is half-rectified to become just the current due to the positive halves of the input voltage (Fig. 7.4).

7.3.2 Thyristors and triacs

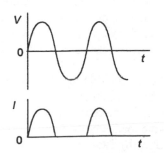

Fig. 7.4   Half-wave rectification

The *thyristor*, or *silicon-controlled rectifier* (SCR), can be regarded as a diode which has a gate controlling the conditions under which the diode can be switched on. Figure 7.5 shows the thyristor characteristic. With the gate current zero, the thyristor passes negligible current when reverse biased (unless sufficiently

Fig. 7.5   Thyristor characteristic

Fig. 7.6   Thyristor circuit

Fig. 7.7   Triac characteristic

Fig. 7.8   Voltage control:
(a) thyristor, (b) triac

reverse biased, hundreds of volts, when it breaks down). When forward biased the current is also negligible until the forward breakdown voltage is exceeded. When this occurs the voltage across the diode falls to a low level, about 1 to 2 V, and the current is then only limited by the external resistance in a circuit. Thus, for example, if the forward breakdown is at 300 V then when this voltage is reached the thyristor switches on and the voltage across it drops to 1 or 2 V. If the thyristor is in series with a resistance of, say, 20 Ω (Fig. 7.6) then before breakdown we have a very high resistance in series with the 20 Ω and so virtually all the 300 V is across the thyristor and there is negligible current. When forward breakdown occurs, the voltage across the thyristor drops to, say, 2 V and so there is now 300 − 2 = 298 V across the 20 Ω resistor, hence the current rises to 298/20 = 14.9 A. Once switched on the thyristor remains on until the forward current is reduced to below a level of a few milliamps. The voltage at which forward breakdown occurs is determined by the current entering the gate, the higher the current the lower the breakdown voltage. The power-handling capability of a thyristor is high and thus it is widely used for switching high power applications. As an example, the Texas Instruments CF106D has a a maximum off-state voltage of 400 V and a maximum gate trigger current of 0.2 mA.

The *triac* is similar to the thyristor and is equivalent to a pair of thyristors connected in reverse parallel on the same chip. The triac can be turned on in either the forward or reverse direction. Figure 7.7 shows the characteristic. As an example, the Motorola MAC212-4 triac has a maximum off-state voltage of 200 V and a maximum on-state current of 12 A r.m.s. Triacs are simple, relatively inexpensive, methods of controlling a.c. power.

Figure 7.8 shows the type of effect that occurs when a sinusoidal alternating voltage is applied across (a) a thyristor and (b) a triac. Forward breakdown occurs when the voltage reaches the breakdown value and then the voltage across the device remains low.

As an example of how such devices can be used for control purposes, Figure 7.9 illustrates how a thyristor could be used to control a steady d.c. voltage V. In this the thyristor is operated as a switch by using the gate to switch the device on or off. By using an alternating signal to the gate, the supply voltage can be chopped and an intermittent voltage produced. The average value of the output d.c. voltage is thus varied and hence controlled by the alternating signal to the gate.

Another example of control is that of alternating current for electric heaters, electric motors or lamp dimmers. Figure 7.10 shows a half-wave, variable resistance, phase-control circuit. The alternating current is applied across the load, e.g. the lamp for the lamp dimming circuit, in series with a thyristor. R_1 is a current-limiting resistor and R_2 is a potentiometer which sets the level at which the thyristor is triggered. The diode is to prevent the

Fig. 7.9   Thyristor d.c. control

Fig. 7.10   Phase-control circuit

negative part of the alternating voltage cycle being applied to the gate. By adjusting R_2 the thyristor can be made to trigger at any point between 0° and 90° in the positive half-cycle of the applied alternating voltage. When the thyristor is triggered near the beginning of the cycle, i.e. 0°, it conducts for the entire positive half-cycle and the maximum power is delivered to the load. As the triggering of the thyristor is delayed to later in the cycle so the power delivered to the load is reduced.

Fig. 7.11   Snubber circuit

When a source voltage is suddenly applied to a thyristor, or a triac, with the gate off, the thyristor may switch from off to on. A typical rate of voltage change that would produce this effect is of the order of 50 V/μs. If the source is a d.c. voltage the thyristor can remain in this conducting state until there is a circuit interruption. In order to prevent this sudden change in source voltage producing this effect, the rate at which the voltage changes with time, i.e. dV/dt, is controlled by using a *snubber circuit*. This is a resistor in series with a capacitor and is placed in parallel with the thyristor (Fig. 7.11). The snubber capacitance C is given by:

$$C = \frac{(V_A)^2_{max}}{L_L (dV/dt)^2_{max}}$$

and its resistance R by:

$$R + R_L = 2\sqrt{\frac{L_L}{C}}$$

where R_L is the load resistance and L_L its inductance.

Collector   Emitter

Base   Base
current   current

Base   Base

Emitter   Collector

(a)   (b)

Fig. 7.12   Transistor symbols:
(a) npn, (b) pnp

(a)

(b)

(c)   0

Fig. 7.13   Transistor switch

7.3.3 Bipolar transistors

Bipolar transistors come in two forms, the npn and the pnp. Figure 7.12 shows the symbol for each. For the npn transistor, the main current flows in at the collector and out at the emitter, a controlling signal being applied to the base. The pnp transistor has the main current flowing in at the emitter and out at the collector, a controlling signal being applied to the base.

For a npn transistor connected as shown in Figure 7.13(a), the so-termed common-emitter circuit, the relationship between the collector current I_C and the potential difference between the collector and emitter V_{CE} is described by the series of graphs shown in Figure 7.13(b). When the base current I_B is zero the transistor is cut off; in this state both the base-emitter and the base-collector junctions are reverse biased. When the base current is increased, the collector current increases and V_{CE} decreases as a result of more of the voltage being dropped across R_C. When V_{CE} reaches a value $V_{CE(sat)}$, the base-collector junction becomes forward biased and the collector current can increase no further, even if the base current is further increased. This is termed saturation. By switching the base current between 0 and a value that drives the transistor into saturation, bipolar transistors can be used as switches. When there is no input voltage V_{in} then virtually the entire V_{CC} voltage appears at the output. When the input voltage is made sufficiently high the transistor switches so that very little of the V_{CC} voltage appears at the output (Fig. 7.13(c)).

The relationship between collector current and the base current I_B at values below that which drive the transistor into saturation is

$$I_C = h_{FE}I_B$$

where h_{FE} is the *current gain*. At saturation the collector current $I_{C(sat)}$ is

$$I_{C(sat)} = \frac{V_{CC} - V_{CE(sat)}}{R_C}$$

To ensure that the transistor is driven into saturation the base current must thus rise to at least

$$I_{B(sat)} = \frac{I_{C(sat)}}{h_{FE}}$$

Thus for a transistor with h_{FE} of 50 and $V_{CE(sat)}$ of 1 V, then for a circuit with $R_C = 10\ \Omega$ and $V_{CC} = 5$ V, the base current must rise to at least about 8 mA.

Because the base current needed to drive a bipolar power transistor is fairly large, a second transistor is often needed to enable switching to be obtained with relatively small currents, e.g. that supplied by a microprocessor. Thus the switching circuit

Buffer transistor

Power transistor

Input

Load

Fig. 7.14   Switching a load

Base current

(a)

Base current

(b)

Fig. 7.15   Darlington pairs

can be of the form shown in Figure 7.14. Such a combination of a pair of transistors to enable a high current to be switched with a small input current is termed a *Darlington pair* and they are available as single-chip devices. A *protection diode* is generally connected in parallel with the power transistor to prevent damage to it when the transistor is switched off since it is generally used with inductive loads and large transient voltages can occur. The integrated circuit ULN2001N from SGS-Thompson contains seven separate Darlington pairs, each pair being provided with a protection diode. Each pair is rated as 500 mA continuous and can withstand surges up to 600 mA.

Figure 7.15(a) shows the Darlington connections when a small npn transistor is combined with a large npn transistor, the result being equivalent to a large npn transistor with a large amplification factor. Figure 7.15(b) shows the Darlington connections for a small pnp transistor with a large npn transistor, the result being equivalent to a single large pnp transistor.

In using transistor switched actuators with a microprocessor, attention has to be given to the size of the base current required and its direction. The base current required can be too high and so a buffer might be used. The buffer increases the drive current to the required value. It might also be used to invert. Figure 7.16 illustrates how a buffer might be used when transistor switching is used to control a d.c. motor by on off switching. Type 240 buffer is inverting while types 241 and 244 are non-inverting. Buffer 74LS240 has a high-level maximum output current of 15 mA and a low-level maximum output current of 24 mA.

Bipolar transistor switching is implemented by base currents and higher frequencies of switching are possible than with thyristors. The power handling capability is less than that of thyristors.

12 V

Ouput port of micro- processor

Buffer

Base current

5 V

Motor

Fig. 7.16   Control of d.c. motor

(a)

(b)

Fig. 7.17   MOSFETs:
(a) n-channel, (b) p-channel

7.3.4 MOSFETs

MOSFETs (metal-oxide field-effect transistors) come in two types, the n-channel and the p-channel. Figure 7.17 shows the symbols. The main difference between the use of a MOSFET for switching and a bipolar transistor is that no current flows into the gate to exercise the control. The gate voltage is the controlling signal. Thus drive circuitry can be simplified in that there is no need to be concerned about the size of the current.

Figure 7.18 illustrates the use of a MOSFET as an on-off switch for a motor; compare the circuit with that in Figure 7.16 where bipolar transistors are used. A level shifter buffer is indicated, this being to raise the voltage level to that required for the MOSFET.

Fig. 7.18   Control of d.c. motor

With MOSFETs, very high frequency switching is possible, up to 1 MHz, and interfacing with a microprocessor is simpler than with bipolar transistors.

For more information on solid-state switches, the reader is referred to specialist texts such *Advanced Industrial Electronics* by N. Morris (McGraw-Hill 1974), *Electronics* by D.I. Crecraft, D.A. Gorham and J.J. Sparkes (Chapman and Hall 1993) or *Power Electronics for the Microprocessor Age* by T. Kenjo (Oxford University Press 1990).

7.4 Solenoids

Solenoids can be used to provide electrically operated actuators. *Solenoid valves* are an example of such devices, being used to control fluid flow in hydraulic or pneumatic systems (see Fig. 5.9). When a current passes through a coil a soft iron core is pulled into the coil and, in doing so, can open or close ports to allow the flow of a fluid.

7.5 D.C. motors

Electric motors are frequently used as the final control element in positional or speed-control systems. Motors can be classified into two main categories: d.c. motors and a.c. motors, most motors

Fig. 7.19   Force on a current carrying conductor

Fig. 7.20   Induced e.m.f.

Fig. 7.21   D.C. motor basics

Fig. 7.22   D.C. motor

used in modern control systems being d.c. motors. The basic principles involved in the action of a motor are:

1   A force is exerted on a conductor in a magnetic field when a current passes through it (Fig. 7.19). For a conductor of length L carrying a current I in a magnetic field of flux density B at right angles to the conductor, the force F equals BIL.

2   When a conductor moves in a magnetic field then an e.m.f. is induced across it (Fig. 7.20). The induced e.m.f. e is equal to the rate at which the magnetic flux Φ (flux equals the product of the flux density and the area) swept through by the conductor changes (Faraday's law), i.e. $e = -d\Phi/dt$. The minus sign is because the e.m.f. is in such a direction as to oppose the change producing it (Lenz's law), i.e. the direction of the induced e.m.f. is in such a direction as to produce a current which sets up magnetic fields which tend to neutralise the change in magnetic flux linked by the coil and which was responsible for the e.m.f. For this reason it is often referred to as a back e.m.f.

7.5.1 Basic principles

Figure 7.21 shows the basic principle of the d.c. motor, a loop of wire which is free to rotate in the field of a permanent magnet. When a current is passed through the coil, the resulting forces acting on its sides at right angles to the field cause forces to act on those sides to give rotation. However, for the rotation to continue, when the coil passes through the vertical position the current direction through the coil has to be reversed.

In the conventional d.c. motor, coils of wire are mounted in slots on a cylinder of magnetic material called the *armature*. The armature is mounted on bearings and is free to rotate. It is mounted in the magnetic field produced by *field poles*. These may be, for small motors, permanent magnets or electromagnets with their magnetism produced by a current through the *field coils*. Figure 7.22 shows the basic principle of a four-pole d.c. motor with the magnetic field produced by current carrying coils. The ends of each armature coil are connected to adjacent segments of a segmented ring called the commutator with electrical contacts made to the segments through carbon contacts called brushes. As the armature rotates, the commutator reverses the current in each coil as it moves between the field poles. This is necessary if the forces acting on the coil are to remain acting in the same direction and so the rotation continue. The direction of rotation of the d.c. motor can be reversed by reversing either the armature current or the field current.

Fig. 7.23 Armature

Fig. 7.24 Equivalent circuit

Fig. 7.25 Torque–speed characteristic

7.5.2 Permanent magnet d.c. motor

Consider a permanent magnet d.c. motor, the permanent magnet giving a constant value of flux density. For an armature conductor of length L and carrying a current i the force resulting from a magnetic flux density B at right angles to the conductor is BiL (Fig. 7.23). With N such conductors the force is $NBiL$. The forces result in a torque T about the coil axis of Fb, with b being the breadth of the coil. Thus:

$$\text{torque } T = NBbLi = k_t i$$

where k_t is the torque constant. Since an armature coil is rotating in a magnetic field, electromagnetic induction will occur and a back e.m.f. will be induced. The back e.m.f. v_b is proportional to the rate at which the flux linked by the coil changes and hence, for a constant magnetic field, is proportional to the angular velocity ω of the rotation. Thus:

$$\text{back e.m.f. } v_b = k_v \omega$$

where k_v is the back e.m.f. constant.

We can consider a d.c. motor to have the equivalent circuit shown in Figure 7.24, i.e. the armature coil being represented by a resistor R in series with an inductance L in series with a source of back e.m.f. If we neglect the inductance of the armature coil then the voltage providing the current i through the resistance is the applied voltage V minus the back e.m.f., i.e. $V - v_b$. Hence:

$$i = \frac{V - v_b}{R} = \frac{V - k_v \omega}{R}$$

The torque T is thus:

$$T = k_t i = \frac{k_t}{R}(V - k_v \omega)$$

Graphs of the torque against the rotational speed ω are a series of straight lines for different voltage values (Fig. 7.25). The starting torque, i.e. the torque when $\omega = 0$, is thus proportional to the applied voltage, the no-load speed is proportional to the applied voltage and the torque decreases with increasing speed.

As an example, a small permanent magnet motor S6M41 by PMI Motors has $k_t = 3.01$ N cm/A, $k_v = 3.15$ V/krpm, a terminal resistance of $1.207 \ \Omega$ and an armature resistance of $0.940 \ \Omega$.

7.5.3 D.C. motors with field coils

D.C. motors with field coils are classified as series, shunt, compound and separately excited according to how the field windings and armature windings are connected (Fig. 7.26).

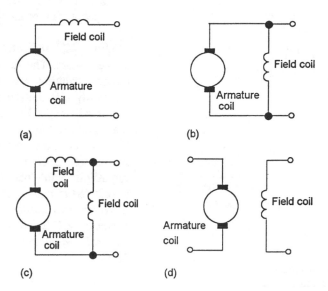

Fig. 7.26   D.C. motors: (a) series,
(b) shunt, (c) compound, (d) separately
wound

1   *Series wound motor*

With the series wound motor the armature and fields coils are in series. Such a motor exerts the highest starting torque and has the greatest no-load speed. With light loads there is a danger that a series wound motor might run at too high a speed. Reversing the polarity of the supply to the coils has no effect on the direction of rotation of the motor; it will continue rotating in the same direction since both the field and armature currents have been reversed.

2   *Shunt wound motor*

With the shunt wound motor the armature and field coils are in parallel. It provides the lowest starting torque, a much lower no-load speed and has good speed regulation. Because of this almost constant speed regardless of load, shunt wound motors are very widely used. To reverse the direction of rotation, either the armature or field supplied must be reversed. For this reason, the separately excited windings are preferable for such a situation.

3   *Compound motor*

The compound motor has two field windings, one in series with the armature and one in parallel. Compound wound motors aim to get the best features of the series and shunt wound motors, namely a high starting torque and good speed regulation.

Fig. 7.27   Torque–speed characteristics

Fig. 7.28   PWM: (a) principle of PWM circuit, (b) varying the average armature voltage by chopping the d.c. voltage

Fig. 7.29   PWM

4   *Separately excited motor*

The separately excited motor has separate control of the armature and field currents and can be considered to be a special case of the shunt wound motor.

Figure 7.27 indicates the torque–speed characteristics of the above motors. The speed of such d.c. motors can be changed by either changing the armature current or the field current. Generally it is the armature current that is varied. The choice of motor will depend on its application. For example, with a robot manipulator, the robot wrist might use a series wound motor because the speed decreases as the load increases. A shunt wound motor would be used where a constant speed was required, regardless of the load. For more details of d.c. motors, the reader is referred to texts such as *Electric Machines and Drives* by J.D. Edwards (Macmillan 1991), *Electrical Machines and Drive Systems* by C.B. Gray (Longman 1989) or *Electric Motors and Control Techniques* by I.M. Gottleib (TAB Books, McGraw-Hill 1994).

7.5.4 Control of d.c. motors

The speed of a permanent magnet motor depends on the current through the armature coil. With a field coil motor the speed can be changed by either varying the armature current or the field current; generally it is the armature current that is varied. Thus speed control can be obtained by controlling the voltage applied to the armature. However, because fixed voltage supplies are often used, a variable voltage is obtained by an electronic circuit.

With an alternating current supply, the thyristor circuit of Figure 7.10 can be used to control the average voltage applied to the armature. However, we are often concerned with the control of d.c. motors by means of control signals emanating from microprocessors. In such cases the technique known as *pulse width modulation* (PWM) is generally used. This basically involves taking a constant d.c. supply voltage and chopping it so that the average value is varied (Fig. 7.28). Figure 7.29 shows how PWM can be obtained by means of a basic transistor circuit. The transistor is switched on or off by means of a signal applied to its base. The diode is to provide a path for current which arises when the transistor is off as a result of the motor acting as a generator. Such a circuit can only be used to drive the motor in one direction; a circuit (Fig. 7.30) involving four transistors, termed an H circuit, can be used to enable the motor to be operated in forward and reverse directions. This circuit can be modified by the use of logic gates so that one input controls the switching and one the direction of rotation (Fig. 7.31).

Fig. 7.30   H circuit

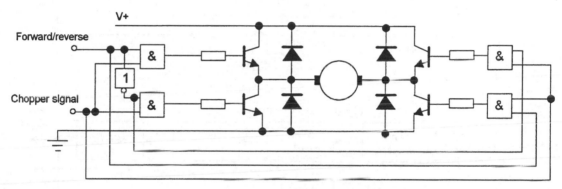

Fig. 7.31   H circuit

The above are examples of open-loop control; this assumes that conditions will remain constant, e.g. the supply voltage and the load driven by the motor. Closed-loop control systems use feedback to modify the motor speed if conditions change. Figure 7.32 shows some of the methods that might be employed.

In Figure 7.32(a) the feedback signal is provided by a tachogenerator, this giving an analogue signal which has to be converted to a digital signal by an ADC for input to the microprocessor. The output from the microprocessor is converted to an analogue signal by an ADC and used to vary the voltage applied to the armature of the d.c. motor. In Figure 7.32(b) the feedback signal is provided by an encoder, this giving a digital signal which after code conversion can be directly inputted to the microprocessor. As in (a) the system shows an analogue voltage being varied to control the motor speed. In Figure 7.32(c) the system is completely digital and PWM is used to control the average voltage applied to the armature.

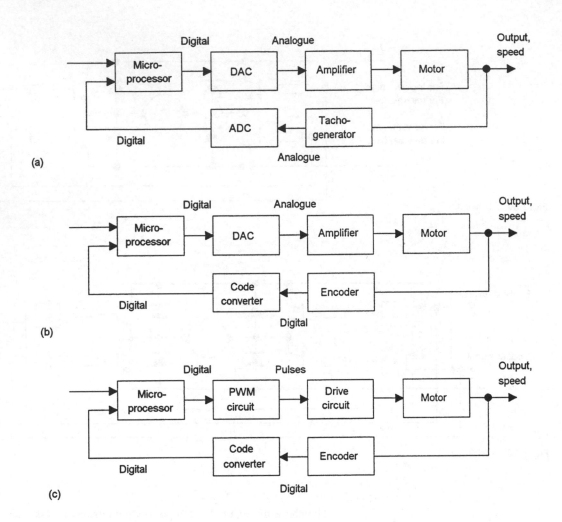

Fig. 7.32 Speed control with feedback

Brush

Commutator
rings connected to
each armature coil

Fig. 7.33 Commutator

7.5.5 Brushless permanent magnet d.c. motors

A problem with d.c. motors is that they require a commutator and brushes (Fig. 7.33) in order to periodically reverse the current through each armature coil. The brushes make sliding contacts with the commutator and as a consequence sparks jump between the two and they suffer wear. Brushes thus have to be periodically changed and the commutator resurfaced. To avoid such problems brushless motors have been designed.

Essentially they consist of a sequence of stator coils and a permanent magnet rotor. A current-carrying conductor in a magnetic field experiences a force; likewise, as a consequence of Newton's third law of motion, the magnet will also experience an opposite and equal force. With the conventional d.c. motor the

Fig. 7.34   Brushless permanent magnet d.c. motor

magnet is fixed and the current-carrying conductors made to move. With the brushless permanent magnet d.c. motor the reverse is the case, the current carrying conductors are fixed and the magnet moves. The rotor is a ferrite or ceramic permanent magnet. Figure 7.34 shows the basic form of such a motor. The current to the stator coils is electronically switched by transistors in sequence round the coils, the switching being controlled by the position of the rotor so that there are always forces acting on the magnet causing it to rotate in the same direction. Hall sensors are generally used to sense the position of the rotor and initiate the switching by the transistors, the sensors being positioned around the stator.

Figure 7.35 shows the transistor switching circuits that might be used with the motor shown in Figure 7.34. To switch the coils in sequence we need to supply signals to switch the transistors on in the right sequence. This is provided by the outputs from the three sensors operating through a decoder circuit to give the appropriate base currents. Thus when the rotor is in the vertical position, i.e. 0°, there is an output from sensor c but none from a and b. This is used to switch on transistors A+ and B−. For the rotor in the 60° position there arc signals from the sensors b and c and transistors A+ and C− are switched on. Table 7.1 shows the entire switching sequence. The entire circuit for controlling such a motor is available on a single integrated circuit.

Fig. 7.35   Transistor switching

Table 7.1 Switching sequence

Rotor position	Sensor signals a	b	c	Transistors on	
0°	0	0	1	A+	B–
60°	0	1	1	A+	C–
120°	0	1	0	B+	C–
180°	1	1	0	B+	A–
240°	1	0	0	C+	A–
360°	1	0	1	C+	B–

Brushless permanent magnet d.c. motors are becoming increasingly used in situations where high performance coupled with reliability and low maintenance are essential. Because of their lack of brushes, they are quiet and capable of high speeds. For more details of brushless d.c. motors, the reader is referred to specialist texts such as *Electric Machines and Drives* by J.D. Edwards (Macmillan 1991) or *Brushless Permanent-magnet and Reluctance Motor Drives* by T.J.E. Miller (Oxford University Press 1989).

7.6 A.C. motors

Alternating current motors can be classified into two groups, single phase and polyphase with each group being further sub-divided into induction and synchronous motors. Single-phase motors tend to be used for low power requirements while poly-phase motors are used for higher powers. Induction motors tend to be cheaper than synchronous motors and are thus very widely used.

The *single-phase squirrel-cage induction motor* consists of a squirrel-cage rotor, this being copper or aluminium bars that fit into slots in end rings to form complete electrical circuits (Fig. 7.36). There are no external electrical connections to the rotor. The basic motor consists of this rotor with a stator having a set of windings. When an alternating current passes through the stator windings an alternating magnetic field is produced. As a result of electromagnetic induction, e.m.f.s are induced in the conductors of the rotor and currents flow in the rotor. Initially, when the rotor is stationary, the forces on the current carrying conductors of the rotor in the magnetic field of the stator are such as to result in no net torque. The motor is not self-starting. A number of methods are used to make the motor self-starting and give this initial impetus to start it; one is to use an auxiliary starting winding to give the rotor an initial push. The rotor rotates at a speed determined by the frequency of the alternating current applied to the stator. For a constant frequency supply to a two-pole single-phase motor the magnetic field will alternate at

Rotor conductors giving the squirrel cage

End rings connecting the ends of all the conductors to give the circuits in which currents are induced

Stator   Rotor

Pole   Pole

End view of squirrel cage

Fig. 7.36   Single-phase induction motor

Fig. 7.37 Three-phase induction motor

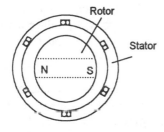

Fig. 7.38 Three-phase synchronous motor

this frequency. This speed of rotation of the magnetic field is termed the *synchronous speed*. The rotor will never quite match this frequency of rotation, typically differing from it by about 1 to 3%. This difference is termed *slip*. Thus for a 50 Hz supply the speed of rotation of the rotor will be almost 50 revolutions per second.

The *three-phase induction motor* (Fig. 7.37) is similar to the single-phase induction motor but has a stator with three windings located 120° apart, each winding being connected to one of the three lines of the supply. Because the three phases reach their maximum currents at different times, the magnetic field can be considered to rotate round the stator poles, completing one rotation in one full cycle of the current. The rotation of the field is much smoother than with the single-phase motor. The three-phase motor has a great advantage over the single-phase motor of being self-starting. The direction of rotation is reversed by interchanging any two of the line connections, this changing the direction of rotation of the magnetic field.

Synchronous motors have stators similar to those described above for induction motors but a rotor which is a permanent magnet (Fig. 7.38). The magnetic field produced by the stator rotates and so the magnet rotates with it. With one pair of poles per phase of the supply, the magnetic field rotates through 360° in one cycle of the supply and so the frequency of rotation with this arrangement is the same as the frequency of the supply. Synchronous motors are used when a precise speed is required. They are not self-starting and some system has to be employed to start them.

A.C. motors have the great advantage over d.c. motors of being cheaper, more rugged, reliable and maintenance free. However speed control is generally more complex than with d.c. motors and as a consequence a speed-controlled d.c. drive generally works out cheaper than a speed-controlled a.c. drive, though the price difference is steadily dropping as a result of technological developments and the reduction in price of solid-state devices. Speed control of a.c. motors is based around the provision of a variable frequency supply, since the speed of such motors is determined by the frequency of the supply. The torque developed by an a.c. motor is constant when the ratio of the applied stator voltage to frequency is constant. Thus to maintain a constant torque at the different speeds when the frequency is varied the voltage applied to the stator has also to be varied. With one method, the a.c. is first rectified to d.c. by a *converter* and then *inverted* back to a.c. again but at a frequency that can be selected (Fig. 7.39). Another method that is often used for operating slow-speed motors is the *cycloconverter*. This converts a.c. at one frequency directly to a.c. at another frequency without the intermediate d.c. conversion.

Fig. 7.39 Variable speed
a.c. motor

For more details of a.c. motors, the reader is referred to texts such as *Electric Machines and Drives* by J.D. Edwards (Macmillan 1991) or *Electrical Machines and Drive Systems* by C.B. Gray (Longman 1989).

7.7 Stepper motors

The *stepper motor* is a device that produces rotation through equal angles, the so-called *steps*, for each digital pulse supplied to its input. Thus, for example, if with such a motor 1 pulse produces a rotation of 6° then 60 pulses will produce a rotation through 360°. There are a number of forms of stepper motor:

1 *Variable reluctance stepper*

Figure 7.40 shows the basic form of the variable reluctance stepper motor. With this form the rotor is made of soft steel and is cylindrical with four poles, i.e. fewer poles than on the stator. When an opposite pair of windings has current switched to them, a magnetic field is produced with lines of force which pass from the stator poles through the nearest set of poles on the rotor. Since lines of force can be considered to be rather like elastic thread and always trying to shorten themselves, the rotor will move until the rotor and stator poles line up. This is termed the position of minimum reluctance. This form of stepper generally gives step angles of 7.5° or 15°.

2 *Permanent magnet stepper*

Figure 7.41 shows the basic form of the *permanent magnet* motor. The motor shown has a stator with four poles. Each pole is wound with a field winding, the coils on opposite pairs of poles being in series. Current is supplied from a d.c. source to the windings through switches. The rotor is a permanent magnet and thus when a pair of stator poles has a current switched to it, the rotor will move to line up with it. Thus for the currents giving the situation shown in the figure the rotor moves to the 45° position. If the current is then switched so that the polarities are reversed, the rotor will move a further 45° in order to line up again. Thus by switching the currents through the coils the rotor rotates in 45° steps. With this type of motor, step angles are commonly 1.8°, 7.5°, 15°, 30°, 34° or 90°.

Rotor   Stator

This pair of poles energised by current being switched to them and rotor rotates to position shown below

This pair of poles energised by current being switched to them to give next step

Fig. 7.40 Variable reluctance stepper motor

Fig. 7.41   Permanent magnet stepper motor

Teeth on   Permanent magnet
end caps

Fig. 7.42   Hybrid motor rotor

3   *Hybrid stepper*
Hybrid stepper motors combine the features of both the variable reluctance and permanent magnet motors, having a permanent magnet encased in iron caps which are cut to have teeth (Fig. 7.42). The rotor sets itself in the minimum reluctance position in response to a pair of stator coils being energised. Typical step angles are 0.9° and 1.8°. Such stepper motors are extensively used in high-accuracy positioning applications, e.g. in computer hard disc drives.

7.7.1 Stepper motor specifications

The following are some of the terms commonly used in specifying stepper motors:

1   *Phase*
This term refers to the number of independent windings on the stator, e.g. a four-phase motor. The current required per phase and its resistance and inductance will be specified so that the controller switching output is specified. Two-phase motors, e.g. Figure 7.41, tend to be used in light-duty applications, three-phase motors tend to be variable reluctance steppers, e.g. Figure 7.40, and four-phase motors tend to be used for higher power applications.

2   *Step angle*
This is the angle through which the rotor rotates for one switching change for the stator coils.

3   *Holding torque*
This is the maximum torque that can be applied to a powered motor without moving it from its rest position and causing spindle rotation.

4   *Pull-in torque*
    This is the maximum torque against which a motor will start, for a given pulse rate, and reach synchronism without losing a step.

5   *Pull-out torque*
    This is the maximum torque that can be applied to a motor, running at a given stepping rate, without losing synchronism.

6   *Pull-in rate*
    This is the maximum switching rate at which a loaded motor can start without losing a step.

7   *Pull-out rate*
    This is the switching rate at which a loaded motor will remain in synchronism as the switching rate is reduced.

8   *Slew range*
    This is the range of switching rates between pull-in and pull-out within which the motor runs in synchronism but cannot start up or reverse.

Figure 7.43 shows the general characteristics of a stepper motor.

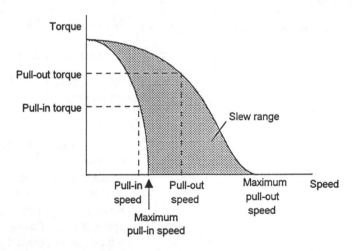

Fig. 7.43   Stepper motor characteristics

7.7.2 Stepper motor control

Solid-state electronics is used to switch the d.c. supply between the pairs of stator windings. Two-phase motors, e.g. Figure 7.41, are termed *bipolar motors* when they have four connecting wires for signals to generate the switching sequence (Fig. 7.44). Such a motor can be driven by H circuits (see Fig. 7.30 and the associated discussion); Figure 7.45 shows the circuit and Table

Phase A   Phase B

Fig. 7.44   Bipolar motor

7.2 shows the switching sequence required for the transistors to carry out the four steps, the sequence then being repeated for further steps. The sequence gives a clockwise rotation; for an anti- clockwise rotation the sequence is reversed.

Table 7.2   Switching sequence for full-stepping bipolar stepper

Step	Transistors			
	1 and 4	2 and 3	5 and 8	6 and 7
1	On	Off	On	Off
2	On	Off	Off	On
3	Off	On	Off	On
4	Off	On	On	Off

Half-steps, and hence finer resolution, are obtainable if instead of the full-stepping sequence needed to implement a pole reversal to get from one step to the next, the coils are switched so that the rotor stops at a position halfway to the next full step. Table 7.3 shows the sequence for half-stepping with the bipolar stepper.

Table 7.3   Half-steps for bipolar stepper

Step	Transistors			
	1 and 4	2 and 3	5 and 8	6 and 7
1	On	Off	On	Off
2	On	Off	Off	Off
3	On	Off	Off	On
4	Off	Off	Off	On
5	Off	On	Off	On
6	Off	On	Off	Off
7	Off	On	On	Off
8	Off	Off	On	Off

Fig. 7.45   H circuit

Fig. 7.46 Unipolar motor

Two-phase motors are termed *unipolar* when they have six connecting wires for the generation of the switching sequence (Fig. 7.46). Each of the coils has a centre-tap. With the centre-taps of the phase coils connected together, such a form of stepper motor can be switched with just four transistors. Table 7.4 gives the switching sequence for the transistors in order to produce the steps for clockwise rotation, the sequence then being repeated for further steps. For anti-clockwise rotation the sequence is reversed. Table 7.5 shows the sequence when the unipolar is half-stepping.

Integrated circuits are available to provide the drive circuitry. Figure 7.47 shows the connections with the integrated circuit SAA 1027 for a four-phase stepper. The three inputs are controlled by applying high or low signals to them. When the set terminal is held high, the output from the integrated circuit changes state each time the trigger terminal goes from low to high. The sequence repeats itself at four-step intervals but can be reset to the zero condition at any time by applying a low signal to the trigger terminal. When the rotation input is held low there is clockwise rotation, when high it is anti-clockwise.

Table 7.4 Switching sequence for full-stepping unipolar stepper

Step	Transistors			
	1	2	3	4
1	On	Off	On	Off
2	On	Off	Off	On
3	Off	On	Off	On
4	Off	On	On	Off

Table 7.5 Half-steps for unipolar stepper

Step	Transistors			
	1	2	3	4
1	On	Off	On	Off
2	On	Off	Off	Off
3	On	Off	Off	On
4	Off	Off	Off	On
5	Off	On	Off	On
6	Off	On	Off	Off
7	Off	On	On	Off
8	Off	Off	On	Off

Internal block diagram of SAA 1027

Typical connections for SAA 1027 to a four-phase stepper

Fig. 7.47 Integrated circuit SAA 1027 for stepper motor

Some applications require very small step angles. While the step angle can be made small by increasing the number of rotor teeth and/or the number of phases, generally more than four phases and 50 to 100 teeth are not used. Instead a technique known as *mini-stepping* is used. This involves dividing each step into a number of equal size sub-steps. This is done by using different currents to the coils so that the rotor moves to intermediate positions between normal step positions. Thus, for example, a step of 1.8° might be subdivided into 10 equal steps.

Stepper motors can be used to give controlled rotational steps but also can give continuous rotation with their rotational speed controlled by controlling the rate at which pulses are applied to it to cause stepping. This gives a very useful controlled variable speed motor which finds many applications.

Because stepper coils have inductance and switched inductive loads can generate large back e.m.f.s when switched, when steppers are connected to microprocessor output ports it is necessary to include protection to avoid damage to the microprocessor. This may take the form of resistors in the lines to limit the current, though these must have values carefully chosen to both provide the protection but also not to limit the value of the current needed to switch the transistors. Diodes across the coils prevent current in the reverse direction and so give protection. An alternative is to use optoisolators (see Section 3.3).

For more details of stepping motors and their drive circuits, the reader is referred to texts such as *Stepping Motors and their Microprocessor Controls* by T. Kenjo (Oxford University Press 1984), *Power Electronics for the Microprocessor Age* by T. Kenjo (Oxford University Press 1990) or *Electrical Machines and Drive Systems* by C.B. Gray (Longman 1989).

Problems

Fig. 7.48 Problem 1

1 Explain how the circuit shown in Figure 7.48 can be used to debounce a switch.
2 Explain how a thyristor can be used to control the level of a d.c. voltage by chopping the output from a constant voltage supply.
3 A d.c. motor is required to have (a) a high torque at low speeds for the movement of large loads, (b) a torque which is almost constant regardless of speed. Suggest suitable forms of motor.
4 Suggest possible motors, d.c. or a.c., which can be considered for applications where (a) cheap, constant torque operation is required, (b) high controlled speeds are required, (c) low speeds are required, (d) maintenance requirements have to be minimised.
5 Explain the principle of the brushless d.c. permanent magnet motor.
6 Explain the principles of operation of the variable reluctance stepper motor.
7 If a stepper motor has a step angle of 7.5°, what digital input rate is required to produce a rotation of 10 rev/s?

8 Basic system models

8.1 Mathematical models

Consider the following situation. A microprocessor switches on a motor. How will the rotation of the motor shaft vary with time? The speed will not immediately assume the full-speed value but will only attain that speed after some time. Consider another situation. A hydraulic system is used to open a valve which allows water into a tank to restore the water level to that required. How will the water level vary with time? The water level will not immediately assume the required level but will only attain that level after some time. This chapter, and Chapters 9, 10 and 11, are about determining how systems behave with time when subject to some disturbance.

In order to understand the behaviour of systems, *mathematical models* are needed. These mathematical models are equations which describe the relationship between the input and output of a system. They can be used to enable forecasts to be made of the behaviour of a system under specific conditions. The basis for any mathematical model is provided by the fundamental physical laws that govern the behaviour of the system. In this chapter a range of systems will be considered, including mechanical, electrical, thermal and fluid examples.

Like a child building houses, cars, cranes, etc., from a number of basic building blocks, systems can be made up from a range of building blocks. Each building block is considered to have a single property or function. Thus to take a simple example, an electrical circuit system may be made up from building blocks which represent the behaviour of resistors, capacitors and inductors. The resistor building block is assumed to have purely the property of resistance, the capacitor purely that of capacitance and the inductor purely that of inductance. By combining these building blocks in different ways a variety of electrical circuit systems can be built up and the overall input–output relationships obtained for the system by combining in an appropriate way the relationships for the building blocks. Thus a mathematical model for the system can be obtained. A system built up in this way is called a *lumped parameter* system. This is because each parameter, i.e. property or function, is considered independently.

There are similarities in the behaviour of building blocks used in mechanical, electrical, thermal and fluid systems. This chapter is about the basic building blocks and their combination to produce mathematical models for physical, real, systems. Chapter 9 looks at more complex models.

A more detailed discussion will be found in texts such as *Dynamic Modelling and Control of Engineering Systems* by J. Lowen Shearer and Bohdan T. Kulakowski (Prentice-Hall 1997) or *Modelling and Analysis of Dynamic Systems* by C. Frederick (Houghton Mifflin 1993).

8.2 Mechanical system building blocks

The models used to represent mechanical systems have the basic building blocks of springs, dashpots and masses. *Springs* represent the stiffness of a system, *dashpots* the forces opposing motion, i.e. frictional or damping effects, and *masses* the inertia or resistance to acceleration. The mechanical system does not have to be really made up of springs, dashpots and masses but have the properties of stiffness, damping and inertia. All these building blocks can be considered to have a force as an input and a displacement as an output.

The stiffness of a *spring* is described by the relationship between the forces F used to extend or compress a spring and the resulting extension or compression x (Fig. 8.1). In the case of a spring where the extension or compression is proportional to the applied forces, i.e. a linear spring,

$$F = kx$$

Fig. 8.1   Spring

Fig. 8.2   Dashpot

where k is a constant. The bigger the value of k the greater the forces have to be to stretch or compress the spring and so the greater the stiffness. The object applying the force to stretch the spring is also acted on by a force, the force being that exerted by the stretched spring (Newton's third law). This force will be in the opposite direction and equal in size to the force used to stretch the spring, i.e. kx.

The *dashpot* building block represents the types of forces experienced when we endeavour to push an object through a fluid or move an object against frictional forces. The faster the object is pushed the greater becomes the opposing forces. The dashpot which is used pictorially to represent these damping forces which slow down moving objects consists of a piston moving in a closed cylinder (Fig. 8.2). Movement of the piston requires the fluid on one side of the piston to flow through or past the piston. This flow produces a resistive force. In the ideal case, the damping or resistive force F is proportional to the velocity v of the piston.

Thus

$$F = cv$$

where c is a constant. The larger the value of c the greater the damping force at a particular velocity. Since velocity is the rate of change of displacement x of the piston, i.e. $v = dx/dt$, then

$$F = c\frac{dx}{dt}$$

Thus the relationship between the displacement x of the piston, i.e. the output, and the force as the input is a relationship depending on the rate of change of the output.

The *mass* building block (Fig. 8.3) exhibits the property that the bigger the mass the greater the force required to give it a specific acceleration. The relationship between the force F and the acceleration a is (Newton's second law) $F = ma$, where the constant of proportionality between the force and the acceleration is the constant called the mass m. Acceleration is the rate of change of velocity, i.e. dv/dt, and velocity v is the rate of change of displacement x, i.e. $v = dx/dt$. Thus

$$F = ma = m\frac{dv}{dt} = m\frac{d(dx/dt)}{dt} = m\frac{d^2x}{dt^2}$$

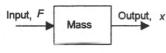

Fig. 8.3   Mass

Energy is needed to stretch the spring, accelerate the mass and move the piston in the dashpot. However, in the case of the spring and the mass we can get the energy back but with the dashpot we cannot. The spring when stretched stores energy, the energy being released when the spring springs back to its original length. The energy stored when there is an extension x is $\frac{1}{2}kx^2$. Since $F = kx$ this can be written as

$$E = \frac{1}{2}\frac{F^2}{k}$$

There is also energy stored in the mass when it is moving with a velocity v, the energy being referred to as kinetic energy, and released when it stops moving.

$$E = \frac{1}{2}mv^2$$

However, there is no energy stored in the dashpot. It does not return to its original position when there is no force input. The dashpot dissipates energy rather than storing it, the power P dissipated depending on the velocity v and being given by

$$P = cv^2$$

8.2.1 Rotational systems

The spring, dashpot and mass are the basic building blocks for mechanical systems where forces and straight line displacements are involved without any rotation. If there is rotation then the equivalent three building blocks are a torsional spring, a rotary damper and the moment of inertia, i.e. the inertia of a rotating mass. With such building blocks the inputs are torque and the outputs angle rotated. With a *torsional spring* the angle θ rotated is proportional to the torque T. Hence

$$T = k\theta$$

With the *rotary damper* a disc is rotated in a fluid and the resistive torque T is proportional to the angular velocity ω, and since angular velocity is the rate at which angle changes, i.e. $d\theta/dt$,

$$T = c\omega = c\frac{d\theta}{dt}$$

The *moment of inertia* building block exhibits the property that the greater the moment of inertia I the greater the torque needed to produce an angular acceleration a.

$$T = Ia$$

Thus, since angular acceleration is the rate of change of angular velocity, i.e. $d\omega/dt$, and angular velocity is the rate of change of angular displacement, then

$$T = I\frac{d\omega}{dt} = I\frac{d(d\theta/dt)}{dt} = I\frac{d^2\theta}{dt^2}$$

The torsional spring and the rotating mass store energy; the rotary damper just dissipates energy. The energy stored by a torsional spring when twisted through an angle θ is $\frac{1}{2}k\theta^2$ and since $T = k\theta$ this can be written as

$$E = \frac{1}{2}\frac{T^2}{k}$$

The energy stored by a mass rotating with an angular velocity ω is the kinetic energy E, where

$$E = \frac{1}{2}I\omega^2$$

The power P dissipated by the rotatory damper when rotating with an angular velocity ω is

$$P = c\omega^2$$

Table 8.1 summarises the equations defining the characteristics of the mechanical building blocks when there is, in the case of straight line displacements (termed translational), a force input F and a displacement x output and, in the case of rotation, a torque T and angular displacement θ.

Table 8.1 Mechanical building blocks

Building block	Describing equation	Energy stored or power dissipated
Translational		
Spring	$F = kx$	$E = \frac{1}{2}\frac{F^2}{k}$
Dashpot	$F = c\dfrac{dx}{dt}$	$P = cv^2$
Mass	$F = m\dfrac{d^2x}{dt^2}$	$E = \frac{1}{2}mv^2$
Rotational		
Spring	$T = k\theta$	$E = \frac{1}{2}\frac{T^2}{k}$
Rotational damper	$T = c\dfrac{d\theta}{dt}$	$P = c\omega^2$
Moment of inertia	$T = I\dfrac{d^2\theta}{dt^2}$	$E = \frac{1}{2}I\omega^2$

Force due to spring

Mass

F

Force due to dashpot

Input, F → Spring–dashpot–mass system → Output, x

Fig. 8.4 Spring–dashpot–mass system

Force due to spring

Mass → F

Force due to dashpot

Fig. 8.5 Free-body diagram

8.2.2 Building up a mechanical system

Many systems can be considered to be essentially a mass, a spring and dashpot combined in the way shown in Figure 8.4. To evaluate the relationship between the force and displacement for the system the procedure to be adopted is to consider just one mass, and just the forces acting on that body. A diagram of the mass and just the forces acting on it is called a *free-body diagram* (Fig. 8.5). When several forces act concurrently on a body, their single equivalent resultant can be found by vector addition. If the forces are all acting along the same line or parallel lines, this means that the resultant or net force acting on the block is the algebraic sum. Thus for the mass in Figure 8.4, if we consider just the forces acting on that block then the net force applied to the mass is the applied force F minus the force resulting from the stretching or compressing of the spring and minus the force from the damper. Thus

Net force applied to mass $m = F - kx - cv$

where v is the velocity with which the piston in the dashpot, and hence the mass, is moving. This net force is the force applied to the mass to cause it to accelerate. Thus

Output, displacement

Mass

Ground

Input, force

Fig. 8.6   Mathematical model for a machine mounted on the ground

Output, displacement

Mass of car

Suspension

Mass of suspension

Tyre

Road

Input, force

Fig. 8.7   Mathematical model of a wheel of a car moving along a road

Net force applied to mass = ma

Hence

$$F - kx - c\frac{dx}{dt} = m\frac{d^2x}{dt^2}$$

or, when rearranged,

$$m\frac{d^2x}{dt^2} + c\frac{dx}{dt} + kx = F$$

This equation, called a *differential equation*, describes the relationship between the input of force F to the system and the output of displacement x. Because of the d^2x/dt^2 term, it is a *second-order* differential equation; a first-order differential equation would only have dx/dt.

There are many systems which can be built up from suitable combinations of the spring, dashpot and mass building blocks. Figure 8.6 shows the model for a machine mounted on the ground and could be used as a basis for studying the effects of ground disturbances on the displacements of a machine bed. Figure 8.7 shows a model for the wheel and its suspension for a car or truck and can be used for the study of the behaviour that could be expected of the vehicle when driven over a rough road and hence as a basis for the design of the vehicle suspension. The procedure to be adopted for the analysis of such models is just the same as outlined above for the simple spring–dashpot–mass model. A free-body diagram is drawn for each mass in the system, such diagrams showing each mass independently and just the forces acting on it. Then for each mass the resultant of the forces acting on it is then equated to the product of the mass and the acceleration of the mass.

Similar models can be constructed for rotating systems. To evaluate the relationship between the torque and angular displacement for the system the procedure to be adopted is to consider just one rotational mass block, and just the torques acting on that body. When several torques act on a body simultaneously, their single equivalent resultant can be found by addition in which the direction of the torques is taken into account. Thus a system involving a torque being used to rotate a mass on the end of a shaft (Fig. 8.8(a)) can be considered to be represented by the rotational building blocks shown in Figure 8.8(b). This is a compar- able situation to that analysed above (Fig. 8.4) for linear dis- placements and yields a similar equation

$$I\frac{d^2\theta}{dt^2} + c\frac{d\theta}{dt} + k\theta = T$$

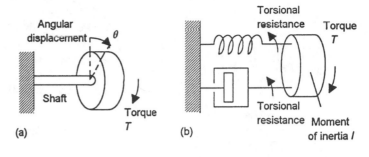

Fig. 8.8 Rotating a mass on the end of a shaft: (a) physical situation, (b) building block model

(a)   (b)

Fig. 8.9 Example 1

To illustrate the above, consider the development of the equations in the following examples.

1 Derive the differential equation describing the relationship between the input of the force F and the output of displacement x for the system shown in Figure 8.9.

The net force applied to the mass is F minus the resisting forces exerted by each of the springs. Since these are $k_1 x$ and $k_2 x$, then

$$\text{Net force} = F - k_1 x - k_2 x$$

Since the net force causes the mass to accelerate, then

$$\text{Net force} = m \frac{\mathrm{d}^2 x}{\mathrm{d} t^2}$$

Hence

$$m \frac{\mathrm{d}^2 x}{\mathrm{d} t^2} + (k_1 + k_2) x = F$$

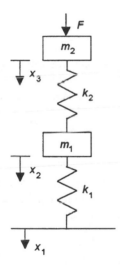

Fig. 8.10 Mass–spring system

2 Derive the differential equation describing the motion of the mass m_1 in Figure 8.10 when a force F is applied.

Consider the free-body diagrams (Fig. 8.11). For mass m_2 these are the force F and the force exerted by the upper spring. The force exerted by the upper spring is due to it being stretched by $(x_2 - x_3)$ and so is $k_2(x_3 - x_2)$. Thus the net force acting on the mass is:

$$\text{Net force} = F - k_2(x_3 - x_2)$$

This force will cause the mass to accelerate and so:

$$F - k_2(x_3 - x_2) = m_2 \frac{\mathrm{d}^2 x_3}{\mathrm{d} t}$$

Fig. 8.11 Free-body diagrams

(a)

Torsional resistance

Torsional resistance

Moment of inertia *I*

(b)

Fig. 8.12   Torsional spring–mass system

For the free-body diagram for mass m_1, the force exerted by the upper spring is $k_2(x_3 - x_2)$ and that by the lower spring is $k_1(x_1 - x_2)$. Thus the net force acting on the mass is:

$$\text{Net force} = k_1(x_2 - x_1) - k_2(x_3 - x_2)$$

This force will cause the mass to accelerate and so:

$$k_1(x_2 - x_1) - k_2(x_3 - x_2) = m_1 \frac{d^2 x_2}{dt}$$

We thus have two simultaneous second-order differential equations to describe the behaviours of the system.

3   A motor is used to rotate a load. Devise a model and obtain the differential equation for it.

The model used can be that described by Figure 8.12, essentially being the same as that in Figure 8.8. The differential equation is thus, as then:

$$I \frac{d^2 \theta}{dt^2} + c \frac{d\theta}{dt} + k\theta = T$$

8.3 Electrical system building blocks

The basic building blocks of electrical systems are inductors, capacitors and resistors. For an *inductor* the potential difference v across it at any instant depends on the rate of change of current (di/dt) through it:

$$v = L \frac{di}{dt}$$

where L is the inductance. The direction of the potential difference is in the opposite direction to the potential difference used to drive the current through the inductor, hence the term back e.m.f. The equation can be rearranged to give

$$i = \frac{1}{L} \int v \, dt$$

For a *capacitor*, the potential difference across it depends on the charge q on the capacitor plates at the instant concerned.

$$v = \frac{q}{C}$$

where C is the capacitance. Since the current i to or from the capacitor is the rate at which charge moves to or from the capacitor plates, i.e. $i = dq/dt$, then the total charge q on the plates is given by

$$q = \int i \, dt$$

and so

$$v = \frac{1}{C} \int i \, dt$$

Alternatively, since $v = q/C$ then

$$\frac{dv}{dt} = \frac{1}{C} \frac{dq}{dt} = \frac{1}{C} i$$

and so

$$i = C \frac{dv}{dt}$$

For a *resistor*, the potential difference v across it at any instant depends on the current i through it.

$$v = Ri$$

where R is the resistance.

Both the inductor and capacitor store energy which can then be released at a later time. A resistor does not store energy but just dissipates it. The energy stored by an inductor when there is a current i is

$$E = \tfrac{1}{2} L i^2$$

The energy stored by a capacitor when there is a potential difference v across it is

$$E = \tfrac{1}{2} C v^2$$

The power P dissipated by a resistor when there is a potential difference v across it is

$$P = iv = \frac{v^2}{R}$$

Table 8.2 summarises the equations defining the characteristics of the electrical building blocks when the input is current and the output is potential difference. Compare them with the equations given in Table 8.1 for the mechanical system building blocks.

8.3.1 Building up a model for an electrical system

The equations describing how the electrical building blocks can be combined are *Kirchhoff's laws*. These can be expressed as:

Table 8.2 Electrical building blocks

Building block	Describing equation	Energy stored or power dissipated
Inductor	$i = \frac{1}{L} \int v \, dt$	$E = \frac{1}{2}Li^2$
Capacitor	$i = C\frac{dv}{dt}$	$E = \frac{1}{2}Cv^2$
Resistor	$i = \frac{v}{R}$	$P = \frac{v^2}{R}$

Law 1: the total current flowing towards a junction is equal to the total current flowing from that junction, i.e. the algebraic sum of the currents at the junction is zero.

Law 2: in a closed circuit or loop, the algebraic sum of the potential differences across each part of the circuit is equal to the applied e.m.f.

A convenient way of using law 1 is called *node analysis* since the law is applied to each principal node of a circuit, a node being a point of connection or junction between building blocks or circuit elements and a principal node being one where three or more branches of the circuit meet. A convenient way of using law 2 is called *mesh analysis* since the law is applied to each mesh, a mesh being a closed path or loop which contains no other loop.

To illustrate the use of these two methods of analysis to generate relationships, consider the circuit shown in Figure 8.13. All the components are resistors for this illustrative example. With node analysis a principal node, point A on the figure, is picked and the voltage at the node given a value v_A with reference to some other principal node that has been picked as the reference. In this case it is convenient to pick node B as the reference. We then consider all the currents entering and leaving node A and thus, according to Kirchhoff's first law,

$$i_1 = i_2 + i_3$$

The current entering through R_1 is i_1 and since the potential difference across R_1 is $(v_A - v)$, then $i_1 R_1 = v_A - v$. The current through R_2 is i_2 and since the potential difference across R_2 is v_A then $i_2 R_2 = v_A$. The current i_3 passes through R_3 in series with R_4 and there is a potential difference of v_A across the combination. Hence $i_3(R_3 + R_4) = v_A$. Thus equating the currents gives

$$\frac{v - v_A}{R_1} = \frac{v_A}{R_2} + \frac{v_A}{R_3 + R_4}$$

Fig. 8.13   Node analysis

To illustrate the use of mesh analysis for the circuit in Figure 8.13 we assume there are currents circulating in each mesh in the way shown in Figure 8.14. Then Kirchhoff's second law is applied to each mesh. Thus for the mesh with current i_1 circulating, since the current through R_1 is i_1 and that through R_2 is $(i_1 - i_2)$, then

$$v = i_1R_1 + (i_1 - i_2)R_2$$

Fig. 8.14   Mesh analysis

Similarly for the mesh with current i_2 circulating, since there is no source of e.m.f., then

$$0 = i_2R_3 + i_2R_4 + (i_2 - i_1)R_2$$

We thus have two simultaneous equations which can be solved to obtain the two mesh currents and hence the currents through each branch of the circuit. In general, when the number of nodes in a circuit is less than the number of meshes it is easier to employ nodal analysis.

Now consider a simple electrical system consisting of a resistor and capacitor in series, as shown in Figure 8.15. Applying Kirchhoff's second law to the circuit loop gives

$$v = v_R + v_C$$

Fig. 8.15   Resistor–capacitor system

where v_R is the potential difference across the resistor and v_C that across the capacitor. Since it is just a single loop the current i through all the circuit elements will be the same. If the output from the circuit is the potential difference across the capacitor, v_C, then since $v_R = iR$ and $i = C(\mathrm{d}v_C/\mathrm{d}t)$,

$$v = RC\frac{\mathrm{d}v_C}{\mathrm{d}t} + v_C$$

This gives the relationship between the output v_C and the input v and is a first-order differential equation.

Figure 8.16 shows a resistor–inductor–capacitor system. If Kirchhoff's second law is applied to this circuit loop,

$$v = v_R + v_L + v_C$$

Fig. 8.16   Resistor–inductor–capacitor system

where v_R is the potential difference across the resistor, v_L that across the inductor and v_C that across the capacitor. Since there is just a single loop the current i will be the same through all circuit elements. If the output from the circuit is the potential difference across the capacitor, v_C, then since $v_R = iR$ and $v_L = L(\mathrm{d}i/\mathrm{d}t)$

$$v = iR + L\frac{\mathrm{d}i}{\mathrm{d}t} + v_C$$

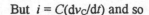

But $i = C(dv_C/dt)$ and so

$$\frac{di}{dt} = C\frac{d(dv_C/dt)}{dt} = C\frac{d^2v_C}{dt^2}$$

Hence

$$v = RC\frac{dv_C}{dt} + LC\frac{d^2v_C}{dt^2} + v_C$$

This is a second-order differential equation.

To illustrate the above, consider the relationship between the output, the potential difference across the inductor of v_L, and the input v for the circuit shown in Figure 8.17. Applying Kirchhoff's second law to the circuit loop gives

$$v = v_R + v_L$$

where v_R is the potential difference across the resistor R and v_L that across the inductor. Since $v_R = iR$,

$$v = iR + v_L$$

Since

$$i = \frac{1}{L}\int v_L\,dt$$

then the relationship between the input and output is:

$$v = \frac{R}{L}\int v_L\,dt + v_L$$

Fig. 8.17   Resistor–inductor system

As another example, consider the relationship between the output, the potential difference v_C across the capacitor, and the input v for the circuit shown in Figure 8.18. Using nodal analysis, node B is taken as the reference node and node A taken to be at a potential of v_A relative to B. Applying Kirchhoff's law 1 to node A gives

$$i_1 = i_2 + i_3$$

But

$$i_1 = \frac{v - v_A}{R}$$

$$i_2 = \frac{1}{L}\int v_A\,dt$$

$$i_3 = C\frac{dv_A}{dt}$$

Fig. 8.18   Resistor–capacitor–
inductor system

Hence

$$\frac{v - v_A}{R} = \frac{1}{L} \int v_A \, dt + C \frac{dv_A}{dt}$$

But $v_C = v_A$. Hence, with some rearrangement,

$$v = RC \frac{dv_C}{dt} + v_C + \frac{R}{L} \int v_C \, dt$$

The same answer could have been obtained by mesh analysis.

8.3.2 Electrical and mechanical analogies

The building blocks for electrical and mechanical systems have many similarities. For example, the electrical resistor does not store energy but dissipates it with the current i through the resistor being given by $i = v/R$, where R is a constant, and the power P dissipated by $P = v^2/R$. The mechanical analogue of the resistor is the dashpot. It also does not store energy but dissipates it with the force F being related to the velocity v by $F = cv$, where c is a constant, and the power P dissipated by $P = cv^2$. Both these sets of equations have similar forms. Comparing them, and taking the current as being analogous to the force, then the potential difference is analogous to the velocity and the dashpot constant c to the reciprocal of the resistance, i.e. $(1/R)$. These analogies between current and force, potential difference and velocity, hold for the other building blocks with the spring being analogous to inductance and mass to capacitance.

The analogy between current and force is the one most often used. However, another set of analogies can be drawn between potential difference and force.

8.4 Fluid system building blocks

Fig. 8.19   Fluid system block

In fluid flow systems there are three basic building blocks which can be considered to be the equivalent of electrical resistance, capacitance and inductance. For such systems (Fig. 8.19) the input, the equivalent of the electrical current, is the volumetric rate of flow q, and the output, the equivalent of electrical potential difference, is pressure difference $(p_1 - p_2)$. Fluid systems can be considered to fall into two categories: *hydraulic*, where the fluid is a liquid and is deemed to be incompressible; and *pneumatic*, where it is a gas which can be compressed and consequently shows a density change.

Hydraulic resistance is the resistance to flow which occurs as a result of a liquid flowing through valves or changes in a pipe diameter (Fig. 8.20). The relationship between the volume rate of flow of liquid q through the resistance element and the resulting pressure difference $(p_1 - p_2)$ is

$$p_1 - p_2 = Rq$$

Fig. 8.20   Hydraulic resistance examples

Fig. 8.21   Hydraulic capacitance

Fig. 8.22   Hydraulic inertance

where R is a constant called the hydraulic resistance. The bigger the resistance the bigger the pressure difference for a given rate of flow. This equation, like that for the electrical resistance and Ohm's law, assumes a linear relationship. Such hydraulic linear resistances occur with orderly flow through capillary tubes and porous plugs but non-linear resistances occur with flow through sharp-edged orifices or if flow is turbulent.

Hydraulic capacitance is the term used to describe energy storage with a liquid where it is stored in the form of potential energy. A height of liquid in a container (Fig. 8.21), i.e. a so-called pressure head, is one form of such a storage. For such a capacitance, the rate of change of volume V in the container, i.e. dV/dt, is equal to the difference between the volumetric rate at which liquid enters the container q_1 and the rate at which it leaves q_2,

$$q_1 - q_2 = \frac{dV}{dt}$$

But $V = Ah$, where A is the cross-sectional area of the container and h the height of liquid in it. Hence

$$q_1 - q_2 = \frac{d(Ah)}{dt} = A \frac{dh}{dt}$$

But the pressure difference between the input and output is p, where $p = h\rho g$ with ρ being the liquid density and g the acceleration due to gravity. Thus, if the liquid is assumed to be incompressible, i.e. its density does not change with pressure,

$$q_1 - q_2 = A \frac{d(p/\rho g)}{dt} = \frac{A}{\rho g} \frac{dp}{dt}$$

The hydraulic capacitance C is defined as being

$$C = \frac{A}{\rho g}$$

Thus

$$q_1 - q_2 = C \frac{dp}{dt}$$

Integration of this equation gives

$$p = \frac{1}{C} \int (q_1 - q_2)\, dt$$

Hydraulic inertance is the equivalent of inductance in electrical systems or a spring in mechanical systems. To accelerate a fluid and so increase its velocity a force is required. Consider a block of liquid of mass m (Fig. 8.22). The net force acting on the liquid is

$$F_1 - F_2 = p_1A - p_2A = (p_1 - p_2)A$$

where $(p_1 - p_2)$ is the pressure difference and A the cross-sectional area. This net force causes the mass to accelerate with an acceleration a, and so

$$(p_1 - p_2)A = ma$$

But a is the rate of change of velocity dv/dt, hence

$$(p_1 - p_2)A = m\frac{dv}{dt}$$

But the mass of liquid concerned has a volume of AL, where L is the length of the block of liquid or the distance between the points in the liquid where the pressures p_1 and p_2 are measured. If the liquid has a density ρ then $m = AL\rho$ and so

$$(p_1 - p_2)A = AL\rho\frac{dv}{dt}$$

But the volume rate of flow $q = Av$, hence

$$(p_1 - p_2)A = L\rho\frac{dq}{dt}$$

$$p_1 - p_2 = I\frac{dq}{dt}$$

where the hydraulic inertance I is defined as

$$I = \frac{L\rho}{A}$$

With *pneumatic systems* the three basic building blocks are, as with hydraulic systems, resistance, capacitance and inertance. However, gases differ from liquids in being compressible, i.e. a change in pressure causes a change in volume and hence density. *Pneumatic resistance* R is defined in terms of the mass rate of flow dm/dt (note that this is often written as an m with a dot above it to indicate that the symbol refers to the mass rate of flow and not just the mass) and the pressure difference $(p_1 - p_2)$ as

$$p_1 - p_2 = R\frac{dm}{dt} = R\dot{m}$$

Pneumatic capacitance C is due to the compressibility of the gas, and is comparable to the way in which the compression of a spring stores energy. If there is a mass rate of flow dm_1/dt entering a container of volume V and a mass rate of flow of dm_2/dt leaving it, then the rate at which the mass in the container is changing is $(dm_1/dt - dm_2/dt)$. If the gas in the container has a density ρ then the rate of change of mass in the container is

Rate of change of mass in container $= \dfrac{\mathrm{d}(\rho V)}{\mathrm{d}t}$

But, because a gas can be compressed, both ρ and V can vary with time. Hence

Rate of change of mass in container $= \rho \dfrac{\mathrm{d}V}{\mathrm{d}t} + V \dfrac{\mathrm{d}\rho}{\mathrm{d}t}$

Since $(\mathrm{d}V/\mathrm{d}t) = (\mathrm{d}V/\mathrm{d}p)(\mathrm{d}p/\mathrm{d}t)$ and, for an ideal gas, $pV = mRT$ with consequently $p = (m/V)RT = \rho RT$ and $\mathrm{d}\rho/\mathrm{d}t = (1/RT)(\mathrm{d}p/\mathrm{d}t)$, then

Rate of change of mass in container $= \rho \dfrac{\mathrm{d}V}{\mathrm{d}p} \dfrac{\mathrm{d}p}{\mathrm{d}t} + \dfrac{V}{RT} \dfrac{\mathrm{d}p}{\mathrm{d}t}$

where R is the gas constant and T the temperature, assumed to be constant, on the Kelvin scale. Thus

$$\frac{\mathrm{d}m_1}{\mathrm{d}t} - \frac{\mathrm{d}m_2}{\mathrm{d}t} = \left(\rho \frac{\mathrm{d}V}{\mathrm{d}p} + \frac{V}{RT} \right) \frac{\mathrm{d}p}{\mathrm{d}t}$$

The pneumatic capacitance due to the change in volume of the container C_1 is defined as

$$C_1 = \rho \frac{\mathrm{d}V}{\mathrm{d}p}$$

and the pneumatic capacitance due to the compressibility of the gas C_2 as

$$C_2 = \frac{V}{RT}$$

Hence

$$\frac{\mathrm{d}m_1}{\mathrm{d}t} - \frac{\mathrm{d}m_2}{\mathrm{d}t} = (C_1 + C_2) \frac{\mathrm{d}p}{\mathrm{d}t}$$

or

$$p_1 - p_2 = \frac{1}{C_1 + C_2} \int (\dot{m}_1 - \dot{m}_2) \, \mathrm{d}t$$

Pneumatic inertance is due to the pressure drop necessary to accelerate a block of gas. According to Newton's second law the net force is $ma = \mathrm{d}(mv)/\mathrm{d}t$. Since the force is provided by the pressure difference $(p_1 - p_2)$, then if A is the cross-sectional area of the block of gas being accelerated

$$(p_1 - p_2)A = \frac{\mathrm{d}(mv)}{\mathrm{d}t}$$

But m, the mass of the gas being accelerated, is ρLA with ρ being the gas density and L the length of the block of gas being accelerated. But the volume rate of flow $q = Av$, where v is the velocity. Thus

$$mv = \rho LA\frac{q}{A} = \rho Lq$$

and so

$$(p_1 - p_2)A = L\frac{d(\rho q)}{dt}$$

But $\dot{m} = \rho q$ and so

$$p_1 - p_2 = \frac{L}{A}\frac{d\dot{m}}{dt}$$

$$p_1 - p_2 = I\frac{d\dot{m}}{dt}$$

with the pneumatic inertance I being

$$I = \frac{L}{A}$$

Table 8.3 shows the basic characteristics of the fluid building blocks, both hydraulic and pneumatic.

Table 8.3   Hydraulic and pneumatic building blocks

Building block	Describing equation	Energy stored or power dissipated
Hydraulic		
Inertance	$q = \frac{1}{L}\int(p_1 - p_2)\,dt$	$E = \frac{1}{2}Iq^2$
Capacitance	$q = C\frac{d(p_1 - p_2)}{dt}$	$E = \frac{1}{2}C(p_1 - p_2)^2$
Resistance	$q = \frac{p_1 - p_2}{R}$	$P = \frac{1}{R}(p_1 - p_2)^2$
Pneumatic		
Inertance	$\dot{m} = \frac{1}{L}\int(p_1 - p_2)\,dt$	$E = \frac{1}{2}I\dot{m}^2$
Capacitance	$\dot{m} = C\frac{d(p_1 - p_2)}{dt}$	$E = \frac{1}{2}C(p_1 - p_2)^2$
Resistance	$\dot{m} = \frac{p_1 - p_2}{R}$	$P = \frac{1}{R}(p_1 - p_2)^2$

For hydraulics the volumetric rate of flow and for pneumatics the mass rate of flow are analogous to the electrical current in an electric system. For both hydraulics and pneumatics the pressure difference is analogous to the potential difference in electrical

Fig. 8.23   A fluid system

systems. Compare Table 8.3 with Table 8.2. Hydraulic and pneumatic inertance and capacitance are both energy storage elements; hydraulic and pneumatic resistance are both energy dissipaters.

8.4.1 Building up a model for a fluid system

Figure 8.23 shows a simple hydraulic system, a liquid entering and leaving a container. Such a system can be considered to consist of a capacitor, the liquid in the container, with a resistor, the valve. Inertance can be neglected since flow rates change only very slowly. For the capacitor we can write

$$q_1 - q_2 = C\frac{\mathrm{d}p}{\mathrm{d}t}$$

The rate at which liquid leaves the container q_2 equals the rate at which it leaves the valve. Thus for the resistor

$$p_1 - p_2 = Rq_2$$

The pressure difference $(p_1 - p_2)$ is the pressure due to the height of liquid in the container and is thus $h\rho g$. Thus $q_2 = h\rho g/R$ and so substituting for q_2 in the first equation gives

$$q_1 - \frac{h\rho g}{R} = C\frac{\mathrm{d}(h\rho g)}{\mathrm{d}t}$$

and, since $C = A/\rho g$,

$$q_1 = A\frac{\mathrm{d}h}{\mathrm{d}t} + \frac{\rho g h}{R}$$

This equation describes how the height of liquid in the container depends on the rate of input of liquid into the container.

A bellows is an example of a simple pneumatic system (Fig. 8.24). Resistance is provided by a constriction which restricts the rate of flow of gas into the bellows and capacitance is provided by the bellows itself. Inertance can be neglected since the flow rate changes only slowly.

The mass flow rate into the bellows is given by

$$p_1 - p_2 = R\dot{m}$$

where p_1 is the pressure prior to the constriction and p_2 the pressure after the constriction, i.e. the pressure in the bellows. All the gas that flows into the bellows remains in the bellows, there being no exit from the bellows. The capacitance of the bellows is given by

Fig. 8.24   A pneumatic system

$$\dot{m}_1 - \dot{m}_2 = (C_1 + C_2)\frac{\mathrm{d}p_2}{\mathrm{d}t}$$

The mass flow rate entering the bellows is given by the equation for the resistance and the mass leaving the bellows is zero. Thus

$$\frac{p_1 - p_2}{R} = (C_1 + C_2)\frac{dp_2}{dt}$$

Hence

$$p_1 = R(C_1 + C_2)\frac{dp_2}{dt} + p_2$$

This equation describes how the pressure in the bellows p_2 varies with time when there is an input of a pressure p_1.

The bellows expands or contracts as a result of pressure changes inside it. Bellows are just a form of spring and so we can write $F = kx$ for the relationship between the force F causing an expansion or contraction and the resulting displacement x, where k is the spring constant for the bellows. But the force F depends on the pressure p_2, with $p_2 = F/A$ where A is the cross-sectional area of the bellows. Thus $p_2A = F = kx$. Hence substituting for p_2 in the above equation gives

$$p_1 = R(C_1 + C_2)\frac{k}{A}\frac{dx}{dt} + \frac{k}{A}x$$

This equation, a first-order differential equation, describes how the extension or contraction x of the bellows changes with time when there is an input of a pressure p_1. The pneumatic capacitance due to the change in volume of the container C_1 is $p\,dV/dp_2$ and since $V = Ax$, C_1 is $pA\,dx/dp_2$. But for the bellows $p_2A = kx$, thus

$$C_1 = \rho A\,\frac{dx}{d(kx/A)} = \frac{\rho A^2}{k}$$

C_2, the pneumatic capacitance due to the compressibility of the air, is $V/RT = Ax/RT$.

The following illustrates how, for the hydraulic system shown in Figure 8.25, relationships can be derived which describe how the heights of the liquids in the two containers will change with time. With this model inertance is neglected.

Container 1 is a capacitor and thus

$$q_1 - q_2 = C_1\frac{dp}{dt}$$

where $p = h_1\rho g$ and $C_1 = A_1/\rho g$ and so

$$q_1 - q_2 = A_1\frac{dh_1}{dt}$$

Fig. 8.25   A fluid system

The rate at which liquid leaves the container q_2 equals the rate at which it leaves the valve R_1. Thus for the resistor,

$$p_1 - p_2 = R_1 q_2$$

The pressures are $h_1 \rho g$ and $h_2 \rho g$. Thus

$$(h_1 - h_2)\rho g = R_1 q_2$$

Using the value of q_2 given by this equation and substituting it into the earlier equation gives

$$q_1 - \frac{(h_1 - h_2)\rho g}{R_1} = A_1 \frac{\mathrm{d}h_1}{\mathrm{d}t}$$

This equation describes how the height of the liquid in container 1 depends on the input rate of flow.

For container 2 a similar set of equations can be derived. Thus for the capacitor C_2

$$q_2 - q_3 = C_2 \frac{\mathrm{d}p}{\mathrm{d}t}$$

where $p = h_2 \rho g$ and $C_2 = A_2 / \rho g$ and so

$$q_2 - q_3 = A_2 \frac{\mathrm{d}h_2}{\mathrm{d}t}$$

The rate at which liquid leaves the container q_3 equals the rate at which it leaves the valve R_2. Thus for the resistor,

$$p_2 - 0 = R_2 q_3$$

This assumes that the liquid exits into the atmosphere. Thus, using the value of q_3 given by this equation and substituting it into the earlier equation gives

$$q_2 - \frac{h_2 \rho g}{R_2} = A_2 \frac{dh_2}{dt}$$

Substituting for q_2 in this equation using the value given by the equation derived for the first container gives

$$\frac{(h_1 - h_2)\rho g}{R_1} - \frac{h_2 \rho g}{R_2} = A_2 \frac{dh_2}{dt}$$

This equation describes how the height of liquid in container 2 changes.

8.5 Thermal system building blocks

There are only two basic building blocks for thermal systems: resistance and capacitance. There is a net flow of heat between two points if there is a temperature difference between them. The electrical equivalent of this is that there is only a net current i between two points if there is a potential difference v between them, the relationship between the current and potential difference being $i = v/R$, where R is the electrical resistance between the points. A similar relationship can be used to define *thermal resistance R*. If q is the rate of flow of heat and $(T_1 - T_2)$ the temperature difference, then

$$q = \frac{T_2 - T_1}{R}$$

The value of the resistance depends on the mode of heat transfer. In the case of conduction through a solid, for unidirectional conduction

$$q = Ak \frac{T_1 - T_2}{L}$$

where A is the cross-sectional area of the material through which the heat is being conducted and L the length of material between the points at which the temperatures are T_1 and T_2. k is the thermal conductivity. Hence, with this mode of heat transfer,

$$R = \frac{L}{Ak}$$

When the mode of heat transfer is convection, as with liquids and gases, then

$$q = Ah(T_2 - T_1)$$

where A is the surface area across which there is the temperature difference and h is the coefficient of heat transfer. Thus, with this mode of heat transfer,

$$R = \frac{1}{Ah}$$

Thermal capacitance is a measure of the store of internal energy in a system. Thus, if the rate of flow of heat into a system is q_1 and the rate of flow out is q_2, then

Rate of change of internal energy = $q_1 - q_2$

An increase in internal energy means an increase in temperature. Since

Internal energy change = mc × change in temperature

where m is the mass and c the specific heat capacity, then

Rate of change of internal energy = mc × rate of change of
temperature

Thus

$$q_1 - q_2 = mc \frac{dT}{dt}$$

where dT/dt is the rate of change of temperature. This equation can be written as

$$q_1 - q_2 = C \frac{dT}{dt}$$

where C is the thermal capacitance and so $C = mc$. Table 8.4 gives a summary of the thermal building blocks.

Table 8.4   Thermal building blocks

Building block	Describing equation	Energy stored
Capacitance	$q_1 - q_2 = C \dfrac{dT}{dt}$	$E = CT$
Resistance	$q = \dfrac{T_1 - T_2}{R}$	

8.5.1 Building up a model for a thermal system

Consider a thermometer at temperature T which has just been inserted into a liquid at temperature T_L (Fig. 8.26). If the thermal resistance to heat flow from the liquid to the thermometer is R then,

$$q = \frac{T_L - T}{R}$$

Fig. 8.26   A thermal system

where q is the net rate of heat flow from liquid to thermometer. The thermal capacitance C of the thermometer is given by the equation

$$q_1 - q_2 = C\frac{dT}{dt}$$

Since there is only a net flow of heat from the liquid to the thermometer, $q_1 = q$ and $q_2 = 0$. Thus

$$q = C\frac{dT}{dt}$$

Substituting this value of q in the earlier equation gives

$$C\frac{dT}{dt} = \frac{T_L - T}{R}$$

Rearranging this equation gives

$$RC\frac{dT}{dt} + T = T_L$$

This equation, a first-order differential equation, describes how the temperature indicated by the thermometer T will vary with time when the thermometer is inserted into a hot liquid.

In the above thermal system the parameters have been considered to be lumped. This means, for example, that there has been assumed to be just one temperature for the thermometer and just one for the liquid, i.e. the temperatures are only functions of time and not position within a body.

To illustrate the above consider Figure 8.27 which shows a thermal system consisting of an electric fire in a room. The fire emits heat at the rate q_1 and the room loses heat at the rate q_2. Assuming that the air in the room is at a uniform temperature T and that there is no heat storage in the walls of the room, derive an equation describing how the room temperature will change with time.

If the air in the room has a thermal capacity C then

$$q_1 - q_2 = C\frac{dT}{dt}$$

If the temperature inside the room is T and that outside the room T_0 then

$$q_2 = \frac{T - T_0}{R}$$

where R is the resistivity of the walls. Substituting for q_2 gives

$$q_1 - \frac{T - T_0}{R} = C\frac{dT}{dt}$$

Fig. 8.27   Thermal system

Hence

$$RC\frac{dT}{dt} + T = Rq_1 + T_0$$

Problems

(a)

(b)

Fig. 8.28 Problem 1

Fig. 8.29 Problem 3

1 Derive an equation relating the input, force F, with the output, displacement x, for the systems described by Figure 8.28.

2 Propose a model for the metal wheel of a railway carriage running on a metal track.

3 Derive an equation relating the input angular displacement θ_i with the output angular displacement θ_o for the rotational system shown in Figure 8.29.

4 Propose a model for a stepped shaft (i.e. a shaft where there is a step change in diameter) used to rotate a mass and derive an equation relating the input torque and the angular rotation. You may neglect damping.

5 Derive the relationship between the output, the potential difference across the resistor R of v_R, and the input v for the circuit shown in Figure 8.30 which has a resistor in series with a capacitor.

6 Derive the relationship between the output, the potential difference across the resistor R of v_R, and the input v for the series LCR circuit shown in Figure 8.31.

7 Derive the relationship between the output, the potential difference across the capacitor C of v_C, and the input v for the circuit shown in Figure 8.32.

8 Derive the relationship between the height h_2 and time for the hydraulic system shown in Figure 8.33. Neglect inertance.

9 A hot object, capacitance C and temperature T, cools in a large room at temperature T_r. If the thermal system has a resistance R derive an equation describing how the temperature of the hot object changes with time and give an electrical analogue of the system.

Fig. 8.30 Problem 5

Fig. 8.31 Problem 6

Fig. 8.32 Problem 7

Fig. 8.33   Problem 8

Fig. 8.34   Problem 10

10 Figure 8.34 shows a thermal system involving two compartments, with one containing a heater. If the temperature of the
compartment containing the heater is T_1, the temperature of
the other compartment T_2 and the temperature surrounding
the compartments T_3, develop equations describing how the
temperatures T_1 and T_2 will vary with time. All the walls of
the containers have the same resistance and negligible
capacity. The two containers have the same capacity C.

11 Derive the differential equation relating the pressure input p
to a diaphragm actuator (as in Fig. 5.22) to the displacement
x of the stem.

12 Derive the differential equation for a motor driving a load
through a gear system (Fig. 8.35) which relates the angular
displacement of the load with time.

Fig. 8.35   Problem 12

9   System models

9.1 Engineering systems

In Chapter 8 the basic building blocks of translational mech-anical, rotational mechanical, electrical, fluid and thermal systems were separately considered. However, many systems encountered in engineering involve aspects of more than one of these disciplines. For example, an electric motor involves both electrical and mechanical elements. This chapter looks at how single-discipline building blocks can be combined to give models for such multi-discipline systems.

In combining blocks we are assuming that the relationship for each block is linear. The following is a discussion of linearity and how, because many real engineering items are non-linear, we can make a linear approximation for a non-linear item.

9.1.1 Linearity

The relationship between the force F and the extension x produced for an ideal spring is linear, being given by $F = kx$. This means that if force F_1 produces an extension x_1 and force F_2 produces an extension x_2, a force equal to $(F_1 + F_2)$ will produce an extension $(x_1 + x_2)$. This is called the *principle of super-position* and is a necessary condition for a system that can be termed a *linear system*. Another condition for a linear system is that if an input F_1 produces an extension x_1 then an input cF_1 will produce an output cx_1, where c is a constant multiplier. A graph of the force F plotted against the extension x is a straight line passing through the origin when the relationship is linear (Fig. 9.1(a)).

Real springs, like any other real components, are not perfectly linear (Fig. 9.1(b)). However, there is often a range of operation for which linearity can be assumed. Thus for the spring giving the graph in Fig. 9.1(b), linearity can be assumed provided the spring is only used over the central part of its graph. For many system components, linearity can be assumed for operations within a range of values of the variable about some operating point.

(a)

(b)

Fig. 9.1   Springs: (a) ideal, (b) real

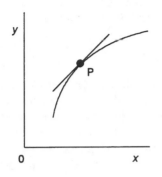

Fig. 9.2   A non-linear relationship

For some system components (Fig. 9.2) the relationship is non-linear. For such components the best that can be done to obtain a linear relationship is to just work with the straight line which is the slope of the graph at the operating point. Thus for the relationship between y and x in Figure 9.2, at the operating point P where the slope has the value m

$$\Delta y = m\, \Delta x$$

where Δy and Δx are small changes in input and output signals at the operating point.

Thus, for example, the rate of flow of liquid q through an orifice is given by

$$q = c_d A \sqrt{\frac{2(p_1 - p_2)}{\rho}}$$

where c_d is a constant called the discharge coefficient, A the cross-sectional area of the orifice, ρ the fluid density and $(p_1 - p_2)$ the pressure difference. For a constant cross-sectional area and density the equation can be written as

$$q = C\sqrt{p_1 - p_2}$$

where C is a constant. This is a non-linear relationship between the rate of flow and the pressure difference. We can obtain a linear relationship by considering the straight line representing the slope of the rate of flow/pressure difference graph (Fig. 9.3) at the operating point. The slope m is $dq/d(p_1 - p_2)$ and has the value

Fig. 9.3   Flow through an orifice

$$m = \frac{dq}{d(p_1 - p_2)} = \frac{C}{2\sqrt{p_{o1} - p_{o2}}}$$

where $(p_{o1} - p_{o2})$ is the value at the operating point. For small changes about the operating point we will assume that we can replace the non-linear graph by the straight line of slope m and therefore can write $m = \Delta q/\Delta(p_1 - p_2)$ and hence

$$\Delta q = m\, \Delta(p_1 - p_2)$$

Hence, if we had $C = 2$ m³/s per kPa, i.e. $q = 2(p_1 - p_2)$, then for an operating point of $(p_1 - p_2) = 4$ kPa with $m = 2/(2\sqrt{4}) = 0.5$, the linearised version of the equation would be

$$\Delta q = 0.5\, \Delta(p_1 - p_2)$$

In the above discussion it was assumed that the flow was through an orifice of constant cross-sectional area. If the orifice is a flow control valve then this is not the case, the cross-sectional area being adjusted to vary the flow rate. In such a situation,

$$q = CA \sqrt{p_1 - p_2}$$

Since both A and $(p_1 - p_2)$ can change, then we have to obtain the linearised equation when either or both these variables can change. Because of the principle of superposition we can consider each of these variables changing independently and then add the two results to obtain the equation for when both change. Thus for changes about the operating point the slopes of a graph of q against A would be

$$m_1 = \frac{dq}{dA} = C \sqrt{p_{o1} - p_{o2}}$$

and thus $\Delta q = m_1 \Delta A$. The subscript o is used to indicate values at the operating point. For a graph of q against $(p_1 - p_2)$

$$m_2 = \frac{dq}{d(p_1 - p_2)} = \frac{CA_o}{2 \sqrt{p_{o1} - p_{o2}}}$$

and thus $\Delta q = m_2 \Delta(p_1 - p_2)$. The linearised version when both variables can change is thus

$$\Delta q = m_1 \Delta A + m_2 \Delta(p_1 - p_2)$$

with m_1 and m_2 having the values indicated above.

Linearised mathematical models are used because most of the techniques of control systems are based on there being linear relationships for the elements of such systems. Also, because most control systems are maintaining an output equal to some reference value, the variations from this value tend to be rather small and so the linearised model is perfectly appropriate.

To illustrate the above, consider a thermistor being used for temperature measurement in a control system. The relationship between the resistance R of the thermistor and its temperature T is given by

$$R = k e^{-cT}$$

We can linearise this equation about an operating point of T_o. The slope m of a graph of R against T at the operating point T_o is given by dR/dT. Thus

$$m = \frac{dR}{dT} = -kc\, e^{-cT_o}$$

Hence

$$\Delta R = m\,\Delta T = (-kc\,e^{-cT_0})\Delta T$$

9.2 Rotational– translational systems

There are many mechanisms which involve the conversion of rotational motion to translational motion or vice versa. For example, there are rack-and-pinion, shafts with lead screws, pulley and cable systems, etc.

To illustrate how such systems can be analysed, consider a rack-and-pinion system (Fig. 9.4). The rotational motion of the pinion is transformed into translational motion of the rack. Consider first the pinion element. The net torque acting on it is $(T_{in} - T_{out})$. Thus, considering the moment of inertia element, and assuming negligible damping,

Input torque T_{in}, Pinion, Output, velocity v, ω, Rack, Torque output T_{out}, Radius r

Fig. 9.4   Rack-and-pinion

$$T_{in} - T_{out} = I\frac{d\omega}{dt}$$

where I is the moment of inertia of the pinion and ω its angular velocity. The rotation of the pinion will result in a translational velocity v of the rack. If the pinion has a radius r, then $v = r\omega$. Hence we can write

$$T_{in} - T_{out} = \frac{I}{r}\frac{dv}{dt}$$

Now consider the rack element. There will be a force of T/r acting on it due to the movement of the pinion. If there is a frictional force of cv then the net force is

$$\frac{T_{out}}{r} - cv = m\frac{dv}{dt}$$

Eliminating T_{out} from the two equations gives

$$T_{in} - rcv = \left(\frac{I}{r} + mr\right)\frac{dv}{dt}$$

and so

$$\frac{dv}{dt} = \left(\frac{r}{I + mr^2}\right)(T_{in} - rcv)$$

The result is a first-order differential equation describing how the output is related to the input.

9.3 Electromechanical systems

Electromechanical devices, such as potentiometers, motors and generators, transform electrical signals to rotational motion or vice versa. This section is a discussion of how we can derive models for such systems. A potentiometer has an input of a rotation and an output of a potential difference. An electric motor has an input of a potential difference and an output of rotation of a shaft. A generator has an input of rotation of a shaft and an output of a potential difference.

9.3.1 Potentiometer

The *rotary potentiometer* (Fig. 9.5) is a potential divider and thus:

$$\frac{v_o}{V} = \frac{\theta}{\theta_{max}}$$

where V is the potential difference across the full length of the potentiometer track and θ_{max} is the total angle swept out by the slider in being rotated from one end of the track to the other. The output is v_o for the input θ.

Fig. 9.5   Rotary potentiometer

9.3.2 D.C. motor

The d.c. motor is used to convert an electrical input signal into a mechanical output signal, a current through the armature coil of the motor resulting in a shaft being rotated and hence the load rotated (Fig. 9.6). The motor basically consists of a coil, the armature coil, which is free to rotate. This coil is located in the magnetic field provided by a current through field coils or a permanent magnet. When a current i_a flows through the armature coil then, because it is in a magnetic field, forces act on the coil and cause it to rotate (Fig. 9.7). The force F acting on a wire carrying a current i_a and of length L in a magnetic field of flux density B at right angles to the wire is given by $F = Bi_aL$ and with N wires is $F = Nbi_aL$. The forces on the armature coil wires result in a torque T, where $T = Fb$, with b being the breadth of the coil. Thus

Fig. 9.6   Motor driving a load

$$T = NBi_aLb$$

The resulting torque is thus proportional to (Bi_a), the other factors all being constants. Hence we can write

$$T = k_1Bi_a$$

Since the armature is a coil rotating in a magnetic field, a voltage will be induced in it as a consequence of electromagnetic induction. This voltage will be in such a direction as to oppose the change producing it and is called the back e.m.f. This back e.m.f. v_b is proportional to the rate or rotation of the armature and the flux linked by the coil, hence the flux density B. Thus

Fig. 9.7   One wire of armature coil

$$v_b = k_2B\omega$$

where ω is the shaft angular velocity and k_2 a constant.

Consider a d.c. motor which has the armature and field coils separately excited (see Fig. 7.22(d) and associated discussion). With a so-called *armature-controlled motor* the field current i_f is held constant and the motor controlled by adjusting the armature voltage v_a. A constant field current means a constant magnetic flux density B for the armature coil. Thus

$$v_b = k_2 B \omega = k_3 \omega$$

where k_3 is a constant. The armature circuit can be considered to be a resistance R_a in series with an inductance L_a (Fig. 9.8). If v_a is the voltage applied to the armature circuit then, since there is a back e.m.f. of v_b, we have

$$v_a - v_b = L_a \frac{di_a}{dt} + R_a i_a$$

We can think of this equation in terms of the block diagram shown in Fig. 9.9(a).

Fig. 9.8   D.C. motor circuits

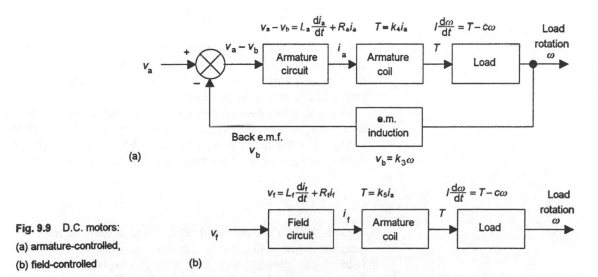

Fig. 9.9   D.C. motors:

(a) armature-controlled,

(b) field-controlled

The input to the motor part of the system is v_a and this is summed with the feedback signal of the back e.m.f. v_b to give an error signal which is the input to the armature circuit. The above equation thus describes the relationship between the input of the error signal to the armature coil and the output of the armature current i_a. Substituting for v_b:

$$v_a - k_3\omega = L_a\frac{di_a}{dt} + R_ai_a$$

The current i_a in the armature generates a torque T. Since, for the armature-controlled motor, B is constant we have

$$T = k_1Bi_a = k_4i_a$$

where k_4 is a constant. This torque then becomes the input to the load system. The net torque acting on the load will be

Net torque $= T -$ damping torque

The damping torque is $c\omega$, where c is a constant. Hence, if any effects due to the torsional springiness of the shaft are neglected,

Net torque $= k_4i_a - c\omega$

This will cause an angular acceleration of $d\omega/dt$, hence

$$I\frac{d\omega}{dt} = k_4i_a - c\omega$$

We thus have two equations that describe the conditions occurring for an armature-controlled motor, namely

$$v_a - k_3\omega = L_a\frac{di_a}{dt} + R_ai_a \quad\text{and}\quad I\frac{d\omega}{dt} = k_4i_a - c\omega$$

We can thus obtain the equation relating the output ω with the input v_a to the system by eliminating i_a. See the brief discussion of the Laplace transform in Chapter 10, or that in Appendix A, for details of how this might be done.

With a so-called *field-controlled motor* the armature current is held constant and the motor controlled by varying the field voltage. For the field circuit (Fig. 9.8) there is essentially just inductance L_f in series with a resistance R_f. Thus for that circuit

$$v_f = R_fi_f + L_f\frac{di_f}{dt}$$

We can think of the field-controlled motor in terms of the block diagram shown in Fig. 9.9(b). The input to the system is v_f. The field circuit converts this into a current i_f, the relationship between v_f and i_f being the above equation. This current leads to the production of a magnetic field and hence a torque acting on the armature coil, as given by $T = k_1 B i_a$. But the flux density B is proportional to the field current i_f and i_a is constant, hence

$$T = k_1 B i_a = k_5 i_f$$

where k_5 is a constant. This torque output is then converted by the load system into an angular velocity ω. As earlier, the net torque acting on the load will be

Net torque = T – damping torque

The damping torque is $c\omega$, where c is a constant. Hence, if any effects due to the torsional springiness of the shaft are neglected,

Net torque = $k_5 i_f - c\omega$

This will cause an angular acceleration of $d\omega/dt$, hence

$$I\frac{d\omega}{dt} = k_5 i_f - c\omega$$

The conditions occurring for a field-controlled motor are thus described by the equations

$$v_f = R_f i_f + L_f \frac{d i_f}{dt} \quad \text{and} \quad I\frac{d\omega}{dt} = k_5 i_f - c\omega$$

We can thus obtain the equation relating the output ω with the input v_f to the system by eliminating i_f. See the brief discussion of the Laplace transform in Chapter 10, or that in the Appendix, for details of how this might be done.

9.4 Hydraulic–mechanical systems

Hydraulic–mechanical converters involve the transformation of hydraulic signals to translational or rotational motion, or vice versa. Thus, for example, the movement of a piston in a cylinder as a result of hydraulic pressure involves the transformation of a hydraulic pressure input to the system to a translational motion output.

Figure 9.10 shows a hydraulic system in which an input of displacement x_i is, after passing through the system, transformed into a displacement x_o of a load. The system consists of a *spool valve* and a *cylinder*. The input displacement x_i to the left results in the hydraulic fluid supply pressure p_s causing fluid to flow into the left-hand side of the cylinder. This pushes the piston in the cylinder to the right and expels the fluid in the right-hand side of the chamber through the exit port at the right-hand end of the spool valve.

Fig. 9.10 Hydraulic system and load

The rate of flow of fluid to and from the chamber depends on the extent to which the input motion has uncovered the ports allowing the fluid to enter or leave the spool valve. When the input displacement x_i is to the right the spool valve allows fluid to move to the right-hand end of the cylinder and so results in a movement of the piston in the cylinder to the left.

The rate of flow of fluid q through an orifice, which is what the ports in the spool valve are, is a non-linear relationship (see Section 9.1.1) depending on the pressure difference between the two sides of the orifice and its cross-sectional area A. However, a linearised version of the equation can be used (see Section 9.1.1 for its derivation)

$$\Delta q = m_1 \Delta A + m_2 \Delta(\text{pressure difference})$$

where m_1 and m_2 are constants at the operating point. For the fluid entering the chamber the pressure difference is $(p_s - p_1)$ and for the exit $(p_2 - p_o)$. If the operating point about which the equation is linearised is taken to be the point at which the spool valve is central and the ports connecting it to the cylinder are both closed, then for this condition q is zero, and so $\Delta q = q$, A is proportional to x_s if x_s is measured from this central position, and the change in pressure on the inlet side of the piston is $-\Delta p_1$ relative to p_s and on the exit side Δp_2 relative to p_o. Thus, for the inlet port the equation can be written as

$$q = m_1 x_i + m_2(-\Delta p_1)$$

and for the exit port

$$q = m_1 x_i + m_2 \Delta p_2$$

Adding the two equations gives

$$2q = 2m_1x_i - m_2(\Delta p_1 - \Delta p_2)$$

$$q = m_1x_i - m_3(\Delta p_1 - \Delta p_2)$$

where $m_3 = m_2/2$.

For the cylinder the change in the volume of fluid entering the left-hand side of the chamber, or leaving the right-hand side, when the piston moves a distance x_o is Ax_o, where A is the cross-sectional area of the piston. Thus the rate at which the volume is changing is $A(dx_o/dt)$. The rate at which fluid is entering the left-hand side of the cylinder is q. However, since there is some leakage flow of fluid from one side of the piston to the other

$$q = A\frac{dx_o}{dt} + q_L$$

where q_L is the rate of leakage. Substituting for q, gives

$$m_1x_i - m_3(\Delta p_1 - \Delta p_2) = A\frac{dx_o}{dt} + q_L$$

The rate of leakage flow q_L is a flow through an orifice, the gap between the piston and the cylinder. This is of constant cross-section and has a pressure difference $(\Delta p_1 - \Delta p_2)$. Hence, using the linearised equation for such a flow,

$$q_L = m_4(\Delta p_1 - \Delta p_2)$$

Thus, using this equation to substitute for q_L

$$m_1x_i - m_3(\Delta p_1 - \Delta p_2) = A\frac{dx_o}{dt} + m_4(\Delta p_1 - \Delta p_2)$$

$$m_1x_i - (m_3 + m_4)(\Delta p_1 - \Delta p_2) = A\frac{dx_o}{dt}$$

The pressure difference across the piston results in a force being exerted on the load, the force exerted being $(\Delta p_1 - \Delta p_2)A$. There is, however, some damping of motion, i.e. friction, of the mass. This is proportional to the velocity of the mass, i.e. (dx_o/dt). Hence the net force acting on the load is

$$\text{net force} = (\Delta p_1 - \Delta p_2)A - c\frac{dx_o}{dt}$$

This net force causes the mass to accelerate, the acceleration being (d^2x_o/dt^2). Hence

$$m\frac{d^2x_o}{dt^2} = (\Delta p_1 - \Delta p_2)A - c\frac{dx_o}{dt}$$

Rearranging this equation gives

$$\Delta p_1 - \Delta p_2 = \frac{m}{A}\frac{d^2x_o}{dt^2} + \frac{c}{A}\frac{dx_o}{dt}$$

Using this equation to substitute for the pressure difference in the earlier equation,

$$m_1x_i - (m_3 + m_4)\left(\frac{m}{A}\frac{d^2x_o}{dt^2} + \frac{c}{A}\frac{dx_o}{dt}\right) = A\frac{dx_o}{dt}$$

Rearranging gives

$$\frac{(m_3+m_4)m}{A}\frac{d^2x_o}{dt^2} + \left(A + \frac{c(m_3+m_4)}{A}\right)\frac{dx_o}{dt} = m_1x_i$$

and rearranging this equation leads to

$$\frac{(m_3+m_4)m}{A^2+c(m_3+m_4)}\frac{d^2x_o}{dt^2} + \frac{dx_o}{dt} = \frac{Am_1}{A^2+c(m_3+m_4)}x_i$$

This equation can be simplified by introducing two constants k and τ, the latter constant being called the time constant (see Chapter 10). Hence

$$\tau\frac{d^2x_o}{dt^2} + \frac{dx_o}{dt} = kx_i$$

Thus the relationship between input and output is described by a second-order differential equation.

Problems

1   The relationship between the force F used to stretch a spring and its extension x is given by

$$F = kx^2$$

where k is a constant. Linearise this equation for an operating point of x_o.

2   The relationship between the e.m.f. E generated by a thermocouple and the temperature T is given by

$$E = aT + bT^2$$

where a and b are constants. Linearise this equation for an operating point of temperature T_o.

3   The relationship between the torque T applied to a simple pendulum and the angular deflection (Fig. 9.11) is given by

$$T = mgL \sin \theta$$

Fig. 9.11   Problem 3

where m is the mass of the pendulum bob, L the length of the pendulum and g the acceleration due to gravity. Linearise this equation for the equilibrium angle θ of $0°$.

4 Derive a differential equation relating the input voltage to a d.c. servo motor and the output angular velocity, assuming that the motor is armature controlled and the equivalent circuit for the motor has an armature with just resistance, its inductance being neglected.

5 Derive differential equations for a d.c. generator. The generator may be assumed to have a constant magnetic field. The armature circuit has the armature coil, having both resistance and inductance, in series with the load. Assume that the load has both resistance and inductance.

6 Derive differential equations for a permanent magnet d.c. motor.

7 Consider a solenoid actuator in which a current passing through the solenoid results in movement of a rod actuator into or out of the solenoid. Propose models for the arrangement which could then be used to develop a differential equation relating the input of current to the output of displacement.

10 Dynamic responses of systems

10.1 Modelling dynamic systems

The most important function of a model devised for measurement or control systems is to be able to predict what the output will be for a particular input. We are not just concerned with a static situation, i.e. that after some time when the steady state has been reached an output of x corresponds to an input of y. We have to consider how the output will change with time when there is a change of input or when the input changes with time. For example, how will the temperature of a temperature-controlled system change with time when the thermostat is set to a new temperature? For a control system, how will the output of the system change with time when the set value is set to a new value or perhaps increased at a steady rate?

Chapters 8 and 9 were concerned with models of systems when the inputs varied with time, with the results being expressed in terms of differential equations. This chapter is about how we can use such models to make predictions about how outputs will change with time when the input changes with time.

10.1.1 Differential equations

To describe the relationship between the input to a system and its output we must describe the relationship between inputs and outputs which are both possible functions of time. We thus need a form of equation which will indicate how the system output will vary with time when the input is varying with time. This can be done by the use of a *differential equation*. Such an equation includes derivatives with respect to time and so gives information about how the response of a system varies with time. A derivative dx/dt describes the rate at which x varies with time, the derivative d^2x/dt^2 states how dx/dt varies with time. Differential equations can be classed as *first-order*, *second-order*, *third-order*, etc. according to the highest order of the derivative in the equation.

For a first-order equation the highest order will be dx/dt, with a second-order d^2x/dt^2, with a third-order d^3x/dt^3, with nth-order d^nx/dt^n.

This chapter is about the types of responses we can expect from first-order and second-order systems and the solution of such differential equations in order that the response of the system to different types of input can be obtained. This chapter uses the 'try a solution' approach in order to find a solution; the Laplace transformation method is introduced in Chapter 11. A more detailed discussion of the transform is given in Appendix A. For a more detailed consideration of differential equations, the reader is referred to *Mathematics for Engineers and Technologists* by H. Fox and W. Bolton (Butterworth-Heinemann 2002) and of the Laplace transform to *Laplace and z-Transforms* by W. Bolton (Longman 1994).

10.1.2 Natural and forced responses

An example of a first-order system is water flowing out of a tank (Fig. 10.1). For such a system we have

$$p_1 - p_2 = Rq$$

where R is the hydraulic resistance. But $p_1 - p_2 = h\rho g$, where ρ is the density of the water, and q is the rate at which water leaves the tank and so is $-dV/dt$, with V being the volume of water in the tank and so being Ah. Thus $q = -d(Ah)/dt = -Adh/dt$. Thus the above equation can be written as:

$$h\rho g = -RA\frac{dh}{dt}$$

The range of change of the variable h is proportional to the variable. This might be termed a *natural response* in that there is no input to the system forcing the variable to change. We can draw attention to this by writing the differential equation with all the output terms, i.e. h, on the same side of the equals sign, i.e.

$$RA\frac{dh}{dt} + (\rho g)h = 0$$

In Section 8.4.1 the differential equation was derived for a water tank from which water was flowing but also into which there was a flow of water (Fig. 10.2). This equation has a forcing function of q_1 and can be written as:

$$RA\frac{dh}{dt} + (\rho g)h = q_1$$

As another example, consider a thermometer being placed in a hot liquid at some temperature T_L. The rate at which the reading

Fig. 10.1   Water flowing out of a tank

Fig. 10.2   Water flowing out of a tank with forcing input

Fig. 10.3 Transient and steady-state responses of a spring system

Fig. 10.4 Step input at time 0

Fig. 10.5 Impulse

Fig. 10.6 Ramp input at time 0

of the thermometer T changes with time and was derived in Section 8.5.1 as being given by the differential equation:

$$RC\frac{dT}{dt} + T = T_L$$

Such a differential equation has a forcing input of T_L.

10.1.3 Transient and steady-state responses

The total response of a control system, or element of a system, can be considered to be made up of two aspects, the steady-state response and the transient response. The *transient response* is that part of a system response which occurs when there is a change in input and which dies away after a short interval of time. The *steady-state response* is the response that remains after all transient responses have died down.

To give a simple illustration of this, consider a vertically suspended spring (Fig. 10.3) and what happens when a weight is suddenly suspended from it. The deflection of the spring abruptly increases and then may well oscillate until after some time it settles down to a steady value. The steady value is the steady-state response of the spring system, the oscillation that occurs prior to this steady state is the transient response.

The input to the spring system, the weight, is a quantity which varies with time. Up to some particular time there is no added weight, i.e. no input, then after that time there is an input which remains constant for the rest of the time. This type of input is known as a *step input* and has a graph of the form shown in Figure 10.4.

The input signal to systems can take other forms, e.g. impulse, ramp, and sinusoidal signals. An impulse is a very short duration input (Fig. 10.5); a ramp is a steadily increasing input (Fig. 10.6) and can be described by an equation of the form $y = kt$, where k is a constant and a sinusoidal input can be described by an equation of the form $y = k \sin \omega t$, with ω being the so-called angular frequency and equal to $2\pi f$ where f is the frequency.

Both the input and the output are functions of time. One way of indicating this is to write them in the form $f(t)$, where f is the function and (t) indicates that its value depends on time t. Thus for the weight W input to the spring system we could write $W(t)$ and for the deflection d output $d(t)$. $y(t)$ is commonly used for an input and $x(t)$ for an output.

10.2 First-order systems

Consider a first-order system (Fig. 10.7) with $y(t)$ as the input to the system and $x(t)$ the output and which has a forcing input b_0y and can be described by a differential equation of the form

$$a_1\frac{dx}{dt} + a_0x = b_0y$$

Fig. 10.7   System

Fig. 10.8   Natural response

where a_1, a_0 and b_0 are constants.

A method that we can use to solve a first-order equation and give an equation which directly indicates how the output varies with time involves the recognition of the type of solution that would fit the differential equation and then establishing that such a solution is valid.

The input $y(t)$ can take many forms. Consider first the situation when the input is 0. Because there is no input to the system we have no signal forcing the system to respond in any way other than its natural response with no input. The differential equation is then

$$a_1 \frac{dx}{dt} + a_0 x = 0$$

Let us try a solution of the form $x = A\,e^{st}$, where A and s are constants. We then have $dx/dt = sA\,e^{st}$ and so when these values are substituted in the differential equation we obtain

$$a_1 s A\,e^{st} + a_0 A\,e^{st} = 0$$

and so $a_1 s + a_0 = 0$ and $s = -a_0/a_1$. Thus the solution is

$$x = A\,e^{-a_0 t/a_1}$$

This is termed the *natural response* since there is no forcing function. We can determine the value of the constant A given some initial (boundary) condition. Thus if $x = 1$ when $t = 0$ then $A = 1$. Figure 10.8 shows the natural response, i.e. an exponential decay.

Now consider the differential equation when there is a *forcing function*, i.e.

$$a_1 \frac{dx}{dt} + a_0 x = b_0 y$$

Consider the solution to this equation to be made up of two parts, i.e. $x = u + v$. One part represents the transient part of the solution and the other the steady-state part. Substituting this into the differential equation gives

$$a_1 \frac{d(u+v)}{dt} + a_0(u+v) = b_0 y$$

Rearranging this gives

$$\left(a_1 \frac{du}{dt} + a_0 u \right) + \left(a_1 \frac{dv}{dt} + a_0 v \right) = b_0 y$$

If we let

$$a_1 \frac{dv}{dt} + a_0 v = b_0 y$$

then we must have

$$a_1 \frac{du}{dt} + a_0 u = 0$$

and so two differential equations, one of which contains a forcing function and one which is just the natural response equation. This last equation is just the natural equation which we solved earlier in this section and so will have a solution of the form

$$u = A\, e^{-a_0 t/a_1}$$

The other differential equation contains the forcing function y. For this differential equation the form of solution we try depends on the form of the input signal y. For a step input when y is constant for all times greater than 0, i.e. $y = k$, we can try a solution $v = A$, where A is a constant. If we have an input signal of the form $y = a + bt + ct^2 + ...$, where a, b and c are constants which can be zero, then we can try a solution which is of the form $v = A + Bt + Ct^2 + ...$. For a sinusoidal signal we can try a solution of the form $v = A \cos \omega t + B \sin \omega t$.

To illustrate this, assume there is a step input at a time of $t = 0$ with the size of the step being k. Then we try a solution of the form $v = A$. Differentiating a constant gives 0; thus when this solution is substituted into the differential equation we obtain $a_0 A = b_0 k$ and so $v = (b_0/a_0)k$.

The full solution will be given by $y = u + v$ and so will be

$$y = A\, e^{-a_0 t/a_1} + \frac{b_0}{a_0} k$$

We can determine the value of the constant A given some initial (boundary) conditions. Thus if the output $y = 0$ when $t = 0$ then

$$0 = A + \frac{b_0}{a_0} k$$

Thus $A = -(b_0/a_0)k$. The solution then becomes

$$x = \frac{b_0}{a_0} k(1 - e^{-a_0 t/a_1})$$

When $t \to \infty$ the exponential term tends to 0. The exponential term thus gives that part of the response which is the transient solution. The steady-state response is the value of x when $t \to \infty$ and so is $(b_0/a_0)k$. Thus the equation can be written as

$$x = \text{steady-state value} \times (1 - e^{-a_0 t/a_1})$$

Fig. 10.9   (a) Step input,
(b) resulting output

Figure 10.9 shows a graph of how the output x varies with time for the step input.

As a further illustration of the above, consider the following examples.

1   An electrical transducer system which consists of a resistance in series with a capacitor. When subject to a step input of size V it is found to give an output of a potential difference across the capacitor v which is given by the differential equation

$$RC\frac{dv}{dt} + v = V$$

What is the solution of the differential equation, i.e. what is the response of the system and how does v vary with time?

Comparing the differential equation with the equation solved earlier: $a_1 = RC$, $a_0 = 1$, and $b_0 = 1$. Then the solution is of the form

$$v = V(1 - e^{-t/RC})$$

2   Consider an electrical circuit consisting of a 1 MΩ resistance in series with a 2 μF capacitance. At a time $t = 0$ the circuit is subject to a ramp voltage of $4t$ V, i.e. the voltage increases at the rate of 4 V every 1 s. Determine how the voltage across the capacitor will vary with time.

The differential equation will be of a similar form to that given in the previous example but with the step voltage V of that example replaced by the ramp voltage $4t$, i.e.

$$RC\frac{dv}{dt} + v = 4t$$

Thus, using the values given in the question,

$$2\frac{dv}{dt} + v = 4t$$

Taking $v = v_n + v_f$, i.e. the sum of the natural and forced responses, we have for the natural response

$$2\frac{dv_n}{dt} + v_n = 0$$

and for the forced response

$$2\frac{dv_f}{dt} + v_f = 4t$$

For the natural response differential equation we can try a solution of the form $v_n = A\,e^{st}$. Hence, using this value

$$2As\, e^{st} + A\, e^{st} = 0$$

Thus $s = -\frac{1}{2}$ and so $v_n = A\, e^{-t/2}$.

For the forced response differential equation, since the right-hand side of the equation is $4t$ we can try a solution of the form $v_f = A + Bt$. Using this value gives $2B + A + Bt = 4t$. Thus we must have $B = 4$ and $A = -2B = -8$. Hence the solution is $v_f = -8 + 4t$. Thus the full solution is

$$v = v_n + v_f = A\, e^{-t/2} - 8 + 4t$$

Since $v = 0$ when $t = 0$ we must have $A = 8$. Hence

$$v = 8\, e^{-t/2} - 8 + 4t$$

3 Consider a motor when the relationship between the output angular velocity ω and the input voltage v for a motor is given by

$$\frac{IR}{k_1 k_2} \frac{d\omega}{dt} + \omega = \frac{1}{k_1} v$$

What will be the steady-state value of the angular velocity when the input is a step of size 1 V?

Comparing the differential equation with the equation solved earlier, then $a_1 = IR/k_1 k_2$, $a_0 = 1$ and $b_0 = 1/k_1$. The steady-state value for a step input is thus $(b_0/a_0) = 1/k_1$.

10.2.1 The time constant

For a first-order system subject to a step input of size k we have an output y which varies with time t according to

$$x = \frac{b_0}{a_0} k(1 - e^{-a_0 t/a_1})$$

or

$$x = \text{steady-state value} \times (1 - e^{-a_0 t/a_1})$$

When the time $t = (a_1/a_0)$ then the exponential term has the value $e^{-1} = 0.37$ and

$$x = \text{steady-state value} \times (1 - 0.37)$$

In this time the output has risen to 0.63 of its steady-state value. This time is called the *time constant* τ.

$$\tau = \frac{a_1}{a_0}$$

Fig. 10.10 Response of a first-order system to a step input

In a time of $2(a_1/a_0) = 2\tau$, the exponential term becomes $e^{-2} = 0.14$ and so

$$x = \text{steady-state value} \times (1 - 0.14)$$

In this time the output has risen to 0.86 of its steady-state value. In a similar way, values can be calculated for the output after 3τ, 4τ, 5τ, etc. Table 10.1 shows the results of such calculations and Figure 10.10 the graph of how the output varies with time for a unit step input.

Table 10.1 Response of a first-order system to a step input

Time t	Fraction of steady-state output
0	0
1τ	0.63
2τ	0.86
3τ	0.95
4τ	0.98
5τ	0.99
∞	1

In terms of the time constant τ, we can write the equation describing the response of a first-order system as

$$x = \text{steady-state value} \times (1 - e^{-t/\tau})$$

The time constant τ is (a_1/a_0), thus we can write our general form of the first-order differential equation

$$a_1 \frac{dx}{dt} + a_0 x = b_0 y$$

as

$$\tau \frac{dx}{dt} + x = \frac{b_0}{a_0} y$$

But b_0/a_0 is the factor by which the input y is multiplied to give the steady-state value. We can term it the *steady-state gain* since it is the factor stating by how much bigger the output is than the input under steady-state conditions. Thus if we denote this by G_{ss} then the differential equation can be written in the form

$$\tau \frac{dx}{dt} + x = G_{ss} y$$

Fig. 10.11 Example

To illustrate this consider Figure 10.11 which shows how the output v_o of a first-order system varies with time when subject to a step input of 5 V. The time constant is the time taken for a first-order system output to change from 0 to 0.63 of its final steady-state value. In this case this time is about 3 s. We can check this value, and that the system is first order, by finding the value at 2, i.e. 6 s. With a first-order system it should be 0.86 of the steady-state value. In this case it is. The steady-state output is 10 V. Thus the steady-state gain G_{ss} is (steady-state output/input) = 10/5 = 2. The differential equation for a first-order system can be written as

$$\tau \frac{dx}{dt} + x = G_{ss}y$$

Thus, for this system, we have

$$3\frac{dv_o}{dt} + v_o = 2v_i$$

10.3 Second-order systems

Fig. 10.12 Spring–dashpot–mass system

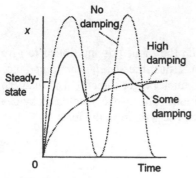

Fig. 10.13 Effect of damping

Many second-order systems can be considered to be essentially just a stretched spring with a mass and some means of providing damping. Figure 10.12 shows the basis of such a system. Such a system was analysed in Section 8.2.2. The equation describing the relationship between the input of force F and the output of a displacement x is

$$m\frac{d^2x}{dt^2} + c\frac{dx}{dt} + kx = F$$

where m is the mass, c the damping constant and k the spring constant.

The way in which the resulting displacement x will vary with time will depend on the amount of damping in the system. Thus if the force was applied as a step input and there was no damping at all then the mass would freely oscillate on the spring and the oscillations would continue indefinitely. No damping means $c = 0$ and so the dx/dt term is 0. However, damping will cause the oscillations to die away until a steady displacement of the mass is obtained. If the damping is high enough there will be no oscillations and the displacement of the mass will just slowly increase with time and gradually the mass will move towards its steady displacement position. Figure 10.13 shows the general way the displacements, for a step input, vary with time with different degrees of damping.

10.3.1 The second-order differential equation

Consider a mass on the end of a spring. In the absence of any damping and left to freely oscillate without being forced the output of the second-order system is a continuous oscillation (simple harmonic motion). Thus, suppose we describe this oscillation by the equation

$$x = A \sin \omega_n t$$

where x is the displacement at a time t, A the amplitude of the oscillation and ω_n the angular frequency of the free undamped oscillations. Differentiating this gives

$$\frac{dx}{dt} = \omega_n A \cos \omega_n t$$

Differentiating a second time gives

$$\frac{d^2 x}{dt^2} = -\omega_n^2 A \sin \omega_n t = -\omega_n^2 x$$

This can be reorganised to give the differential equation

$$\frac{d^2 x}{dt^2} + \omega_n^2 x = 0$$

But for a mass m on a spring of stiffness k we have a restoring force of kx and thus

$$m \frac{d^2 x}{dt^2} = -kx$$

This can be written as

$$\frac{d^2 x}{dt^2} + \frac{k}{m} x = 0$$

Thus, comparing the two differential equations, we must have

$$\omega_n^2 = \frac{k}{m}$$

and $x = A \sin \omega_n t$ is the solution to the differential equation.

Now consider when we have damping. The motion of the mass is then described by

$$m \frac{d^2 x}{dt^2} + c \frac{dx}{dt} + kx = F$$

We can solve this second-order differential equation by the same method used earlier for the first-order differential equation and consider the solution to be made up of two elements, a transient response and a forced response, i.e. $x = x_n + x_f$. Substituting for x in the above equation then gives

$$m \frac{d^2 (x_n + x_f)}{dt^2} + c \frac{d(x_n + x_f)}{dt} + k(x_n + x_f) = F$$

If we let

$$m\frac{d^2x_n}{dt^2} + c\frac{dx_n}{dt} + kx_n = 0$$

then we must have

$$m\frac{d^2x_f}{dt^2} + c\frac{dx_f}{dt} + kx_f = F$$

To solve the transient equation we can try a solution of the form $x_n = A\ e^{st}$. This gives $dx_n/dt = As\ e^{st}$ and $d^2x_n/dt^2 = As^2\ e^{st}$. Thus, substituting these values in the differential equation gives

$$mAs^2\ e^{st} + cAs\ e^{st} + kA\ e^{st} = 0$$

$$ms^2 + cs + k = 0$$

Thus $x_n = A\ e^{st}$ can only be a solution provided the above equation equals 0. This equation is called the *auxiliary equation*. The roots of the equation can be obtained by factoring or using the formula for the roots of a quadratic equation. Thus

$$s = \frac{-c \pm \sqrt{c^2 - 4mk}}{2m} = -\frac{c}{2m} \pm \sqrt{\left(\frac{c}{2m}\right)^2 - \frac{k}{m}}$$

$$= -\frac{c}{2m} \pm \sqrt{\frac{k}{m}\left(\frac{c^2}{4mk}\right) - \frac{k}{m}}$$

But $\omega_n^2 = k/m$ and so, if we let $\zeta^2 = c^2/4mk$ we can write the above equation as

$$s = -\zeta\omega_n \pm \omega_n\sqrt{\zeta^2 - 1}$$

ζ is termed the *damping factor*.

The value of s obtained from the above equation depends very much on the value of the square root term. Thus when ζ^2 is greater than 1 the square root term gives a square root of a positive number, and when ζ^2 is less than 1 we have the square root of a negative number. The damping factor determines whether the square root term is that of a positive or negative number and so the form of the output from the system.

With $\zeta > 1$ there are two different real roots s_1 and s_2:

$$s_1 = -\zeta\omega_n + \omega_n\sqrt{\zeta^2 - 1}$$

$$s_2 = -\zeta\omega_n - \omega_n\sqrt{\zeta^2 - 1}$$

and so the general solution for x_n is

$$x_n = A\,e^{s_1 t} + B\,e^{s_2 t}$$

For such conditions the system is said to be *over-damped*.

When $\zeta = 1$ there are two equal roots with $s_1 = s_2 = -\omega_n$. For this condition, which is called *critically damped*,

$$x_n = (At + B)\,e^{-\omega_n t}$$

It may seem that the solution for this case should be $x_n = A\,e^{st}$, but two constants are required and so the solution is of this form (see *Ordinary Differential Equations* by W. Bolton (Longman 1994) for a discussion of this).

With $\zeta < 1$ there are two complex roots since the roots both involve the square root of (-1).

$$s = -\zeta\omega_n \pm \omega_n\sqrt{\zeta^2 - 1} = -\zeta\omega_n \pm \omega_n\sqrt{-1}\,\sqrt{1 - \zeta^2}$$

and so writing j for $\sqrt{-1}$,

$$s = -\zeta\omega_n \pm j\omega_n\sqrt{1 - \zeta^2}$$

If we let

$$\omega = \omega_n\sqrt{1 - \zeta^2}$$

then we can write $s = -\zeta\omega_d \pm j\omega$ and so the two roots are

$$s_1 = -\zeta\omega_d + j\omega \quad \text{and} \quad s_2 = -\zeta\omega_d - j\omega$$

The term ω is the angular frequency of the motion when it is in the damped condition specified by ζ. The solution under these conditions is thus

$$x_n = A\,e^{(-\zeta\omega_n + j\omega)t} + B\,e^{(-\zeta\omega_n - j\omega)t} = e^{-\zeta\omega_n t}(A\,e^{j\omega t} + B\,e^{-j\omega t})$$

But $e^{j\omega t} = \cos\omega t + j\sin\omega t$ and $e^{-j\omega t} = \cos\omega t - j\sin\omega t$. Hence

$$x_n = e^{-\zeta\omega_n t}(A\cos\omega t + jA\sin\omega t + B\cos\omega t - jB\sin\omega t)$$

$$= e^{-\zeta\omega_n t}[(A + B)\cos\omega t + j(A - B)\sin\omega t]$$

If we substitute constants P and Q for $(A + B)$ and $j(A - B)$, then

$$x_n = e^{-\zeta\omega_n t}(P\cos\omega t + Q\sin\omega t)$$

For such conditions the system is said to be *under-damped*.

The above has thus given the solutions for the natural part of the solution. To solve the forcing equation,

$$m\frac{d^2x_f}{dt^2} + c\frac{dx_f}{dt} + kx_f = F$$

we need to consider a particular form of input signal and then try a solution. Thus for a step input of size F at time $t = 0$ we can try a solution $x_f = A$, where A is a constant (see the discussion of the solution of first-order differential equations for a discussion of the choice of solutions). Then $dx_f/dt = 0$ and $d^2x_f/dt^2 = 0$. Thus, when these are substituted in the differential equation $0 + 0 + kA = F$ and so $A = F/k$ and $x_f = F/k$. The complete solution, which is the sum of natural and forced solutions, is thus for the over-damped system

$$x = A\,e^{s_1 t} + B\,e^{s_2 t} + \frac{F}{k}$$

for the critically damped system

$$x = (At + B)\,e^{-\omega_n t} + \frac{F}{k}$$

and for the under-damped system

$$x = e^{-\zeta\omega_n t}(P\cos\omega t + Q\sin\omega t) + \frac{F}{k}$$

When $t \to \infty$ the above three equations all lead to the solution $x = F/k$. This is the steady-state condition.

Thus a second-order differential equation in the form

$$a_2\frac{d^2x}{dt^2} + a_1\frac{dx}{dt} + a_0 x = b_0 y$$

has

$$\omega_n^2 = \frac{a_0}{a_2}$$

and

$$\zeta^2 = \frac{a_1^2}{4a_2 a_0}$$

The following examples are designed to illustrate the points made above.

1   Consider a series RLC circuit (Fig. 10.14) with $R = 100\ \Omega$, $L = 2.0$ H and $C = 20\ \mu$F. The current i in the circuit is given by (see the text associated with Fig. 8.16):

$$\frac{d^2i}{dt^2} + \frac{R}{L}\frac{di}{dt} + \frac{1}{LC}i = \frac{V}{LC}$$

Fig. 10.14   *RLC* system

when there is a step input V. If we compare the equation with the general second-order differential equation of:

$$a_2 \frac{d^2x}{dt^2} + a_1 \frac{dx}{dt} + a_0 x = b_0 y$$

then the natural angular frequency is given by

$$\omega_n^2 = \frac{1}{LC} = \frac{1}{2.0 \times 20 \times 10^{-6}}$$

and so $\omega_n = 158$ Hz. Comparison with the general second-order equation also gives

$$\zeta^2 = \frac{(R/L)^2}{4 \times (1/LC)} = \frac{R^2 C}{4L} = \frac{100^2 \times 20 \times 10^{-6}}{4 \times 2.0}$$

Thus $\zeta = 0.16$. Since ζ is less than 1 the system is under-damped. The damped oscillation frequency ω is given by

$$\omega = \omega_n \sqrt{1 - \zeta^2} = 158 \sqrt{1 - 0.16^2} = 156 \text{ Hz}$$

Because the system is under-damped the solution will be of the same form as

$$x = e^{-\zeta \omega_n t}(P \cos \omega t + Q \sin \omega t) + \frac{F}{k}$$

and so

$$i = e^{-0.16 \times 158 t}(P \cos 156t + Q \sin 156t) + V$$

Since $i = 0$ when $t = 0$, then $0 = 1(P + 0) + V$. Thus $P = -V$. Since $di/dt = 0$ when $t = 0$ then differentiating the above equation and equating it to zero gives

$$\frac{di}{dt} = e^{-\zeta \omega_n t}(\omega P \sin \omega t - \omega Q \cos \omega t)$$
$$- \zeta \omega_n e^{-\zeta \omega_n t}(P \cos \omega t + Q \cos \omega t)$$

Thus $0 = 1(0 - \omega Q) - \zeta \omega_n (P + 0)$ and so

$$Q = \frac{\zeta \omega_n P}{\omega} = -\frac{\zeta \omega_n V}{\omega} = -\frac{0.16 \times 158 V}{156} \approx -0.16$$

Thus the solution of the differential equation is

$$i = V - V e^{-25.3t}(\cos 156t + 0.16 \sin 156t)$$

2   Consider the system shown in Figure 10.15. The input, a torque T, is applied to a disc with a moment of inertia I about

Opposing torsional torque

Torque T

Opposing frictional torque

Moment of inertia I

Fig. 10.15   Torsional system

the axis of the shaft. The shaft is free to rotate at the disc end but is fixed at its far end. The shaft rotation is opposed by the torsional stiffness of the shaft, an opposing torque of $k\theta_o$ occurring for an input rotation of θ_o. k is a constant. Frictional forces damp the rotation of the shaft and provide an opposing torque of $c\ d\theta_o/dt$, where c is a constant. What is the condition for this system to be critically damped?

We first need to obtain the differential equation for the system. The net torque is

$$\text{Net torque} = T - c\frac{d\theta_o}{dt} - k\theta_o$$

The net torque is $I\ d^2\theta_o/dt$, hence

$$I\frac{d^2\theta_o}{dt^2} = T - c\frac{d\theta_o}{dt} - k\theta_o$$

$$I\frac{d^2\theta_o}{dt^2} + c\frac{d\theta_o}{dt} + k\theta_o = T$$

The condition for critical damping is given when the damping ratio ζ equals 1. Comparing the above differential equation with the general form of the second-order differential equation, then

$$\zeta^2 = \frac{a_1^2}{4a_2a_0} = \frac{c^2}{4Ik}$$

Thus for critical damping we must have $c = \sqrt{(Ik)}$.

10.4 Performance measures for second-order systems

Figure 10.16 shows the typical form of the response of an under-damped second-order system to a step input. Certain terms are used to specify such a performance.

The *rise time* t_r is the time taken for the response x to rise from 0 to the steady-state value x_{ss} and is a measure of how fast a system responds to the input. This is the time for the oscillating response to complete a quarter of a cycle, i.e. $\frac{1}{2}\pi$. Thus

$$\omega t_r = \tfrac{1}{2}\pi$$

The rise time is sometimes specified as the time taken for the response to rise from some specified percentage of the steady-state value, e.g. 10%, to another specified percentage, e.g. 90%.

The *peak time* t_p is the time taken for the response to rise from 0 to the first peak value. This is the time for the oscillating response to complete one half-cycle, i.e. π. Thus

$$\omega t_p = \pi$$

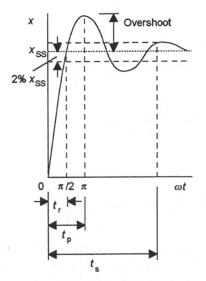

Fig. 10.16   Step response of an under-damped system

The *overshoot* is the maximum amount by which the response overshoots the steady-state value. It is thus the amplitude of the first peak. The overshoot is often written as a percentage of the steady-state value. For the under-damped oscillations of a system we can write

$$x = e^{-\zeta\omega_n t}(P \cos \omega t + Q \sin \omega t) + \text{steady-state value}$$

Since $x = 0$ when $t = 0$ then $0 = 1(P + 0) + x_{SS}$ and so $P = -x_{SS}$. The overshoot occurs at $\omega t = \pi$ and thus

$$x = e^{-\zeta\omega_n \pi/\omega}(P + 0) + x_{SS}$$

The overshoot is the difference between the output at that time and the steady-state value. Hence

$$\text{Overshoot} = x_{SS}\, e^{-\zeta\omega_n \pi/\omega}$$

Since $\omega = \omega_n \sqrt{(1 - \zeta^2)}$ then we can write

$$\text{Overshoot} = x_{SS} \exp\left(\frac{-\zeta\omega_n \pi}{\omega_n \sqrt{1 - \zeta^2}}\right)$$

$$= x_{SS} \exp\left(\frac{-\zeta\pi}{\sqrt{1 - \zeta^2}}\right)$$

Expressed as a percentage of x_{SS},

$$\text{Percentage overshoot} = \exp\left(\frac{-\zeta\pi}{\sqrt{1 - \zeta^2}}\right) \times 100\%$$

Table 10.2 gives values of the percentage overshoot for particular damping ratios.

Table 10.2 Percentage peak overshoot

Damping ratio	Percentage overshoot
0.2	52.7
0.4	25.4
0.6	9.5
0.8	1.5

An indication of how fast oscillations decay is provided by the *subsidence ratio* or *decrement*. This is the amplitude of the second overshoot divided by that of the first overshoot. The first overshoot occurs when we have $\omega t = \pi$, the second overshoot when $\omega t = 3\pi$. Thus,

$$\text{First overshoot} = x_{SS} \exp\left(\frac{-\zeta\pi}{\sqrt{1-\zeta^2}}\right)$$

$$\text{Second overshoot} = x_{SS} \exp\left(\frac{-3\zeta\pi}{\sqrt{1-\zeta^2}}\right)$$

and so

$$\text{Subsidence ratio} = \frac{\text{second overshoot}}{\text{first overshoot}} = \exp\left(\frac{-2\zeta\pi}{\sqrt{1-\zeta^2}}\right)$$

The *settling time* t_s is used as a measure of the time taken for the oscillations to die away. It is the time taken for the response to fall within and remain within some specified percentage, e.g. 2%, of the steady-state value (see Fig. 10.15). This means that the amplitude of the oscillation should be less than 2% of x_{SS}. We have

$$x = e^{-\zeta\omega_n t}(P\cos\omega t + Q\sin\omega t) + \text{steady-state value}$$

and, as derived earlier, $P = -x_{SS}$. The amplitude of the oscillation is $(x - x_{SS})$ when x is a maximum value. The maximum values occur when ωt is some multiple of π and thus we have $\cos \omega t = 1$ and $\sin \omega t = 0$. For the 2% settling time, the settling time t_s is when the maximum amplitude is 2% of x_{SS}, i.e. $0.02x_{SS}$. Thus

$$0.02x_{SS} = e^{-\zeta\omega_n t_s}(x_{SS} \times 1 + 0)$$

Taking logarithms gives $\ln 0.02 = -\zeta\omega_n t_s$ and since $\ln 0.02 = -3.9$ or approximately 4 then

$$t_s = \frac{4}{\zeta\omega_n}$$

The above is the value of the settling time if the specified percentage is 2%. If the percentage is 5% the equation becomes

$$t_s = \frac{3}{\zeta\omega_n}$$

Since the time taken to complete one cycle, i.e. the periodic time, is $1/f$, where f is the frequency, and since $\omega = 2\pi f$ then the time to complete one cycle is $2\pi/f$. In a settling time of t_s the number of oscillations that occur is

$$\text{Number of oscillations} = \frac{\text{settling time}}{\text{periodic time}}$$

and thus for a settling time defined for 2% of the steady-state value,

$$\text{Number of oscillations} = \frac{4/\zeta\omega_n}{2\pi/\omega}$$

Since $\omega = \omega_n\sqrt{(1 - \zeta^2)}$, then

$$\text{Number of oscillations} = \frac{2\omega_n\sqrt{1-\zeta^2}}{\pi\zeta\omega_n} = \frac{2}{\pi}\sqrt{\frac{1}{\zeta^2} - 1}$$

To illustrate the above, consider a second-order system which has a natural angular frequency of 2.0 Hz and a damped frequency of 1.8 Hz. Since $\omega = \omega_n\sqrt{(1 - \zeta^2)}$, then the damping factor is given by

$$1.8 = 2.0\sqrt{1 - \zeta^2}$$

and $\zeta = 0.44$. Since $\omega t_r = \frac{1}{2}\pi$, then the 100% rise time is given by

$$t_r = \frac{\pi}{2 \times 1.8} = 0.87 \text{ s}$$

The percentage overshoot is given by

$$\text{Percentage overshoot} = \exp\left(\frac{-\zeta\pi}{\sqrt{1-\zeta^2}}\right) \times 100\%$$

$$= \exp\left(\frac{-0.44\pi}{\sqrt{1-0.44^2}}\right) \times 100\%$$

The percentage overshoot is thus 21%. The 2% settling time is given by

$$t_s = \frac{4}{\zeta\omega_n} = \frac{4}{0.44 \times 2.0} = 4.5 \text{ s}$$

The number of oscillations occurring within the 2% settling time is given by

$$\text{Number of oscillations} = \frac{2}{\pi}\sqrt{\frac{1}{\zeta^2} - 1}$$

$$= \frac{2}{\pi}\sqrt{\frac{1}{0.44^2} - 1} = 1.3$$

10.5 System identification

In Chapter 9 models were devised for systems by considering them to be made up of simple elements. An alternative way of developing a model for a real system is to use tests to determine its response to some input, e.g. a step input, and then find the model that fits the response. This process of determining a mathematical model is known as *system identification*. Thus if we obtain a response to a step input of the form shown in Figure 10.9 then we might assume that it is a first-order system and determine the time constant from the response curve. For example, suppose the response takes a time of 1.5 s to reach 0.63 of its final height and the final height of the signal is 5 times the size of the step input. Table 10.1 indicates a time constant of 1.5 s and so the differential equation describing the model is:

$$1.5\frac{dx}{dt} + x = 5y$$

An underdamped second-order system will give a response to a step input of the form shown in Figure 10.16. The damping ratio can be determined from measurements of the first and second overshoots with the ratio of these overshoots, i.e. the subsidence ratio, giving the damping ratio. The natural frequency can be determined from the time between successive overshoots. We can then use these values to determine the constants in the second-order differential equation.

Problems

1   A first-order system has a time constant of 4 s and a steady-state transfer function of 6. What is the form of the differential equation for this system?

2   A mercury-in-glass thermometer has a time constant of 10 s. If it is suddenly taken from being at 20°C and plunged into hot water at 80°C, what will be the temperature indicated by the thermometer after (a) 10 s, (b) 20 s?

3   A circuit consists of a resistor R in series with an inductor L. When subject to a step input voltage V at time $t = 0$ the differential equation for the system is

$$\frac{di}{dt} + \frac{R}{L}i = \frac{V}{L}$$

What is (a) the solution for this differential equation, (b) the time constant, (c) the steady-state current i?

4   Describe the form of the output variation with time for a step input to a second-order system with a damping factor of (a) 0, (b) 0.5, (c) 1.0, (d) 1.5.

5   A RLC circuit has a current i which varies with time t when subject to a step input of V and is described by

$$\frac{d^2i}{dt^2} + 10\frac{di}{dt} + 16i = 16V$$

What is (a) the undamped frequency, (b) the damping ratio, (c) the solution to the equation if $i = 0$ when $t = 0$ and $di/dt = 0$ when $t = 0$?

6   A system has an output x which varies with time t when subject to a step input of y and is described by

$$\frac{d^2x}{dt^2} + 10\frac{dx}{dt} + 25x = 50y$$

What is (a) the undamped frequency, (b) the damping ratio, (c) the solution to the equation if $x = 0$ when $t = 0$ and $dx/dt = -2$ when $t = 0$ and there is a step input of size 3 units?

7   An accelerometer (an instrument for measuring acceleration) has an undamped angular frequency of 100 Hz and a damping factor of 0.6. What will be (a) the maximum percentage overshoot and (b) the rise time when there is a sudden change in acceleration?

8   What will be (a) the undamped angular frequency, (b) the damping factor, (c) the damped angular frequency, (d) the rise time, (e) the percentage maximum overshoot and (f) the 0.2% settling time for a system which gave the following differential equation for a step input y?

$$\frac{d^2x}{dt^2} + 5\frac{dx}{dt} + 16x = 16y$$

9   When a voltage of 10 V is suddenly applied to a moving coil voltmeter it is observed that the pointer of the instrument rises to 11 V before eventually settling down to read 10 V. What is (a) the damping factor and (b) the number of oscillations the pointer will make before it is within 0.2% of its steady-state value?

10   A second order system is described by the differential equation:

$$\frac{d^2x}{dt^2} + c\frac{dx}{dt} + 4x = F$$

What value of damping constant c will be needed if the percentage overshoot is to be less than 9.5%?

11   Observation of the oscillations of a damped system when responding to an input indicates that the maximum displacement during the second cycle is 75% of the first displacement. What is the damping factor of the system?

12   A second order system is found to have a time of 1.6 s between the first overshoot and the second overshoot. What is the natural frequency of the system?

11 System transfer functions

11.1 The transfer function

For an amplifier system it is customary to talk of the *gain* of the amplifier. This states how much bigger the output signal will be when compared with the input signal. It enables the output to be determined for specific inputs. Thus, for example, an amplifier with a voltage gain of 10 will give, for an input voltage of 2 mV, an output of 20 mV; or if the input is 1 V an output of 10 V. The gain states the mathematical relationship between the output and the input for the block.

$$\text{Gain} = \frac{\text{output}}{\text{input}}$$

However, for many systems the relationship between the output and the input is in the form of a differential equation and so a statement of the function as just a simple number like the gain of 10 is not possible. We cannot just divide the output by the input because the relationship is a differential equation and not a simple algebraic equation. We can, however, transform a differential equation into an algebraic equation by using what is termed the *Laplace transform*. Differential equations describe how systems behave with time and are transformed by means of the Laplace transform into simple algebraic equations, not involving time, where we can carry out normal algebraic manipulations of the quantities. We talk of behaviour in the *time domain* being transformed to the *s-domain*. Then we can define the relationship between output and input in terms of a *transfer function*. The transfer function states the relationship between the Laplace transform of the output and the Laplace transform of the input, i.e.

$$\text{Transfer function} = \frac{\text{Laplace transform of output}}{\text{Laplace transform of input}}$$

Fig. 11.1 Block diagram

Unit impulse at zero time has
the transform of 1

Unit step at zero time has
the transform of $1/s$

Unit ramp at zero time has
the transform of $1/s^2$

Unitampliture sine wave has
the transform of $\omega/(s^2 + \omega^2)$

Unitampliture cosine wave has
the transform of $s/(s^2 + \omega^2)$

Fig. 11.2 Laplace transforms for
common inputs

We can indicate when a signal is in the time domain, i.e. is a function of time, by writing it as $f(t)$. When in the s-domain a function is written, since it is a function of s, as $F(s)$. It is usual to use a capital letter F for the Laplace transform and a lower-case letter f for the time-varying function $f(t)$.

Suppose that the input to a linear system has a Laplace transform of $Y(s)$ and the Laplace transform of the output is $X(s)$. The *transfer function* $G(s)$ of the system is then defined as

$$G(s) = \frac{X(s)}{Y(s)}$$

with all the initial conditions being zero, i.e. we assume zero output when zero input, zero rate of change of output with time when zero rate of change of input with time. Thus the output transform is $X(s) = G(s)Y(s)$, i.e. the product of the input transform and the transfer function. If we represent a system by a block diagram (Fig. 11.1) then $G(s)$ is the function in the box which takes an input of $Y(s)$ and converts it to an output of $X(s)$.

This chapter gives an indication of how Laplace transforms can be used in relation to the transfer functions of systems. For more details the reader is referred to Appendix A or *Laplace and z-Transforms* by W. Bolton (Longman 1994), part of the Mathematics for Engineers series.

11.1.1 Laplace transforms

To obtain the Laplace transform of a differential equation which includes quantities which are functions of time we can use tables coupled with a few basic rules (Appendix A includes such a table and gives details of the rules). Figure 11.2 shows basic transforms for common forms of inputs.

The following are some of the basic rules involved in working with Laplace transforms:

1   If a function of time is multiplied by a constant then the Laplace transform is multiplied by the same constant, i.e.

$af(t)$ has the transform of $aF(s)$

For example, the Laplace transform of a step input of 6 V to an electrical system is just 6 times the transform for a unit step and thus $6s$.

2   If an equation includes the sum of, say, two separate quantities which are functions of time, then the transform of the equation is the sum of the two separate Laplace transforms, i.e.

$f(t) + g(t)$ has the transform $F(s) + G(s)$

3   The Laplace transform of a first derivative of a function is

$$\text{Transform of } \left\{ \frac{\text{d}}{\text{d}t} f(t) \right\} = sF(s) - f(0)$$

where $f(0)$ is the initial value of $f(t)$ when $t = 0$. However, when we are dealing with a transfer function we have all initial conditions zero.

4   The Laplace transform for the second derivative of a function is

$$\text{Transform of } \left\{ \frac{\text{d}^2}{\text{d}t^2} f(t) \right\} = s^2 F(s) - sf(0) - \frac{\text{d}}{\text{d}t} f(0)$$

where $\text{d}f(0)/\text{d}t$ is the initial value of the first derivative of $f(t)$ when we have $t = 0$. However, when we are dealing with a transfer function we have all initial conditions zero.

5   The Laplace transform of an integral of a function is

$$\text{Transform of } \left\{ \int_0^t f(t)\,\text{d}t \right\} = \frac{1}{s} F(s)$$

Thus, in obtaining the transforms of differential or integral equations when all the initial conditions are zero we:

Replace a function of time f(t) by F(s),
replace a first derivative df(t)/dt by sF(s),
replace a second derivative d²f(t)/dt² by s²F(s),
replace an integral ∫ f(t) dt by F(s)/s.

When algebraic manipulations have occurred in the s-domain, then the outcome can be transformed back to the time domain by using the table of transforms in the inverse manner, i.e. finding the time domain function which fits the s-domain result. Often the transform has to be rearranged to be put into a form given in the table. The following are some useful such inversions; for more inversions see the table given in Appendix A.

	Laplace transform	*Function of time*
1	$\dfrac{1}{s+a}$	e^{-at}
2	$\dfrac{a}{s(s+a)}$	$(1 - e^{-at})$
3	$\dfrac{b-a}{(s+a)(s+b)}$	$e^{-at} - e^{-bt}$
4	$\dfrac{s}{(s+a)^2}$	$(1 - at)\,e^{-at}$
5	$\dfrac{a}{s^2(s+a)}$	$t - \dfrac{1 - e^{-at}}{a}$

The following sections illustrate the application of the above to first-order and second-order systems.

11.2 First-order systems

Consider a system where the relationship between the input and the output is in the form of a first-order differential equation. The differential equation of a first-order system is of the form

$$a_1 \frac{dx}{dt} + a_0 x = b_0 y$$

where a_1, a_0, b_0 are constants, y is the input and x the output, both being functions of time. The Laplace transform of this, with all initial conditions zero, is

$$a_1 s X(s) + a_0 X(s) = b_0 Y(s)$$

and so we can write the transfer function $G(s)$ as

$$G(s) = \frac{X(s)}{Y(s)} = \frac{b_0}{a_1 s + a_0}$$

This can be rearranged to give

$$G(s) = \frac{b_0/a_0}{(a_1/a_0)s + 1} = \frac{G}{\tau s + 1}$$

where G is the *gain* of the system when there are steady-state conditions, i.e. there is no dx/dt term. (a_1/a_0) is the time constant τ of the system (see Section 10.2.3).

When a first-order system is subject to a unit step input then $Y(s) = 1/s$ and the output transform $X(s)$ is

$$X(s) = G(s)Y(s) = \frac{G}{s(\tau s + 1)} = G\frac{(1/\tau)}{s(s + 1/\tau)}$$

Hence, since we have the transform in the form $a/s(s + a)$, using the inverse transformation listed as item 2 in the previous section gives

$$x = G(1 - e^{-t/\tau})$$

The following examples illustrate the above points in the consideration of the transfer function of a first-order system and its behaviour when subject to a step input.

1   Consider a circuit which has a resistance R in series with a capacitance C. The input to the circuit is v and the output is the potential difference v_C across the capacitor. The differential equation relating the input and output is

$$v = RC\frac{dv_C}{dt} + v_C$$

Determine the transfer function.

Taking the Laplace transform, with all initial conditions zero, then

$$V(s) = RCsV_C(s) + V_C(s)$$

Hence the transfer function is

$$G(s) = \frac{V_C(s)}{V(s)} = \frac{1}{RCs+1}$$

2   Consider a thermocouple which has a transfer function linking its voltage output V and temperature input of

$$G(s) = \frac{30 \times 10^{-6}}{10s+1} \text{ V/°C}$$

Determine the response of the system when subject to a step input of size 100°C and hence the time taken to reach 95% of the steady-state value.

Since the transform of the output is equal to the product of the transfer function and the transform of the input, then

$$V(s) = G(s) \times \text{input } (s)$$

A step input of size 100°C, i.e. the temperature of the thermocouple is abruptly increased by 100°C, is 100/s. Thus

$$V(s) = \frac{30 \times 10^{-6}}{10s+1} \times \frac{100}{s} = \frac{30 \times 10^{-4}}{10s(s+0.1)}$$

$$= 30 \times 10^{-4}\frac{0.1}{s(s+0.1)}$$

The fraction element is of the form $a/s(s + a)$ and so the inverse transform is

$$V = 30 \times 10^{-4} (1 - e^{-0.1t}) \text{ V}$$

The final value, i.e. the steady-state value, is when $t \to \infty$ and so is when the exponential term is zero. The final value is therefore 30×10^{-4} V. Thus the time taken to reach, say, 95% of this is given by

$$0.95 \times 30 \times 10^{-4} = 30 \times 10^{-4} (1 - e^{-0.1t})$$

Thus $0.05 = e^{-0.1t}$ and $\ln 0.05 = -0.1t$. The time is thus 30 s.

3 Consider a ramp input to the above thermocouple system of $5t$ °C/s, i.e. the temperature is raised by 5°C every second. Determine how the voltage of the thermocouple varies with time and hence the voltage after 12 s.

The transform of the ramp signal is $5/s^2$. Thus

$$V(s) = \frac{30 \times 10^{-6}}{10s + 1} \times \frac{5}{s^2} = 150 \times 10^{-6} \frac{0.1}{s^2(s + 0.1)}$$

The inverse transform can be obtained using item 5 in the list given in the previous section. Thus

$$V = 150 \times 10^{-6} \left(t - \frac{1 - e^{-0.1t}}{0.1} \right)$$

After a time of 12 s we would have $V = 7.5 \times 10^{-4}$ V.

4 Consider an impulse input of size 100°C, i.e. the thermocouple is subject to a momentary temperature increase of 100°C. Determine how the voltage of the thermocouple varies with time and hence the voltage after 2 s.

The impulse has a transform of 100. Hence

$$V(s) = \frac{30 \times 10^{-6}}{10s + 1} \times 100 = 3 \times 10^{-4} \frac{1}{s + 0.1}$$

Hence $V = 3 \times 10^{-4} e^{-0.1t}$ V. After 2 s, the thermocouple voltage $V = 1.8 \times 10^{-4}$ V.

11.3 Second-order systems

For a second-order system, the relationship between the input y and the output x is described by a differential equation of the form

$$a_2 \frac{d^2x}{dt^2} + a_1 \frac{dx}{dt} + a_0 x = b_0 y$$

where a_2, a_1, a_0 and b_0 are constants. The Laplace transform of this equation, with all initial conditions zero, is

$$a_2 s^2 X(s) + a_1 s X(s) + a_0 X(s) = b_0 Y(s)$$

Hence

$$G(s) = \frac{X(s)}{Y(s)} = \frac{b_0}{a_2 s^2 + a_1 s + a_0}$$

An alternative way of writing the differential equation for a second-order system is

$$\frac{d^2x}{dt^2} + 2\zeta \omega_n \frac{dx}{dt} + \omega_n^2 x = b_0 \omega_n^2 y$$

where ω_n is the natural angular frequency with which the system oscillates and ζ is the damping ratio. The Laplace transform of this equation gives

$$G(s) = \frac{X(s)}{Y(s)} = \frac{b_0\omega_n^2}{s^2 + 2\zeta\omega_n s + \omega_n^2}$$

The above are the general forms taken by the transfer function for a second-order system.

When a second-order system is subject to a unit step input, i.e. $Y(s) = 1/s$, then the output transform is

$$X(s) = G(s)Y(s) = \frac{b_0\omega_n^2}{s(s^2 + 2\zeta\omega_n s + \omega_n)}$$

This can be rearranged as

$$X(s) = \frac{b_0\omega_n^2}{s(s + p_1)(s + p_2)}$$

where p_1 and p_2 are the roots of the equation

$$s^2 + 2\zeta\omega_n s + \omega_n^2 = 0$$

Hence, using the equation for the roots of a quadratic equation,

$$p = \frac{-2\zeta\omega_n \pm \sqrt{4\zeta^2\omega_n^2 - 4\omega_n^2}}{2}$$

and so the two roots p_1 and p_2 are:

$$p_1 = -\zeta\omega_n + \omega_n\sqrt{\zeta^2 - 1}$$

$$p_2 = -\zeta\omega_n - \omega_n\sqrt{\zeta^2 - 1}$$

With $\zeta > 1$ the square root term is real and the system is over-damped. To find the inverse transform we can either use partial fractions (see Appendix A) to break the expression down into a number of simple fractions or use item 14 in the table of transforms in Appendix A; the result in either case is:

$$x = \frac{b_0\omega_n^2}{p_1 p_2}\left[1 - \frac{p_2}{p_2 - p_1}\,e^{-p_2 t} + \frac{p_1}{p_2 - p_1}\,e^{-p_1 t}\right]$$

With $\zeta = 1$ the square root term is zero and so $p_1 = p_2 = -\omega_n$. The system is critically damped. The equation then becomes

$$X(s) = \frac{b_0\omega_n^2}{s(s + \omega_n)^2}$$

This equation can be expanded by means of partial fractions (see Appendix A) to give

$$Y(s) = b_0 \left[\frac{1}{s} - \frac{1}{s+\omega_n} - \frac{\omega_n}{(s+\omega_n)^2} \right]$$

Hence

$$x = b_0 [1 - e^{-\omega_n t} - \omega_n t e^{-\omega_n t}]$$

With $\zeta < 1$, using item 28 in the table in Appendix A, gives

$$x = b_0 \left[1 - \frac{e^{-\zeta \omega_n t}}{\sqrt{1-\zeta^2}} \sin\left(\omega_n \sqrt{(1-\zeta^2)}\, t + \phi \right) \right]$$

where $\cos \phi = \zeta$. This is an under-damped oscillation.
The following examples illustrate the above:

1 What will be the state of damping of a system having the following transfer function and subject to a unit step input?

$$G(s) = \frac{1}{s^2 + 8s + 16}$$

For a unit step input $Y(s) = 1/s$ and so the output transform is

$$X(s) = G(s)Y(s) = \frac{1}{s(s^2 + 8s + 16)}$$

$$= \frac{1}{s(s+4)(s+4)}$$

The roots of $s^2 + 8s + 16$ are thus $p_1 = p_2 = -4$. Both the roots are real and the same and so the system is critically damped.

2 A robot arm having the following transfer function is subject to a unit ramp input. What will be the output?

$$G(s) = \frac{K}{(s+3)^2}$$

The output transform $X(s)$ is:

$$X(s) = G(s)Y(s) = \frac{K}{(s+3)^2} \times \frac{1}{s^2}$$

Using partial fractions (see Appendix A) this becomes

$$X(s) = \frac{K}{9s^2} - \frac{2K}{9(s+3)} + \frac{K}{9(s+3)^2}$$

Hence the inverse transform is

$$x = \tfrac{1}{9}Kt - \tfrac{2}{9}Ke^{-3t} + \tfrac{1}{9}Kte^{-3t}$$

11.4 Systems in series

If a system consists of a number of subsystems in series, as in Figure 11.3, then the transfer function $G(s)$ of the system is given by

$$G(s) = \frac{X(s)}{Y(s)} = \frac{X_1(s)}{Y(s)} \times \frac{X_2(s)}{X_1(s)} \times \frac{X(s)}{X_2(s)}$$

$$= G_1(s) \times G_2(s) \times G_3(s)$$

Fig. 11.3   Systems in series

The transfer function of the system as a whole is the product of the transfer functions of the series elements. The following examples illustrate this. It has been assumed that when subsystems are linked together that no interaction occurs between the blocks which would result in changes in their transfer functions, e.g. with electrical circuits there can be problems when subsystem circuits interact and load each other.

1   What will be the transfer function for a system consisting of three elements in series, the transfer functions of the elements being 10, 2/s, and 4/(s + 3)?
    Using the equation developed above

$$G(s) = 10 \times \frac{2}{s} \times \frac{4}{s+3} = \frac{80}{s(s+3)}$$

2   A field-controlled d.c. motor consists of three subsystems in series, the field circuit, the armature coil and the load. Figure 11.4 illustrates the arrangement and the transfer functions of the subsystems. Determine the overall transfer function.

Fig. 11.4   Field-controlled d.c. motor

The overall transfer function is the product of the transfer functions of the series elements. Thus

$$G(s) = \frac{1}{Ls+R} \times k \times \frac{1}{Is+c} = \frac{k}{(Ls+R)(Is+c)}$$

11.5 Systems with feedback loops

Figure 11.5 shows a simple system having negative feedback. With *negative feedback* the system input and the feedback signals are subtracted at the summing point. The term *forward path* is

Fig. 11.5   Negative feedback system

used for the path having the transfer function $G(s)$ in the figure and *feedback path* for the one having $H(s)$. The entire system is referred to as a *closed-loop system*.

For the negative feedback system, the input to the subsystem having the forward-path transfer function $G(s)$ is $Y(s)$ minus the feedback signal. The feedback loop has a transfer function of $H(s)$ and has as its input $X(s)$, thus the feedback signal is $H(s)X(s)$. Thus the $G(s)$ element has an input of $Y(s) - H(s)X(s)$ and an output of $X(s)$ and so

$$G(s) = \frac{X(s)}{Y(s) - H(s)X(s)}$$

This can be rearranged to give

$$\frac{X(s)}{Y(s)} = \frac{G(s)}{1 + G(s)H(s)}$$

Hence the overall transfer function for the negative feedback system $T(s)$ is

$$T(s) = \frac{X(s)}{Y(s)} = \frac{G(s)}{1 + G(s)H(s)}$$

The following examples illustrate the above:

1   What will be the overall transfer function for a closed-loop system having a forward-path transfer function of $2/(s + 1)$ and a negative feedback-path transfer function of $5s$?
   Using the equation developed above

$$T(s) = \frac{G(s)}{1 + G(s)H(s)}$$

$$= \frac{2/(s+1)}{1 + [2/(s+1)]5s} = \frac{2}{11s+1}$$

2   Consider an armature-controlled d.c. motor (Fig. 11.6). This has a forward path consisting of three elements: the armature circuit with a transfer function $1/(Ls + R)$, the armature coil with a transfer function k and the load with a transfer function $1/(Is + c)$. There is a negative feedback path with a transfer function K. Determine the overall transfer function for the system.
   The forward-path transfer function for the series elements is the product of the transfer functions of the series elements, i.e.

$$G(s) = \frac{1}{Ls+R} \times k \times \frac{1}{Is+c} = \frac{k}{(Ls+R)(Is+c)}$$

Fig. 11.6   Armature-controlled d.c. motor

The feedback path has a transfer function of K. Thus the overall transfer function is

$$T(s) = \frac{G(s)}{1+G(s)H(s)} = \frac{\dfrac{k}{(Ls+R)(Is+c)}}{1+\dfrac{kK}{(Ls+R)(Is+c)}}$$

$$= \frac{k}{(Ls+R)(Is+c)+kK}$$

11.6 Effect of pole location on transient response

Consider a first-order system with a transfer function of $1/(s+1)$ and subject to an input of a unit impulse. The system output $X(s) = [1/(s+1)] \times 1$ and thus $x = e^{-t}$. As the time t increases so the output dies away to eventually become zero. Now consider the unit impulse input to a system with the transfer function $1/(s-1)$. The output is then $x = e^{t}$. As t increases so the output increases with time. Thus a momentary impulse to the system has resulted in an ever increasing output; the system with this pole value is unstable. Thus, in general, for a first-order system with transfer function $1/(s+p)$, the system is *stable* if we have $(s+p)$, i.e. the pole is negative, and unstable if we have $(s-p)$, i.e. the pole is positive (Fig. 11.7).

A system is stable if the real part of all its poles is negative.
A system is unstable if the real part of any of its poles is positive.

For a second-order system with transfer function

$$G(s) = \frac{b_0\omega_n^2}{s^2+2\zeta\omega_n s+\omega_n^2}$$

when subject to a unit impulse input:

$$X(s) = \frac{b_0\omega_n^2}{(s+p_1)(s+p_2)}$$

where p_1 and p_2 are the roots of the equation

Fig. 11.7   First-order systems: (a) root is negative, (b) root is positive

$$s^2 + 2\zeta\omega_n s + \omega_n = 0$$

Using the equation for the roots of a quadratic equation,

$$p = \frac{-2\zeta\omega_n \pm \sqrt{4\zeta^2\omega_n^2 - 4\omega_n^2}}{2} = -\zeta\omega_n \pm \omega_n\sqrt{\zeta^2 - 1}$$

Depending on the value of the damping factor so the term under the square root sign can be real or imaginary. When there is an imaginary term the output involves an oscillation. For example, suppose we have a second-order system with transfer function

$$G(s) = \frac{1}{[s - (-2 + j1)][s - (-2 - j1)]}$$

i.e. $p = -2 \pm j1$. When subject to a unit impulse input the output is $e^{-2t} \sin t$. The amplitude of the oscillation, i.e. e^{-2t}, dies away as the time increases and so the effect of the impulse is a gradually decaying oscillation (Fig. 11.8(a)). The system is stable.

Suppose, however, we have a system with transfer function

$$G(s) = \frac{1}{[s - (2 + j1)][s - (2 - j1)]}$$

i.e. $p = +2 \pm j1$. When subject to a unit impulse input the output is $e^{2t} \sin t$. The amplitude of the oscillation, i.e. e^{2t}, increases as the time increases (Fig. 11.8(b)). The system is unstable.

In general, when an impulse is applied to a system, the output is in the form of the sum of a number of exponential terms. If just one of these terms is of exponential growth then the output continues to grow and the system is unstable. When there are pairs of poles involving ± imaginary terms then the output is an oscillation.

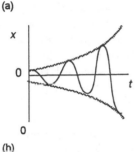

Fig. 11.8   Second-order systems

11.6.1 Compensation

The output from a system might be unstable or perhaps the response is too slow or there is to much overshoot. Systems can have their responses to inputs altered by including *compensators*. A compensator is a block which is incorporated in the system so that it alters the overall transfer function of the system in such a way as to obtain the required characteristics.

As an illustration of the use of a compensator, consider a position control system which has a negative feedback path with a transfer function of 1 and two subsystems in its forward path: a compensator with a transfer function of K and a motor/drive system with a transfer function of $1/s(s + 1)$. What value of K is necessary for the system to be critically damped? The forward path has a transfer function of $K/s(s + 1)$ and the feedback path a transfer function of 1. Thus the overall transfer function of the system is

$$T(s) = \frac{G(s)}{1 + G(s)H(s)} = \frac{\dfrac{K}{s(s+1)}}{1 + \dfrac{K}{s(s+1)}} = \frac{K}{s(s+1)+K}$$

The denominator is thus $s^2 + s + K$. This will have the roots

$$s = \frac{-1 \pm \sqrt{1 - 4K}}{2}$$

To be critically damped we must have $1 - 4K = 0$ and hence the compensator must have the proportional gain of $K = \frac{1}{4}$.

11.7 MATLAB and SIMULINK

Computer software can be used to aid computation and modelling of systems; a program that is often used is MATLAB. The following is a brief introduction to MATLAB (registered trademark of the Mathworks Inc.) version 4.0 or later. For additional information you are referred to the user guide or textbooks such as *The MATLAB Handbook* by Eva Pärt-Enander, Anders Sjöberg, Bo Melin and Pernilla Isaksson (Addison-Wesley 1996) and *Using MATLAB to Analyze and Design Control Systems*, 2nd edition, by Naomi Ehrich Leonard and William S. Levine (Addison-Wesley 1995).

Commands are entered by typing them in after the prompt and then pressing the enter or return key in order that the command can be executed. In the discussion of the commands that follow, this pressing of the enter or return key will not be repeated but should be assumed in all cases. To start MATLAB, in Windows or the Macintosh systems, click on the MATLAB icon, otherwise type matlab. The screen will then produce the MATLAB prompt >>. To quit MATLAB type quit or exit after the prompt. Because MATLAB is case sensitive, lower-case letters should be used throughout for commands.

Typing help after the prompt, or selecting help from the menu bar at the top of the MATLAB window, displays a list of MATLAB broad help topics. To get help on a particular topic in the list, e.g. exponentials, type help exp. Typing lookfor plus some topic will instruct MATLAB to search for information on that topic, e.g. lookfor integ will display a number of commands which could be considered for integration.

In general, mathematical operations are entered into MATLAB in the same way as they would be written on paper. For example:

>> a = 4/2

yields the response:

a =
   2

and:

>> a = 3*2

yields the response:

a =
   6

Operations are carried in the following order: ^ power operation, * multiplication, / division, + addition, − subtraction. Precedence of operators is from left to right but parentheses ( ) can be used to affect the order. For example:

>> a = 1+2^3/4*5

yields the response:

a =
   11

because we have $2^3/4$ which is multiplied by 5 and then added to 1, whereas:

>> a = 1+2^3/(4*5)

yields the response:

a =
   1.4

because we have 2^3 divided by the product of 4 and 5, and then added to 1.

The following are some of the mathematical functions available with MATLAB:

abs(x)	Gives the absolute value of x, i.e. $|x|$
exp(x)	Gives the exponential of x, i.e. e^x
log(x)	Gives the natural logarithm of x, i.e. $\ln x$
log10(x)	Gives the base 10 logarithm of x, i.e. $\log_{10} x$
sqrt(x)	Gives the square root of x, i.e. $\sqrt{x}$
sin(x)	Gives $\sin x$ where x is in radians
cos(x)	Gives $\cos x$ where x is in radians
tan(x)	Gives $\tan x$ where x is in radians
asin(x)	Gives $\arcsin x$, i.e. $\sin^{-1} x$
acos(x)	Gives $\arccos x$, i.e. $\cos^{-1} x$
atan(x)	Gives $\arctan x$, i.e. $\tan^{-1} x$
csc(x)	Gives $1/\sin x$

sec(x)   Gives 1/cos x

cot(x)   Gives 1/tan x

π is entered by typing pi.

Instead of writing a series of commands at the prompt, a text file can be written and then the commands executed by referring MATLAB to that file. The term M-file is used since these text files, containing a number of consecutive MATLAB commands, have the suffix .m. In writing such a file, the first line must begin with the word function followed by a statement identifying the name of the function and the input and output in the form:

function [output] = function name [input]

e.g. function y=cotan(x) which is the file used to determine the value of y given by cotan x. Such a file can be called up in some MATLAB sequence of commands by writing the name followed by the input, e.g. cotan(x). It is in fact already included in MATLAB and is used when the cotangent of x is required. However, the file could have been user written. A function that has multiple inputs should list all of them in the function statement. Likewise, a function that is to return more than one value should list all the outputs.

Lines that start with % are comment lines; they are not interpreted by MATLAB as commands. For example, suppose we write a program to determine the root-mean-square values of a single column of data points, the program might look like:

```
function y=rms(x)
% rms Root mean square
% rms(x) gives the root mean square value of the
% elements of column vector x.
xs=x^2;
s=size(x);
y=sqrt(sum(xs)/s);
```

We have let xs be the square values of each x value. The command s=size(x) obtains the size, i.e. number of entries, in the column of data. The command y=sqrt(sum(xs)/s(1)) obtains the square root of the sum of all the xs values divided by s. The ; command is used at the end of each program line.

MATLAB supplies a number of toolboxes containing collections of M-files. Of particular relevance to this book is the Control System toolbox. It can be used to carry out time responses of systems to impulses, steps, ramps, etc., along with Bode and Nyquist analysis, root locus, etc.

11.7.1 Plotting

Two-dimensional linear plots can be produced by using the plot(x,y) command; this plots the values of x and y. For example, we might have:

```
x=[0 1 2 3 4 5];
y=[0 1 4 9 16 25];
plot(x,y)
```

To plot a function, whether standard or user defined, we use the command fplot(function name,lim), where lim determines the plotting interval, i.e. the minimum and maximum values of x.

The command semilogx(x,y) generates a plot of the values of x and y using a logarithmic scale for x and a linear scale for y. The command semilogy(x,y) generates a plot of the values of x and y using a linear scale for x and a logarithmic scale for y. The command loglog(x,y) generates a plot of the values of x and y using logarithmic scales for both x and y. The command polar(theta,r) plots in polar coordinates with theta being the argument in radians and r the magnitude.

The subplot command enables the graph window to be split into subwindows and plots to be placed in each. For example, we might have:

```
x=(0 1 2 3 4 5 6 7);
y=expx;
subplot(2,1,1);plot(x,y);
subplot(2,1,2);semilogy(x,y);
```

Three integers m, n, p are given with the subplot command; the digits m and n indicate the graph window is to be split into an $m \times n$ grid of smaller windows, where m is the number of rows and n is the number of columns, and the digit p specifies the window to be used for the plot. The subwindows are numbered by row from left to right and top to bottom. Thus the above sequence of commands divides the window into two, with one plot above the other; the top plot is a linear plot and the lower plot is a semilogarithmic plot.

The number and style of grid lines, the plot colour and the adding of text to a plot can all be selected. The command print is used to print a hard copy of a plot, either to a file or a printer. This can be done by selecting the file menu-bar item in the figure window and then selecting the print option.

11.7.2 Transfer functions

The following lines in a MATLAB program illustrate how a transfer function can be entered and displayed on screen:

```
% G(s)=4(s+10)/(s+5)(s+15)
num=4*[1 10];
den=conv([1 5],[1 15]);
printsys(num,den,'s')
```

The command num is used to indicate the numerator of the transfer function, in descending powers of s. The command den is used to indicate the denominator in descending powers of s for each of the two polynomials in the denominator. The command conv multiplies two polynomials, in this case they are $(s + 5)$ and $(s + 15)$. The printsys command displays the transfer function with the numerator and denominator specified and written in the s-domain.

Sometimes we may be presented with a transfer function as the ratio of two polynomials and need to find the poles and zeros. For this we can use:

```
% Finding poles and zeros for the transfer function
% G(s)=(5s^2 + 3s + 4)/(s^3 + 2s^2 + 4s + 7)
num=[5 3 4];
den=[1 2 4 7];
[z,p,k]=tf2zp(num,den)
```

[z,p,k]=tf2zp(num,den) is the command to determine and display the zeros (z), poles (p) and gain (k) of the zero-pole-gain transfer function entered.

MATLAB can be used to give graphs showing the response of a system to different inputs. For example, the following program will give the response of the system to a unit step input $u(t)$ with a specified transfer function:

```
% Display of response to a step input for a system with
% transfer function G(s) = 5/(s^2 + 3s + 12)
num=5;
den=[1 3 12];
step(num,den)
```

11.7.3 Block diagrams

Control systems are often represented as a series of inter-connected blocks, each block having specific characteristics. MATLAB allows systems to be built up from interconnected blocks. The commands used are cloop when a block with a given open-loop transfer function has unity feedback. If the feedback is not unity the command feedback is used, e.g. with Figure 11.9 we have the program:

```
% System with feedback loop
ngo=[1 1];
dgo=conv([1 3],[1 4]);
```

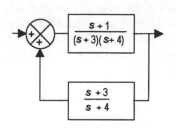

Fig. 11.9   Block diagram

```
nh=[1 3];
dh=[1 4];
[ngc2,dgc2]=feedback(ngo,dgo,nh,dh)
printsys(ngc2,dgc2,'s')
```

ngo and dgo indicate the numerator and denominator of open-loop transfer function $G_0(s)$, nh and dh the numerator and denominator of the feedback loop transfer function $H(s)$. The program results in the display of the transfer function for the system as a whole.

The command series is used to indicate that two blocks are in series in a particular path; the command parallel indicates that they are in parallel.

11.7.4 SIMULINK

SIMULINK is used in conjunction with MATLAB to specify systems by 'connecting' boxes on the screen rather than, as above, writing a series of commands to generate the description of the block diagram. Once MATLAB has been started, SIMULINK is selected using the command >>simulink. This opens the SIMULINK control window with its icons and pull-down menus in its header bar. Click on file, then click on new from the drop-down menu. This opens a window in which a system can be assembled.

To start assembling the blocks required, go back to the control window and double-click on the linear icon. Click and then drag the transfer Fcn icon into the untitled window. If you require a gain block, click and drag the gain icon into the untitled window. Do the same for a sum icon and perhaps an integrator icon. In this way, drag all the required icons into the untitled window. Then double-click on the Sources icon and select the appropriate source from its drop-down menu, e.g. step input, and drag it into the untitled window. Now double-click on the sinks icon and drag the graph icon into the untitled window. To connect the icons, depress the mouse button while the mouse arrow is on the output symbol of an icon and drag to it the input symbol of the icon to which it is to be connected. Repeat this for all the icons until the complete block diagram is assembled.

To give the transfer Fcn box a transfer function, double-click in the box. This will give a dialogue box in which you can use MATLAB commands for numerator and denominator. Click on the numerator and type in [1 1] if $(s + 1)$ is required. Click on the denominator and type in [1 2 3] if $(s^2 + 2s + 3)$ is required. Then click the done icon. Double-click on the gain icon and type in the gain value. Double-click on the sum icon and set the signs to + or − according to whether positive or negative feedback is required. Double-click on the graph icon and set the parameters for the graph. You then have the complete simulation diagram on screen. Figure 11.10 shows the form it might take. To delete any block or connection, select them by clicking and then press the <DEL> key.

To simulate the behaviour of the system, click on Simulation to pull down its menu. Select Parameters and set the start and stop times for the simulation. From the Simulation menu, select Start. SIMULINK will then create a graph window and display the output of the system. The file can be saved by selecting File and clicking on SAVE AS in the drop-down menu. Insert a file name in the dialogue box then click on Done.

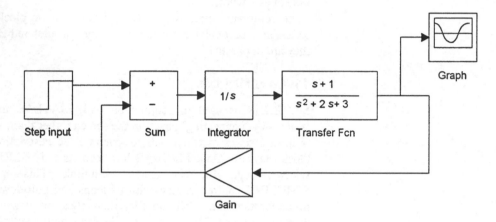

Fig. 11.10 Example of use of SIMULINK

Problems

1 What are the transfer functions for systems giving the following input/output relationships?

(a) A hydraulic system has an input q and an output h where

$$q = A\frac{dh}{dt} + \frac{\rho g h}{R}$$

(b) A spring–dashpot–mass system with an input F and an output x, where

$$m\frac{d^2x}{dt^2} + c\frac{dx}{dt} + kx = F$$

(c) An RLC circuit with an input v and output v_C, where

$$v = RC\frac{dv_C}{dt} + LC\frac{d^2v_C}{dt^2} + v_C$$

2 What are the time constants of the systems giving the transfer functions (a) $G(s) = 5/(3s + 1)$, (b) $G(s) = 2/(2s + 3)$?

3 Determine how the outputs of the following systems vary with time when subject to a unit step input at time $t = 0$: (a) $G(s) = 2/(s + 2)$, (b) $G(s) = 10/(s + 5)$.

4   What is the state of the damping for the systems having the following transfer functions?

(a) $G(s) = \dfrac{5}{s^2 - 6s + 16}$, (b) $G(s) = \dfrac{10}{s^2 + s + 100}$,

(c) $G(s) = \dfrac{2s + 1}{s^2 + 2s + 1}$, (d) $G(s) = \dfrac{3s + 20}{s^2 + 2s + 20}$

5   What is the output of a system with the transfer function $s/(s + 3)^2$ and subject to a unit step input at time $t = 0$?

6   What is the output of a system having the transfer function $G = 2/[(s + 3)(s + 4)]$ and subject to a unit impulse?

7   What are the overall transfer functions of the following negative feedback systems?

Forward path	Feedback path
(a) $G(s) = \dfrac{4}{s(s + 1)}$,	$H(s) = \dfrac{1}{s}$,
(b) $G(s) = \dfrac{2}{s + 1}$,	$H(s) = \dfrac{1}{s + 2}$,
(c) $G(s) = \dfrac{4}{(s + 2)(s + 3)}$,	$H(s) = 5$
(d) two series elements $G_1(s) = 2/(s + 2)$ and $G_2(s) = 1/s$,	$H(s) = 10$

8   What is the overall transfer function for a closed-loop system having a forward-path transfer function of $5/(s + 3)$ and a negative feedback-path transfer function of 10?

9   A closed-loop system has a forward path having two series elements with transfer functions 5 and $1/(s + 1)$. If the feedback path has a transfer function $2/s$, what is the overall transfer function of the system?

10 A closed-loop system has a forward path having two series elements with transfer functions of 2 and $1/(s + 1)$. If the feedback path has a transfer function of s, what is the overall transfer function of the system?

12 Frequency response

12.1 Sinusoidal input

In the previous two chapters, the response of systems to step, impulse and ramp inputs has been considered. This chapter extends this to when there is a sinusoidal input. While for many control systems a sinusoidal input might not be encountered normally, it is a useful testing input since the way the system responds to such an input is a very useful source of information to aid the design and analysis of systems.

Consider a first-order system which is described by the differential equation

$$a_1 \frac{dx}{dt} + a_0 x = b_0 y$$

where y is the input and x the output. Suppose we have the unit amplitude sinusoidal input of $y = \sin \omega t$. What will the output be? Well we must end up with the sinusoid $b_0 \sin \omega t$ when we add $a_1 \, dx/dt$ and $a_0 x$. But sinusoids have the property that when differentiated the result is also a sinusoid and with the same frequency (a cosine is a sinusoid, being just $\sin (\omega t + 90°)$). This applies no matter how many times we carry out the differentiation. Thus we should expect that the steady-state response x will also be sinusoidal and with the same frequency. The output may, however, differ in amplitude and phase from the input.

12.2 Phasors

In discussing sinusoidal signals it is convenient to use *phasors*. Consider a sinusoid described by the equation $v = V \sin (\omega t + \phi)$, where V is the amplitude, ω the angular frequency and ϕ the phase angle. The phasor can be represented by a line of length $|V|$ making an angle of ϕ with the phase reference axis. The $|\ |$ lines are used to indicate that we are only concerned with the magnitude or size of the quantity when specifying its length. A phasor quantity in order to be specified always requires a

Fig. 12.1 Complex representation of a phasor

(a)

(b)

(c)

(d)

Fig. 12.2 Phasor rotation:
(a) 0°, (b) 90°, (c) 180°, (d) 270°

magnitude and angle to be specified. The convention generally adopted is to write a phasor in bold, non-italic, print, e.g. **V**. When such a symbol is seen it implies a quantity having both a magnitude and an angle.

Such a phasor can be described by means of complex number notation. A complex quantity can be represented by $(x + jy)$, where x is the real part and y the imaginary part of the complex number. On a graph with the imaginary component as the y-axis and the real part as the x-axis, x and y are Cartesian coordinates of the point representing the complex number (Fig. 12.1). If we take the line joining this point to the graph origin to represent a phasor, then we have the phase angle ϕ of the phasor represented by

$$\tan \phi = \frac{y}{x}$$

and its length by the use of Pythagoras' theorem as

$$\text{Length of phasor } |V| = \sqrt{x^2 + y^2}$$

Since $x = |V| \cos \phi$ and $y = |V| \sin \phi$, then we can write

$$\mathbf{V} = x + jy = |V| (\cos \theta + j \sin \theta)$$

Thus a specification of the real and imaginary parts of a complex number enables a phasor to be specified.

Consider a phasor of length 1 and phase angle 0° (Fig. 12.2(a)). It will have a complex representation of $1 + j0$. Now consider the same length phasor but with a phase angle of 90° (Fig. 12.2(b)). It will have a complex representation of $0 + j1$. Thus rotation of a phasor anti-clockwise by 90° corresponds to multiplication of the phasor by j. If we now rotate this phasor by a further 90° (Fig. 12.2(c)), then following the same multiplication rule we have the original phasor multiplied by j^2. But the phasor is just the original phasor in the opposite direction, i.e. just multiplied by −1. Hence $j^2 = -1$ and so $j = \sqrt{(-1)}$. Rotation of the original phasor through a total of 270°, i.e. 3 × 90°, is equivalent to multiplying the original phasor by $j^3 = j(j^2) = -j$.

For further discussion of complex numbers and their application in engineering, the reader is referred to *Complex Numbers* by W. Bolton (Longman 1994), part of their Mathematics for Engineers series.

To illustrate the above, consider a voltage v which varies sinusoidally with time according to the equation

$$v = 10 \sin (\omega t + 30°) \text{ V}$$

When represented by a phasor, what are (a) its length, (b) its angle relative to the reference axis, (c) its real and imaginary parts when represented by a complex number?

(a) The phasor will have a length scaled to represent the amplitude of the sinusoid and so is 10 V.
(b) The angle of the phasor relative to the reference axis is equal to the phase angle and so is 30°.
(c) The real part is given by the equation $x = 10 \cos 30° = 8.7$ V and the imaginary part by $y = 10 \sin 30° = 5.0$ V. Thus the phasor is specified by $8.7 + j5.0$ V.

12.2.1 Phasor equations

Consider a phasor to represent the unit amplitude sinusoid of $x = \sin \omega t$. Differentiation of the sinusoid gives $dx/dt = \omega \cos \omega t$. But we can also write this as $dx/dt = \omega \sin (\omega t + 90°)$. In other words, differentiation just results in a phasor with a length increased by a factor of ω and which is rotated round by 90° from the original phasor. Thus, in complex notation, we have multiplied the original phasor by $j\omega$, since multiplication by j is equivalent to a rotation through 90°.

Thus the differential equation

$$a_1 \frac{dx}{dt} + a_0 x = b_0 y$$

can be written, in complex notation, as a *phasor equation*

$$j\omega a_1 \mathbf{X} + a_0 \mathbf{X} = b_0 \mathbf{Y}$$

where the bold, non-italic, letters indicate that the data refers to phasors. We can say that the differential equation, which was an equation in the time domain, has been transformed into an equation in the *frequency domain*. The frequency domain equation can be rewritten as

$$(j\omega a_1 + a_0) \mathbf{X} = b_0 \mathbf{Y}$$

$$\frac{\mathbf{X}}{\mathbf{Y}} = \frac{b_0}{j\omega a_1 + a_0}$$

But, in Section 11.2, when the same differential equation was written in the s-domain, we had

$$G(s) = \frac{X(s)}{Y(s)} = \frac{b_0}{a_1 s + a_0}$$

If we replace s by $j\omega$ we have the same equation. It turns out that we can always do this to convert from the s-domain to the frequency domain. This thus leads to a definition of a *frequency-*

response function or *frequency transfer function* $G(j\omega)$), for the steady state, as

$$G(j\omega) = \frac{\text{output phasor}}{\text{input phasor}}$$

To illustrate the above consider the determination of the frequency-response function for a system having a transfer function of

$$G(s) = \frac{1}{s+1}$$

The frequency-response function is obtained by replacing s by $j\omega$. Thus

$$G(j\omega) = \frac{1}{j\omega + 1}$$

12.3 Frequency response

A first-order system has a transfer function which can be written as

$$G(s) = \frac{1}{1+\tau s}$$

where τ is the time constant of the system (see Section 11.2). The frequency-response function $G(j\omega)$ can be obtained by replacing s by $j\omega$. Hence

$$G(j\omega) = \frac{1}{1+j\omega\tau}$$

We can put this into a more convenient form by multiplying the top and bottom of the expression by $(1 - j\omega\tau)$ to give

$$G(j\omega) = \frac{1}{1+j\omega\tau} \times \frac{1-j\omega\tau}{1-j\omega\tau} = \frac{1-j\omega\tau}{1+j^2\omega^2\tau^2}$$

But $j^2 = -1$, thus

$$G(j\omega) = \frac{1}{1+\omega^2\tau^2} - j\frac{\omega\tau}{1+\omega^2\tau^2}$$

This is of the form $x + jy$ and so, since $G(j\omega)$ is the output phasor divided by the input phasor, we have the size of the output phasor bigger than that of the input phasor by a factor which can be written as $|G(j\omega)|$, with

$$|G(j\omega)| = \sqrt{x^2+y^2} = \sqrt{\left(\frac{1}{1+\omega^2\tau^2}\right)^2 + \left(\frac{\omega\tau}{1+\omega^2\tau^2}\right)^2}$$

$$= \frac{1}{\sqrt{1+\omega^2\tau^2}}$$

$|G(j\omega)|$ tells us how much bigger the amplitude of the output is than the amplitude of the input. It is generally referred to as the *magnitude* or *gain*. The phase difference ϕ between the output phasor and the input phasor is given by

$$\tan\phi = \frac{y}{x} = -\omega\tau$$

The negative sign indicates that the output phasor lags behind the input phasor by this angle.

The following examples illustrate the above:

1   Determine the frequency-response function, the magnitude and phase of a system (an electrical circuit with a resistor in series with a capacitor across which the output is taken) that has a transfer function of

$$G(s) = \frac{1}{RCs+1}$$

The frequency-response function can be obtained by substituting $j\omega$ for s and so give

$$G(j\omega) = \frac{1}{j\omega RC+1}$$

We can multiply the top and bottom of the above equation by $1 - j\omega RC$ and then rearrange the result to give

$$G(j\omega) = \frac{1}{1+\omega^2(RC)^2} - j\frac{\omega(RC)}{1+\omega^2(RC)^2}$$

Hence

$$|G(j\omega)| = \frac{1}{\sqrt{1+\omega^2(RC)^2}}$$

and $\tan\phi = -\omega RC$.

2   Determine the magnitude and phase of the output from a system when subject to a sinusoidal input of $2\sin(3t + 60°)$ if it has a transfer function of

$$G(s) = \frac{4}{s+1}$$

The frequency-response function is obtained by replacing s by $j\omega$. Thus

$$G(j\omega) = \frac{4}{j\omega + 1}$$

Multiplying the top and bottom of the equation by $(-j\omega + 1)$:

$$G(j\omega) = \frac{-j4\omega + 4}{\omega^2 + 1} = \frac{4}{\omega^2 + 1} - j\frac{4\omega}{\omega^2 + 1}$$

The magnitude is thus

$$|G(j\omega)| = \sqrt{x^2 + y^2} = \sqrt{\frac{4^2}{(\omega^2 + 1)^2} + \frac{4^2\omega^2}{(\omega^2 + 1)^2}}$$

$$= \frac{4}{\sqrt{\omega^2 + 1}}$$

and the phase angle is given by $\tan\phi = y/x$ and so

$$\tan\phi = -\omega$$

For the specified input we have $\omega = 3$ rad/s. The magnitude is thus

$$|G(j\omega)| = \frac{4}{\sqrt{3^2 + 1}} = 1.3$$

and the phase is given by $\tan\phi = -3$. Thus $\phi = -72°$. This is the phase angle between the input and the output. Thus the output is $2.6 \sin(3t - 12°)$.

12.3.1 Frequency response for a second-order system

Consider a second-order system with the transfer function (see Section 11.3)

$$G(s) = \frac{\omega_n^2}{s^2 + 2\zeta\omega_n s + \omega_n^2}$$

where ω_n is the natural angular frequency and ζ the damping ratio. The frequency-response function is obtained by replacing s by $j\omega$. Thus

$$G(j\omega) = \frac{\omega_n^2}{-\omega^2 + j2\zeta\omega\omega_n + \omega_n^2} = \frac{\omega_n^2}{(\omega_n^2 - \omega^2) + j2\zeta\omega_n}$$

$$= \frac{1}{\left[1 - \left(\frac{\omega}{\omega_n}\right)^2\right] + j2\zeta\left(\frac{\omega}{\omega_n}\right)}$$

Multiplying the top and bottom of the expression by

$$\left[1-\left(\frac{\omega}{\omega_n}\right)^2\right] - j2\zeta\left(\frac{\omega}{\omega_n}\right)$$

gives

$$G(j\omega) = \frac{\left[1-\left(\frac{\omega}{\omega_n}\right)^2\right] - j2\zeta\left(\frac{\omega}{\omega_n}\right)}{\left[1-\left(\frac{\omega}{\omega_n}\right)^2\right]^2 + \left[2\zeta\left(\frac{\omega}{\omega_n}\right)\right]^2}$$

This is of the form $x + jy$ and so, since $G(j\omega)$ is the output phasor divided by the input phasor, we have the size or magnitude of the output phasor bigger than that of the input phasor by a factor which is given by $\sqrt{(x^2 + y^2)}$ as

$$|G(j\omega)| = \frac{1}{\sqrt{\left[1-\left(\frac{\omega}{\omega_n}\right)^2\right]^2 + \left[2\zeta\left(\frac{\omega}{\omega_n}\right)\right]^2}}$$

The phase ϕ difference between the input and output is given by $\tan\phi = x/y$ and so

$$\tan\phi = -\frac{2\zeta\left(\frac{\omega}{\omega_n}\right)}{1-\left(\frac{\omega}{\omega_n}\right)^2}$$

The minus sign is because the output phase lags behind the input.

12.4 Bode plots

The frequency response of a system is the set of values of the magnitude $|G(j\omega)|$ and phase angle ϕ that occur when a sinusoidal input signal is varied over a range of frequencies. This can be expressed as two graphs, one of the magnitude $|G(j\omega)|$ plotted against the angular frequency ω and the other of the phase ϕ plotted against ω. The magnitude and angular frequency are plotted using logarithmic scales. Such a pair of graphs is referred to as a *Bode plot*.

The magnitude is expressed in decibel units (dB).

$$|G(j\omega)| \text{ in dB} = 20 \lg_{10} |G(j\omega)|$$

Thus, for example, a magnitude of 20 dB means that

$$20 = 20 \lg_{10} |G(j\omega)|$$

so $1 = \lg_{10} |G(j\omega)|$ and $10^1 = |G(j\omega)|$. Thus a magnitude of 20 dB means the magnitude is 10, therefore the amplitude of the output is 10 times that of the input. A magnitude of 40 dB would mean a magnitude of 100 and so the amplitude of the output is 100 times that of the input.

12.4.1 Examples of Bode plots

Consider the Bode plot for a system having the transfer function $G(s) = K$, where K is a constant. The frequency-response function is thus $G(j\omega) = K$. The magnitude $|G(j\omega)| = K$ and so, in decibels, $|G(j\omega)| = 20 \lg K$. The magnitude plot is thus a line of constant magnitude, changing K merely shifts the magnitude line up or down by a certain number of decibels. The phase is zero. Figure 12.3 shows the Bode plot.

Consider the Bode plot for a system having a transfer function $G(s) = 1/s$. The frequency-response function $G(j\omega)$ is thus $1/j\omega$. Multiplying this by j/j gives $G(j\omega) = -j/\omega$. The magnitude $|G(j\omega)|$ is thus $1/\omega$. In decibels this is $20 \lg (1/\omega) = -20 \lg \omega$. When $\omega = 1$ rad/s the magnitude is 0. When $\omega = 10$ rad/s it is -20 dB. When $\omega = 100$ rad/s it is -40 dB. For each tenfold increase in angular frequency the magnitude drops by -20 dB. The magnitude plot is thus a straight line of slope -20 dB per decade of frequency which passes through 0 dB at $\omega = 1$ rad/s. The phase of such a system is given by

Fig. 12.3   Bode plot for $G(s) = K$

$$\tan \phi = \frac{-\frac{1}{\omega}}{0} = -\infty$$

Hence $\phi = -90°$ for all frequencies. Figure 12.4 shows the Bode plot.

Consider the Bode plot for a first-order system for which the transfer function is given by

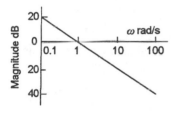

$$G(s) = \frac{1}{\tau s + 1}$$

The frequency-response function is then

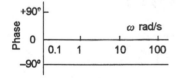

$$G(j\omega) = \frac{1}{j\omega\tau + 1}$$

Fig. 12.4   Bode plot for $G(s) = 1/s$

The magnitude (see Section 12.2.1) is then

$$|G(j\omega)| = \frac{1}{\sqrt{1 + \omega^2\tau^2}}$$

In decibels this is

$$20 \lg \left(\frac{1}{\sqrt{1 + \omega^2\tau^2}} \right)$$

When $\omega \ll 1/\tau$ then $\omega^2\tau^2$ is negligible compared with 1 and so the magnitude is $20 \lg 1 = 0$ dB. Hence at low frequencies there is a straight line magnitude plot at a constant value of 0 dB. For higher frequencies, when $\omega \gg 1/\tau$, $\omega^2\tau^2$ is much greater than 1 and so the 1 can be neglected. The magnitude is then 20 lg

Fig. 12.5 Bode plot for first-order system

$(1/\omega\tau)$, i.e. $-20 \lg \omega\tau$. This is a straight line of slope -20 dB per decade of frequency which intersects the 0 dB line when $\omega\tau = 1$, i.e. when $\omega = 1/\tau$. Figure 12.5 shows these lines for low and high frequencies with their intersection, or so-called *break point* or *corner frequency*, at $\omega = 1/\tau$. The two straight lines are called the asymptotic approximation to the true plot. The true plot rounds off the intersection of the two lines. The difference between the true plot and the approximation is a maximum of 3 dB at the break point.

The phase for the first-order system (see Section 12.2.1) is given by $\tan \phi = -\omega\tau$. At low frequencies, when ω is less than about $0.1/\tau$, the phase is virtually $0°$. At high frequencies, when ω is more than about $10/\tau$, the phase is virtually $-90°$. Between these two extremes the phase angle can be considered to give a reasonable straight line on the Bode plot (Fig. 12.5). The maximum error in assuming a straight line is $5.5°$.

Consider a second-order system with a transfer function of

$$G(s) = \frac{\omega_n^2}{s^2 + 2\zeta\omega_n s + \omega_n^2}$$

The frequency-response function is obtained by replacing s by $j\omega$.

$$G(j\omega) = \frac{\omega_n^2}{-\omega^2 + j2\zeta\omega_n\omega + \omega_n^2}$$

The magnitude is then (see Section 12.2.2)

$$|G(j\omega)| = \frac{1}{\sqrt{\left[1 - \left(\frac{\omega}{\omega_n}\right)^2\right]^2 + \left[2\zeta\left(\frac{\omega}{\omega_n}\right)\right]^2}}$$

Thus, in decibels, the magnitude is

$$20 \lg \frac{1}{\sqrt{\left[1 - \left(\frac{\omega}{\omega_n}\right)^2\right]^2 + \left[2\zeta\left(\frac{\omega}{\omega_n}\right)\right]^2}}$$

$$= -20 \lg \sqrt{\left[1 - \left(\frac{\omega}{\omega_n}\right)^2\right]^2 + \left[2\zeta\left(\frac{\omega}{\omega_n}\right)\right]^2}$$

For $(\omega/\omega_n) \ll 1$ the magnitude approximates to $-20 \lg 1$ or 0 dB and for $(\omega/\omega_n) \gg 1$ the magnitude approximates to $-20 \lg (\omega/\omega_n)^2$. Thus when ω increases by a factor of 10 the magnitude increases by a factor of $-20 \lg 100$ or -40 dB. Thus at low frequencies the magnitude plot is a straight line at 0 dB, while at high frequencies it is a straight line of -40 dB per decade of frequency. The intersection of these two lines, i.e. the break point, is at $\omega = \omega_n$. The magnitude plot is thus approximately given by these two asymptotic lines. The true value, however, depends on the damping ratio ζ. Figure 12.6 shows the two asymptotic lines and the true plots for a number of damping ratios.

The phase is given by (see Section 12.2.2)

$$\tan\phi = -\frac{2\zeta\left(\dfrac{\omega}{\omega_n}\right)}{1-\left(\dfrac{\omega}{\omega_n}\right)^2}$$

For $(\omega/\omega_n) \ll 1$, e.g. $(\omega/\omega_n) = 0.2$, then $\tan\phi$ is approximately 0 and so $\phi = 0°$. For $(\omega/\omega_n) \gg 1$, e.g. $(\omega/\omega_n) = 5$, $\tan\phi$ is approximately $-(-\infty)$ and so $\phi = -180°$. When $\omega = \omega_n$ then we have $\tan\phi = -\infty$ and so $\phi = -90°$. A reasonable approximation is given by a straight line drawn through $-90°$ at $\omega = \omega_n$ and the points $0°$ at $(\omega/\omega_n) = 0.2$ and $-180°$ at $(\omega/\omega_n) = 5$. Figure 12.6 shows the graph.

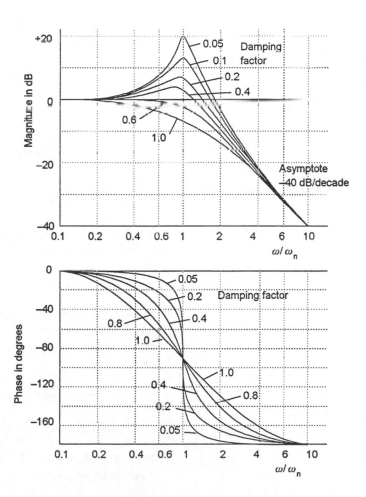

Fig. 12.6 Bode plot for a second-order system

12.4.2 Building up Bode plots

Consider a system involving a number of elements in series. The transfer function of the system as a whole is given by (see Section 11.4)

$$G(s) = G_1(s)G_2(s)G_3(s) \dots \text{etc.}$$

Hence the frequency-response function for a two-element system, when s is replaced by $j\omega$, is

$$G(j\omega) = G_1(j\omega)G_2(j\omega)$$

We can write the transfer function $G_1(j\omega)$ as a complex number (see Section 12.2), i.e.

$$x + jy = |G_1(j\omega)| \, (\cos \phi_1 + j \sin \phi_1)$$

where $|G(j\omega)|$ is the magnitude and ϕ_1 the phase of the frequency-response function. Similarly we can write $G_2(j\omega)$ as

$$|G_2(j\omega)| \, (\cos \phi_2 + j \sin \phi_2)$$

Thus

$$\begin{aligned}
G(j\omega) =\ & |G_1(j\omega)| \, (\cos \phi_1 + j \sin \phi_1) \\
& \times |G_2(j\omega)| \, (\cos \phi_2 + j \sin \phi_2)
\end{aligned}$$

$$\begin{aligned}
=\ & |G_1(j\omega)| \, |G_2(j\omega)| \, [\cos \phi_1 \cos \phi_2 \\
& + j(\sin \phi_1 \cos \phi_2 + \cos \phi_1 \sin \phi_2) + j^2 \sin \phi_1 \sin \phi_2]
\end{aligned}$$

But $j^2 = -1$ and, since $\cos \phi_1 \cos \phi_2 - \sin \phi_1 \sin \phi_2 = \cos (\phi_1 + \phi_2)$ and $\sin \phi_1 \cos \phi_2 + \cos \phi_1 \sin \phi_2 = \sin (\phi_1 + \phi_2)$, thus

$$G(j\omega) = |G_1(j\omega)| \, |G_2(j\omega)| \, [\cos (\phi_1 + \phi_2) + j \sin (\phi_1 + \phi_2)]$$

The frequency-response function of the system has a magnitude which is the product of the magnitudes of the separate elements and a phase which is the sum of the phases of the separate elements, i.e.

$$|G(j\omega)| = |G_1(j\omega)| \, |G_2(j\omega)| \, |G_3(j\omega)| \dots \text{etc.}$$

$$\phi = \phi_1 + \phi_2 + \phi_3 + \dots \text{etc.}$$

Now, considering the Bode plot where the logarithms of the magnitudes are plotted,

$$\lg |G(j\omega)| = \lg |G_1(j\omega)| + \lg |G_2(j\omega)| + \lg |G_3(j\omega)| + \dots \text{etc.}$$

Thus we can obtain the Bode plot of a system by adding together the Bode plots of the magnitudes of the constituent elements. Likewise the phase plot is obtained by adding together the phases of the constituent elements.

By using a number of basic elements, the Bode plots for a wide range of systems can be readily obtained. The basic elements used are:

1 $G(s) = K$
   This gives the Bode plot shown in Figure 12.3.

2 $G(s) = 1/s$
   This gives the Bode plot shown in Figure 12.4.

3 $G(s) = s$
   This gives a Bode plot which is a mirror image of that in Figure 12.4. $|G(j\omega)| = 20$ dB per decade of frequency, passing through 0 dB at $\omega = 1$ rad/s. ϕ is constant at 90°.

4 $G(s) = 1/(\tau s + 1)$
   This gives the Bode plot shown in Figure 12.5.

5 $G(s) = \tau s + 1$
   This gives a Bode plot which is a mirror image of that in Figure 12.5. For the magnitude plot, the break point is at $1/\tau$ with the line prior to it being at 0 dB and after it at a slope of 20 dB per decade of frequency. The phase is zero at $0.1/\tau$ and rises to +90° at $10/\tau$.

6 $G(s) = \omega_n^2/(s^2 + 2\zeta\omega_n s + \omega_n^2)$
   This gives the Bode plot shown in Figure 12.6.

7 $G(s) = (s^2 + 2\zeta\omega_n s + \omega_n^2)/\omega_n^2$
   This gives a Bode plot which is a mirror image of that in Figure 12.6.

To illustrate the above, consider the drawing of the asymptotes of the Bode plot for a system having a transfer function of

$$G(s) = \frac{10}{2s + 1}$$

The transfer function is made up of two elements, one with a transfer function of 10 and one with transfer function $1/(2s + 1)$. The Bode plots can be drawn for each of these and then added together to give the required plot. The Bode plot for transfer function 10 will be of the form given in Figure 12.3 with $K = 10$ and that for $1/(2s + 1)$ like that given in Figure 12.5 with $\tau = 2$. The result is shown in Figure 12.7.

Fig. 12.7 Building up a Bode diagram

As another example, consider the drawing of the asymptotes of the Bode plot for a system having a transfer function of

$$G(s) = \frac{2.5}{s(s^2 + 3s + 25)}$$

The transfer function is made up of three components, one with a transfer function of 0.1, one with transfer function $1/s$ and one with transfer function $25/(s^2 + 3s + 25)$. The transfer function of 0.1 will give a Bode plot like that of Figure 12.3 with $K = 0.1$. The transfer function of $1/s$ will give a Bode plot like that of Figure 12.4. The transfer function of $25/(s^2 + 3s + 25)$ can be represented as $\omega_n^2/(s^2 + 2\zeta\omega_n s + \omega_n^2)$ with $\omega_n = 5$ rad/s and $\zeta = 0.3$. The break point will be when $\omega = \omega_n = 5$ rad/s. The asymptote for the phase passes through $-90°$ at the break point, and is $0°$ when we have $(\omega/\omega_n) = 0.2$ and $-180°$ when $(\omega/\omega_n) = 5$. Figure 12.8 shows the resulting Bode plot.

12.4.3 Bode plots with MATLAB

MATLAB can be used to produce Bode plots (see Section 11.7 for a preliminary discussion of MATLAB). To carry out a Bode plot of a system described by a transfer function $4/(s^2 + 2s +3)$ the program is:

```
% Generate Bode plot for G(s)=4/(s^2 + 2s + 3)
num=4;
```

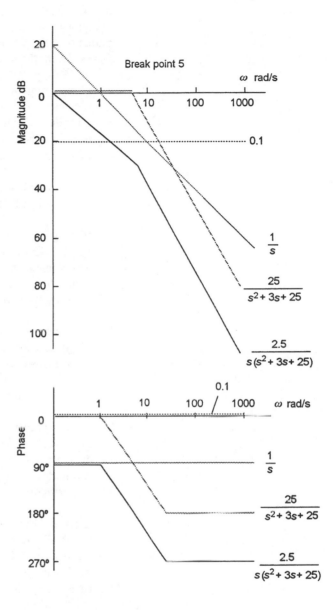

Fig. 12.8 Building up a Bode diagram

```
den=[1 2 3];
bode(num,den)
```

The command bode(num,den) produces the Bode plot of gain in dB against frequency in rad/s on a log scale, and phase in degrees against frequency in rad/s on a log scale.

12.4.4 System identification

If we experimentally determine the Bode diagram for a system by considering its response to a sinusoidal input, then we can obtain the transfer function for the system. Basically we draw the asymptotes on the magnitude Bode plot and consider their gradients. The phase angle curve is used to check the results obtained from the magnitude analysis.

1   If the gradient at low frequencies prior to the first corner frequency is zero then there is no s or $1/s$ element in the transfer function. The K element in the numerator of the transfer function can be obtained from the value of the low frequency magnitude; the magnitude in dB = 20 lg K.

2   If the initial gradient at low frequencies is –20 dB/decade then the transfer function has a $1/s$ element.

3   If the gradient becomes more negative at a corner frequency by 20 dB/decade, there is a $(1 + s/\omega_c)$ term in the denominator of the transfer function, with ω_c being the corner frequency at which the change occurs. Such terms can occur for more than one corner frequency.

4   If the gradient becomes more positive at a corner frequency by 20 dB/decade, there is a $(1 + s/\omega_c)$ term in the numerator of the transfer function, with ω_c being the frequency at which the change occurs. Such terms can occur for more than one corner frequency.

5   If the gradient at a corner frequency becomes more negative by 40 dB/decade, there is a $(s^2/\omega_c^2 + 2\zeta s/\omega_c + 1)$ term in the denominator of the transfer function. The damping ratio ζ can be found by considering the detail of the Bode plot at a corner frequency, as in Figure 12.6.

6   If the gradient at a corner frequency becomes more positive by 40 dB/decade, there is a $(s^2/\omega_c^2 + 2\zeta s/\omega_c + 1)$ term in the numerator of the transfer function. The damping ratio ζ can be found by considering the detail of the Bode plot at a corner frequency, as in Figure 12.6.

7   If the low-frequency gradient is not zero, the K term in the numerator of the transfer function can be determined by considering the value of the low-frequency asymptote. At low frequencies, many terms in transfer functions can be neglected and the gain in dB approximates to 20 lg (K/ω^2). Thus, at $\omega = 1$ the gain in dB approximates to 20 lg K.

As an illustration of the above, consider the Bode magnitude plot shown in Figure 12.9. The initial gradient is 0 and so there is no $1/s$ or s term in the transfer function. The initial gain is 20 and so 20 = 20 lg K and $K = 10$. The gradient changes by –20 dB/decade at a frequency of 10 rad/s. Hence there is a $(1 + s/10)$

Fig. 12.9   Bode plot

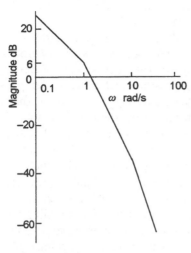

Fig. 12.10   Bode plot

term in the denominator. The transfer function is thus $10/(1 + 0.1s)$.

As a further illustration, consider Figure 12.10. There is an initial slope of -20 dB/decade and so a $1/s$ term. At the corner frequency 1.0 rad/s there is a -20 dB/decade change in gradient and so a $1/(1 + s/1)$ term. At the corner frequency 10 rad/s there is a further -20 dB/decade change in gradient and so a $1/(1 + s/10)$ term. At $z = 1$ the magnitude is 6 dB and so $6 = 20 \lg K$ and $K = 10^{6/20} = 2.0$. The transfer function is thus $2.0/s(1 + s)(1 + 0.1s)$.

As a further illustration, Figure 12.11 shows a Bode plot which has an initial zero gradient which changes by -40 dB/decade at 10 rad/s. The initial magnitude is 10 dB and so $10 = 20 \lg K$ and $K = 10^{0.5} = 3.2$. The change of -40 dB/decade at 10 rad/s means there is $(s^2/10^2 + 2\zeta s/10 + 1)$ term in the denominator. The transfer function is thus $3.2/(0.01s^2 + 0.2\zeta s + 1)$. The damping factor can be obtained by comparison of the Bode plot at the corner frequencies with Figure 12.6. It rises by about 6 dB above the corner and this corresponds to a damping factor of about 0.2. The transfer function is thus $3.2/(0.01s^2 + 0.04s + 1)$.

12.5 Performance specifications

Fig. 12.11   Bode plot

Fig. 12.12   Performance specifications

The terms used to describe the performance of a system when subject to a sinusoidal input are peak resonance and bandwidth. The *peak resonance* M_p is defined as being the maximum value of the magnitude (Fig. 12.12). A large value of the peak resonance corresponds to a large value of the maximum overshoot of a system. For a second-order system it can be directly related to the damping ratio by comparison of the response with the Bode plot of Figure 12.6, a low damping ratio corresponding to a high peak resonance.

The *bandwidth* is defined as the frequency band between which the magnitude does not fall below -3 dB. For the system giving the Bode plot in Figure 12.12, the bandwidth is the spread between zero frequency and the frequency at which the magnitude drops below -3 dB.

12.6 Stability

Fig. 12.13   Self-sustaining oscillations

When there is a sinusoidal input to a system, the output from that system is sinusoidal with the same angular frequency but can have an output with an amplitude and phase which differ from that of the input. Consider a closed-loop system with negative feedback (Fig. 12.13) and no input to the system. Suppose, somehow, we have a half-rectified sinusoidal pulse as the error signal in the system and that it passes through to the output and is fed back to arrive at the comparator element with amplitude unchanged but delayed by just half a cycle, i.e. a phase change of 180° as shown in the figure. When this signal is subtracted from the input signal we have a resulting error signal which just continues the initial half-rectified pulse. This pulse then goes back round the feedback loop and once again arrives just in time to continue the signal. Thus we have a self-sustaining oscillation.

For self-sustained oscillations to occur we must have a system which has a frequency-response function with a magnitude of 1 and a phase of −180°. The system through which the signal passes is $G(s)$ in series with $H(s)$. If the magnitude is less than 1 then each succeeding half-wave pulse is smaller in size and so the oscillation dies away. If the magnitude is greater than 1 then each succeeding pulse is larger than the previous one and so the wave builds up and the system is unstable.

1   A control system will oscillate with a constant amplitude if the magnitude resulting from the system $G(s)$ in series with $H(s)$ is 1 and the phase is −180°.

2   A control system will oscillate with a diminishing amplitude if the magnitude resulting from the system $G(s)$ in series with $H(s)$ is less than 1 and the phase is −180°.

3   A control system will oscillate with an increasing amplitude, and so is unstable, if the magnitude resulting from the system $G(s)$ in series with $H(s)$ is greater than 1 and the phase is −180°.

A good, stable control system usually requires that the magnitude of $G(s)H(s)$ should be significantly less than 1. Typically a value between 0.4 and 0.5 is used. In addition, the phase angle should be between about −115° and −125°. Such values produce a slightly under-damped control system which gives, with a step input, about a 20 to 30% overshoot with a subsidence ratio of about 3 to 1 (see Section 10.4 for an explanation of these terms).

A concern with a control system is how stable it is and thus not likely to oscillate as a result of some small disturbance. The term *gain margin* is used for the factor by which the magnitude ratio must be multiplied when the phase is −180° to make it have the value 1 and so give instability. The term *phase margin* is used for the number of degrees by which the phase angle is numerically smaller than −180° when the magnitude is 1. These rules mean a gain margin of between 2 and 2.5 and a phase margin between 45° and 65° for a good, stable control system.

Problems

1   What are the magnitudes and phases of the systems having the following transfer functions?

(a) $\dfrac{5}{s+2}$, (b) $\dfrac{2}{s(s+1)}$, (c) $\dfrac{1}{(2s+1)(s^2+s+1)}$

2   What will be the steady-state response of a system with a transfer function $1/(s+2)$ when subject to the sinusoidal input $3 \sin(5t + 30°)$?

3   What will be the steady-state response of a system with a transfer function $5/(s^2 + 3s + 10)$ when subject to the input $2 \sin(2t + 70°)$?

4   Determine the values of the magnitudes and phase at angular frequencies of (i) 0 rad/s, (ii) 1 rad/s, (iii) 2 rad/s, (iv) ∞ rad/s for systems with the transfer functions (a) $1/[s(2s + 1)$, (b) $1/(3s + 1)$.

5   Draw Bode plot asymptotes for systems having the transfer functions (a) $10/[s(0.1s + 1)]$, (b) $1/[(2s + 1)(0.5s + 1)]$.

6   Obtain the transfer functions of the systems giving the Bode plots in Figure 12.14.

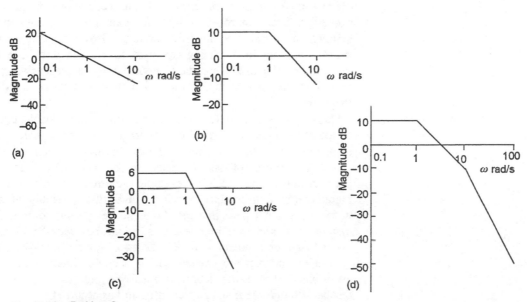

Fig. 12.14   Problem 6

13   Closed-loop controllers

13.1 Continuous and discrete processes

Open-loop control is essentially just a switch on–switch off form of control, e.g. an electric fire is either switched on or off in order to heat a room. With closed-loop control systems, a controller is used to compare the output of a system with the required condition and convert the error into a control action designed to reduce the error. The error might arise as a result of some change in the conditions being controlled or because the set value is changed, e.g. there is a step input to the system to change the set value to a new value. In this chapter we are concerned with the ways in which controllers can react to error signals, i.e. the *control modes* as they are termed, which occur with continuous processes. Such controllers might, for example, be pneumatic systems or operational amplifier systems. However, computer systems are rapidly replacing many of these. The term *direct digital control* is used when the computer is in the feedback loop and exercising control in this way. This chapter is about closed-loop control.

Many processes not only involve controlling some variable, e.g. temperature, to a required value but also involve the sequencing of operations. A domestic washing machine (see Section 1.4.3) where a number of actions have to be carried out in a predetermined sequence is an example. Another example is the manufacture of a product which involves the assembly of a number of discrete parts in a specific sequence by some controlled system. The sequence of operations might be *clock-based* or *event-based* or a combination of the two. With a clock-based system the actions are carried out at specific times; with an event-based system the actions are carried out when there is feedback to indicate that a particular event has occurred.

The term *programmable logic controller* (PLC) is used for a simple controller based on a microprocessor and operates by examining the input signals from sensors and carrying out logic instructions which have been programmed into its memory. The output after such processing is signals which feed into correcting/actuator units. Thus it can carry out sequences of operations. The main difference between a PLC and a computer is that programming is predominantly concerned with logic and switching operations, and the interfacing for input and output devices is inside the controller. Such controllers are discussed in more detail in Chapter 18.

In many processes there can be a mixture of continuous and discrete control. For example, in the domestic washing machine there will be sequence control for the various parts of the washing cycle with feedback loop control of the temperature of the hot water and the level of the water.

13.1.1 Lag

In any control system there are lags. Thus, for example, a change in the condition being controlled does not immediately produce a correcting response from the control system. This is because time is required for the system to make the necessary responses. For example, in the control of the temperature in a room by means of a central heating system, a lag will occur between the room temperature falling below the required temperature and the control system responding and switching on the heater. This is not the only lag. Even when the control system has responded there is a lag in the room temperature responding as time is taken for the heat to transfer from the heater to the air in the room.

13.1.2 Steady-state error

We might get an error signal to the controller occurring as a result of the controlled variable changing or a change in the set value input. For example, we might have a ramp input to the system with the aim that the controlled variable increases steadily with time. The term *steady-state error* is used for the difference between the set value input and the output after all transients have died away. It is thus a measure of the accuracy of the control system in tracking the set value input.

Consider a control system which has unity feedback (Fig. 13.1). When there is a reference input of $R(s)$ there is an output of $X(s)$. The feedback signal is $X(s)$ and so the error signal is $E(s) = R(s) - X(s)$. If $G(s)$ is the forward-path transfer function, then for the unity feedback system as a whole:

$$\frac{X(s)}{R(s)} = \frac{G(s)}{1 + G(s)H(s)} = \frac{G(s)}{1 + G(s)}$$

Hence

$$E(s) = R(s) - X(s) = R(s) - \frac{G(s)R(s)}{1 + G(s)} = \frac{1}{1 + G(s)}R(s)$$

Unity feedback

Fig. 13.1 Unity feedback system

The error thus depends on $G(s)$.

In order to determine the steady-state error we can determine the error e as a function of time and then determine the value of the error when all transients have died down and so the error as the time t tends to an infinite value. While we could determine the inverse of $E(s)$ and then determine its value when $t \to \infty$, there is a simpler method using the *final-value theorem* (see Appendix A); this involves finding the value of $sE(s)$ as s tends to a zero value.

$$e_{SS} = \lim_{t \to \infty} e(t) = \lim_{s \to 0} sE(s)$$

To illustrate the above, consider a unity feedback system with a forward-path transfer function of $k/(\tau s + 1)$ and subject to a unit step input of $1/s$.

$$e_{SS} = \lim_{s \to 0} sE(s) = \lim_{s \to 0} \left[s \frac{1}{1 + k/(\tau s + 1)} \frac{1}{s} \right] = \frac{1}{1 + k}$$

There is thus a steady-state error; the output from the system will never attain the set value. By increasing the gain k of the system then the steady-state error can be reduced.

However, if the unity-feedback system had a forward-path transfer function of $k/s(\tau s + 1)$ and was subject to a step input, then the steady-state error would be:

$$e_{SS} = \lim_{s \to 0} sE(s) = \lim_{s \to 0} \left[s \frac{1}{1 + k/s(\tau s + 1)} \frac{1}{s} \right] = 0$$

There is no steady-state error with this system.

13.2 Control modes

There are a number of ways by which a control unit can react to an error signal and supply an output for correcting elements:

1   The *two-step mode* in which the controller is essentially just a switch which is activated by the error signal and supplies just an on-off correcting signal.

2   The *proportional mode* (P) which produces a control action that is proportional to the error. The correcting signal thus becomes bigger the bigger the error. Thus as the error is reduced the amount of correction is reduced and the correcting process slows down.

3   The *derivative mode* (D) which produces a control action that is proportional to the rate at which the error is changing. When there is a sudden change in the error signal the controller gives a large correcting signal; when there is a gradual change only a small correcting signal is produced.

Derivative control can be considered to be a form of anticipatory control in that the existing rate of change of error is measured, a coming larger error is anticipated and correction applied before the larger error has arrived. Derivative control is not used alone but always in conjunction with proportional control and, often, integral control.

4   The *integral mode* (I) which produces a control action that is proportional to the integral of the error with time. Thus a constant error signal will produce an increasing correcting signal. The correction continues to increase as long the error persists. The integral controller can be considered to be 'looking-back', summing all the errors and thus responding to changes that have occurred.

5   Combinations of modes: proportional plus derivative modes (PD), proportional plus integral modes (PI), proportional plus integral plus derivative modes (PID). The term *three-term controller* is used for PID control.

These five modes of control are discussed in the following sections of the chapter. The controller can achieve these modes by means of pneumatic circuits, analogue electronic circuits involving operational amplifiers or by the programming of a microprocessor or computer.

13.3 Two-step mode

An example of the *two-step mode* of control is the bimetallic thermostat (see Fig. 2.49) that might be used with a simple temperature control system. This is just a switch which is switched on or off according to the temperature. If the room temperature is above the required temperature then the bimetallic strip is in an off position and the heater is off. If the room temperature falls below the required temperature then the bimetallic strip moves into an on position and the heater is switched fully on. The controller in this case can be in only two positions, on or off, as indicated by Figure 13.2.

With the two-step mode the control action is discontinuous. A consequence of this is that oscillations of the controlled variable occur about the required condition. This is because of lags in the time the control system and the process take to respond. For example, in the case of the temperature control for a domestic central heating system, when the room temperature drops below the required level the time that elapses before the control system responds and switches the heater on might be very small in comparison with the time that elapses before the heater begins to have an effect on the room temperature. In the meantime the temperature has fallen even more. The reverse situation occurs when the temperature has risen to the required temperature. Since time elapses before the control system reacts and switches the

Fig. 13.2   Two-step control

Fig. 13.3   Oscillations with two-step control

Fig. 13.4   Two-step control with two controller switch points

13.4 Proportional mode

Fig. 13.5   Proportional band

heater off, and yet more time while the heater cools and stops heating the room, the room temperature goes beyond the required value. The result is that the room temperature oscillates above and below the required temperature (Fig. 13.3).

With the simple two-step system described above there is the problem that when the room temperature is hovering about the set value the thermostat might be almost continually switching on or off, reacting to very slight changes in temperature. This can be avoided if, instead of just a single temperature value at which the controller switches the heater on or off, two values are used and the heater is switched on at a lower temperature than the one at which it is switched off (Fig. 13.4). The term *dead band* is used for the values between the on and off values. A large dead band results in large fluctuations of the temperature about the set temperature; a small dead band will result in an increased frequency of switching. The bimetallic element shown in Figure 2.49 has a permanent magnet for a switch contact; this has the effect of producing a dead band.

Two-step control action tends to be used where changes are taking place very slowly, i.e. with a process with a large capacitance. Thus, in the case of heating a room, the effect of switching the heater on or off on the room temperature is only a slow change. The result of this is an oscillation with a long periodic time. Two-step control is thus not very precise, but it does involve simple devices and is thus fairly cheap. On–off control is not restricted to mechanical switches such as bimetallic strips or relays; rapid switching can be achieved with the use of thyristor circuits (see Section 7.2.2); such a circuit might be used for controlling the speed of a motor, and operational amplifiers.

With the two-step method of control, the controller output is either an on or an off signal, regardless of the magnitude of the error. With the *proportional mode*, the size of the controller output is proportional to the size of the error. This means the correction element of the control system, e.g. a valve, will receive a signal which is proportional to the size of the correction required.

Figure 13.5 shows how the output of such a controller varies with the size and sign of the error. The linear relationship between controller output and error tends to exist only over a certain range of errors, this range being called the *proportional band*. Within the proportional band the equation of the straight line can be represented by

$$\text{Change in controller output from set point} = K_P e$$

where e is the error and K_P a constant. K_P is thus the gradient of the straight line in Figure 13.5.

The controller output is generally expressed as a percentage of the full range of possible outputs within the proportional band. This output can then correspond to, say, a correction valve changing from fully closed to fully open. Similarly, the error is expressed as a percentage of the full-range value, i.e. the error range corresponding to the 0 to 100% controller output. Thus

% change in controller output from set point
$= K_P \times$ % change in error

Hence, since 100% controller output corresponds to an error percentage equal to the proportional band

$$K_P = \frac{100}{\text{proportional band}}$$

We can rewrite the equation as

change in output $= I_{out} - I_0 = K_P e$

where I_0 is the controller output percentage at zero error, I_{out} the output percentage at percentage error e. Thus taking Laplace transforms:

Change in output $(s) = K_P E(s)$

and so, since

$$\text{Transfer function} = \frac{\text{change in output } (s)}{E(s)}$$

K_P is, within the proportional band, the transfer function of the controller.

Generally a 50% controller output is chosen to be the output when the error is zero. Thus, in the case of the controller being used to control a valve which allows water into a tank, when the error is zero the valve will be half (50%) open. This will give the normal flow rate. Any error will then increase or reduce the flow rate at a value which depends on the size of the error. The result will be to return the error to its zero value and the controller to a 50% output. Suppose the process has the flow of liquid into a tank being controlled and for some reason a new set point is required for the flow rate. We can talk of this change in terms of there being a step input to the control system. This new set value could require the correcting valve to be kept open at a higher percentage, say 60%. This cannot be achieved by the zero error setting but requires a permanent error setting called an *offset* (Fig. 13.6). The size of this offset is directly proportional to the size of the load changes and inversely proportional to the K_P, a

Fig. 13.6 Offset

higher value of K_P giving a steeper gradient in Figure 13.6 and so a smaller error change needed to accommodate a load change.

The proportional mode of control tends to be used in processes where the transfer function K_P can be made large enough to reduce the offset to an acceptable level. However, the larger the transfer function the greater the chance of the system oscillating and becoming unstable.

To illustrate the above discussion of proportional control, consider a proportional controller which is to be used to control the height of water in a tank where the water level can vary from zero to 9.0 m. What proportional band and transfer function will be required if the required height of water is 5.0 m and the controller is to fully close a valve when the water rises to 5.5 m and fully open it when the water falls to 4.5 m? When the error is −0.5 m the controller output must be 100% open and when +0.5 m it must be 0% open. The proportional band must therefore extend from a height error of −0.5 m to one of +0.5 m. Expressed as a percentage, the proportional band extends from

$$-(0.5/9.0) \times 100 = -5.6\% \text{ to } +(0.5/9.0) \times 100 = +5.6\%$$

The proportional band is thus 11.2%. Note that if we work in percentages for the controller, we must work in percentages for the error. This value of proportional band will thus mean a transfer function K_P of $(100\%)/(11.2\%) = 8.9$.

As an illustration of offset error, consider a proportional controller which has a transfer function of 15 and a set point of 50% output. It outputs to a valve which at the set point allows a flow of 200 m³/s. The valve changes its output by 4 m³/s for each per cent change in controller output. What will be the controller output and the offset error when the flow has to be changed to 240 m³/s? The new controller setting, as a percentage, for a flow change from 200 to 240 m³/s is 40/4 = 10% change from 50 to 60%. Hence

$$K_P = 15 = \frac{60 - 50}{e}$$

Thus the offset is $e = 0.67\%$.

13.4.1 Electronic proportional controller

A summing operational amplifier with an inverter can be used as a proportional controller (Fig. 13.7). For a summing amplifier we have (see Section 3.2.3)

$$V_{out} = -R_f\left(\frac{V_0}{R_2} + \frac{V_e}{R_1}\right)$$

Fig. 13.7   Proportional controller

The input to the summing amplifier through R_2 is the zero error voltage value V_0, i.e. the set value, and the input through R_1 is the error signal V_e. But when the feedback resistor $R_f = R_2$, then the equation becomes

$$V_{out} = -\frac{R_2}{R_1}V_e - V_0$$

If the output from the summing amplifier is then passed through an inverter, i.e. an operational amplifier with a feedback resistance equal to the input resistance, then

$$V_{out} = \frac{R_2}{R_1}V_e + V_0$$

$$V_{out} = K_P V_e + V_0$$

where K_P is the proportionality constant. The result is a proportional controller.

As an illustration, Figure 13.8 shows an example of a proportional control system for the control of the temperature of the liquid in a container as liquid is pumped through it.

Fig. 13.8   Proportional controller
for temperature control

13.4.2 System response

With proportional control we have a gain element with transfer function K_P in series with the forward-path element $G(s)$ (Fig. 13.9). The error is thus:

$$E(s) = \frac{K_P G(s)}{1 + K_P G(s)} R(s)$$

and so, for a step input, the steady-state error is:

$$e_{SS} = \lim_{s \to 0} sE(s) = \lim_{s \to 0} \left[s \frac{1}{1 + 1/K_P G(s)} \frac{1}{s} \right]$$

Fig. 13.9 System with proportional control

Unity feedback

This will have a finite value and so there is always a steady-state error. Low values of K_P give large steady-state errors but stable responses. High values of K_P give smaller steady-state errors but a greater tendency to instability.

13.5 Derivative control

With the *derivative mode* of control the change in controller output from the set point value is proportional to the rate of change with time of the error signal. This can be represented by the equation

$$I_{out} - I_0 = K_D \frac{de}{dt}$$

where I_0 is the set point output value, I_{out} the output value that will occur when the error e is changing at the rate de/dt. It is usual to express these controller outputs as a percentage of the full range of output and the error as a percentage of full range. K_D is the constant of proportionality.

The transfer function is obtained by taking Laplace transforms, thus

$$(I_{out} - I_0)(s) = K_D sE(s)$$

Hence the transfer function is $K_D s$.

With derivative control, as soon as the error signal begins to change there can be quite a large controller output since it is proportional to the rate of change of the error signal and not its value. Rapid initial responses to error signals thus occur. Figure

Fig. 13.10   Derivative control

13.10 shows the controller output that results when there is a constant rate of change of error signal with time. The controller output is constant because the rate of change is constant and occurs immediately the deviation occurs. Derivative controllers do not, however, respond to steady-state error signals, since with a steady error the rate of change of error with time is zero. Because of this derivative control is always combined with proportional control; the proportional part gives a response to all error signals, including steady signals, while the derivative part responds to the rate of change.

To illustrate the above, consider a derivative controller which has a set point of 50% and derivative constant K_D of 0.4 s. What will be the controller output when the error (a) changes at 1%/s, (b) is constant at 4%? Using the equation given above

$$I_{out} = K_D \frac{de}{dt} + I_0 = 0.4 \times 1 + 50 = 50.4\%$$

With de/dt zero, then I_{out} equals I_0, i.e. 50%. The output only differs from the set point value when the error is changing.

Figure 13.11 shows the form of an electronic derivative controller circuit, the circuit involving an operational amplifier connected as a differentiator circuit followed by another operational amplifier connected as an inverter. The derivative time K_D is R_2C.

13.5.1 Proportional plus derivative control

Derivative control is never used alone because it is not capable of giving an output when there is a steady error signal and so no correction is possible. It is thus invariably used in conjunction with proportional control so that this problem can be resolved.

With proportional plus derivative control the change in controller output from the set point value is given by

$$\text{Change in output from set point} = K_P e + K_D \frac{de}{dt}$$

Fig. 13.11   Derivative controller

Hence

$$I_{\text{out}} = K_{\text{P}}e + K_{\text{D}}\frac{de}{dt} + I_0$$

where I_0 is the output at the set point, I_{out} the output when the error is e, K_{P} is the proportionality constant and K_{D} the derivative constant, de/dt is the rate of change of error. The system has a transfer function given by

$$(I_{\text{out}} - I_0)(s) = K_{\text{P}}E(s) + K_{\text{D}}sE(s)$$

Hence the transfer function is $K_{\text{P}} + K_{\text{D}}s$. This is often written as:

$$\text{transfer function} = K_{\text{D}}\left(s + \frac{1}{T_{\text{D}}}\right)$$

Fig. 13.12 PD control

where $T_{\text{D}} = K_{\text{D}}/K_{\text{P}}$ and is the *derivative time constant*.

Figure 13.12 shows how the controller output can vary when there is a constantly changing error. There is an initial quick change in controller output because of the derivative action followed by the gradual change due to proportional action. This form of control can thus deal with fast process changes; however, a change in set value will require an offset error (see earlier discussion of proportional control).

To illustrate the above, consider what the controller output will be for a proportional plus derivative controller (a) initially and (b) 2 s after the error begins to change from the zero error at the rate of 1.2%/s. The controller has a set point of 50%, $K_{\text{P}} = 4$ and $K_{\text{D}} = 0.4$ s. Using the equation given above

$$I_{\text{out}} = K_{\text{P}}e + K_{\text{D}}\frac{de}{dt} + I_0$$

Initially the error e is zero. Hence, initially when the error begins to change

$$I_{\text{out}} = 0 + 0.4 \times 1.2 + 50 = 50.48\%$$

Because the rate of change is constant, after 2 s the error will have become 2.4%. Hence, then

$$I_{\text{out}} = 4 \times 2.4 + 0.4 \times 1.2 + 50 = 59.08\%$$

13.6 Integral control

The *integral mode* of control is one where the rate of change of the control output I is proportional to the input error signal e.

$$\frac{dI}{dt} = K_{\text{I}}e$$

K_I is the constant of proportionality and, when the controller output is expressed as a percentage and the error as a percentage, has units of s^{-1}. Integrating the above equation gives

$$\int_{I_0}^{I_{out}} dI = \int_0^t K_I e \, dt$$

$$I_{out} - I_0 = \int_0^t K_I e \, dt$$

I_0 is the controller output at zero time, I_{out} is the output at time t.

The transfer function is obtained by taking the Laplace transform. Thus

$$(I_{out} - I_0)(s) = \frac{1}{s} K_I E(s)$$

and so

$$\text{Transfer function} = \frac{1}{s} K_I$$

Figure 13.13 illustrates the action of an integral controller when there is a constant error input to the controller. We can consider the graphs in two ways. When the controller output is constant the error is zero; when the controller output varies at a constant rate the error has a constant value. The alternative way of considering the graphs is in terms of the area under the error graph.

$$\text{Area under the error graph between } t = 0 \text{ and } t = \int_0^t e \, dt$$

Thus up to the time when the error occurs the value of the integral is zero. Hence $I_{out} = I_0$. When the error occurs it maintains a constant value. Thus the area under the graph is increasing as the time increases. Since the area increases at a constant rate the controller output increases at a constant rate.

To illustrate the above, consider an integral controller with a value of K_I of 0.10 /s and an output of 40% at the set point. What will be the output after times of (a) 1 s, (b) 2 s, if there is a sudden change to a constant error of 20%? Using the equation developed above

$$I_{out} - I_0 = \int_0^t K_I e \, dt$$

When the error does not vary with time the equation becomes

$$I_{out} = K_I e t + I_0$$

Thus for (a) when $t = 1$ s,

$$I_{out} = 0.10 \times 20 \times 1 + 40 = 42\%$$

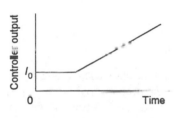

Fig. 13.13   Integral control

For (b) when $t = 2$ s,

$I_{out} = 0.10 \times 20 \times 2 + 40 = 44\%$

Figure 13.14 shows the form of the circuit used for an electronic integral controller. It consists of an operational amplifier connected as an integrator and followed by another operational amplifier connected as a summer to add the integrator output to that of the controller output at zero time. K_I is $1/R_1C$.

Integrator Summing amplifier

V_e

V_0

V_{out}

Fig. 13.14 Integral controller

13.6.1 Proportional plus integral control

The integral mode of control is not usually used alone but is frequently used in conjunction with the proportional mode. When integral action is added to a proportional control system the controller output I_{out} is given by

$$I_{out} = K_P e + K_I \int e \, dt + I_0$$

where K_P is the proportional control constant, K_I the integral control constant, I_{out} the output when there is an error e and I_0 the output at the set point when the error is zero. The transfer function is thus:

$$\text{transfer function} = K_P + \frac{K_I}{s} = \frac{K_P}{s}\left(s + \frac{1}{T_I}\right)$$

where $T_I = K_P/K_I$ and is the *integral time constant*.

Figure 13.15 shows how the system reacts when there is an abrupt change to a constant error. The error gives rise to a proportional controller output which remains constant since the error does not change. There is then superimposed on this a steadily increasing controller output due to the integral action.

Suppose there is a change in the controller set point from, say, 50 to 60%. With just a proportional controller this can only be done by having an offset error, i.e. an error value other than zero for the set point value. However, with the combination of integral

Elements
due to

I

P

Fig. 13.15 PI control

and proportional control this is not the case. The integral part of the control can provide a change in controller output without any offset error. The controller can be said to reset its set point. Figure 13.16 shows the effects of the proportional action and the integral action if we create an error signal which is increased from the zero value and then decreased back to it again. With proportional action alone the controller mirrors the change and ends up back at its original set point value. With the integral action the controller output increases in proportion to the way the area under the error–time graph increases and since, even when the error has reverted back to zero, there is still a value for the area there is a change in controller output which continues after the error has ceased. The result of combining the proportional and integral actions, i.e. adding the two separate graphs, is thus a change in controller output without an offset error. A step input to the control system can thus give a steady-state value with no error.

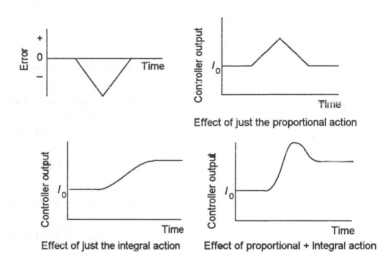

Fig. 13.16   PI control

Effect of just the proportional action

Effect of just the integral action

Effect of proportional + integral action

Because of the lack of an offset error, this type of controller can be used where there are large changes in the process variable. However, because the integration part of the control takes time the changes must be relatively slow to prevent oscillations. Another disadvantage of this form of control is that when the process is started up with controller output at 100%, e.g. with a liquid level control the initial condition may be an empty tank and so the error is so large that the controller has to give a 100% output to fully open a valve, the integral action causes a considerable overshoot of the error before finally settling down.

13.7 PID controller

Combining all three modes of control (proportional, integral and derivative) enables a controller to be produced which has no offset error and reduces the tendency for oscillations. Such a controller is known as a *three-mode controller* or *PID controller*. The equation describing its action can be written as

$$I_{out} = K_Pe + K_I \int e\, dt + K_D \frac{de}{dt} + I_0$$

where I_{out} is the output from the controller when there is an error e which is changing with time t, I_0 is the set point output when there is no error, K_P is the proportionality constant, K_I the integral constant and K_D the derivative constant. One way of considering a three-mode controller is as a proportional controller which has integral control to eliminate the offset error and derivative control to reduce time lags. Taking the Laplace transform gives:

$$(I_{out} - I_0)(s) = K_PE(s) + \frac{1}{s}K_IE(s) + sK_D(s)$$

and so:

$$\text{Transfer function} = K_Pe + \frac{1}{s}K_I + sK_D = K_p\left(1 + \frac{1}{T_{I}s} + T_Ds\right)$$

To illustrate the above, consider what will be the controller output of a three-mode controller having K_P as 4, K_I as 0.6 /s, K_D as 0.5 s, a set point output of 50% and, subject to the error change shown in Figure 13.17, (a) immediately the change starts to occur and (b) 2 s after it starts. Using the equation given above for I_{out}

$$I_{out} = K_Pe + K_I \int e\, dt + K_D \frac{de}{dt} + I_0$$

we have for (a) $e = 0$, $de/dt = 1$ /s, and $\int e\, dt = 0$. Thus

$$I_{out} = 0 + 0 + 0.5 \times 1 + 50 = 50.5\%$$

For (b) we have, at 2 s, $e = 1\%$, $\int e\, dt = 1.5$ s since the integral is the area under the error–time graph up to 2 s, and $de/dt = 0$. Thus

$$I_{out} = 4 \times 1 + 0.6 \times 1.5 + 0 + 50 = 54.9\%$$

13.7.1 Operational amplifier PID circuits

A three-mode controller can be produced by combining the various circuits described earlier in this chapter for the separate proportional, derivative and integral modes. A more practical controller can, however, be produced with a single operational amplifier. Figure 13.18 shows one such circuit. The proportional constant K_p is $R_I/(R + R_D)$, the derivative constant K_D is R_DC_D and the integral constant K_I is $1/R_IC_I$.

Fig. 13.17   Error signal

Fig. 13.18   PID circuit

Fig. 13.19   Digital closed-loop control system

13.8 Digital controllers

Figure 13.19 shows the basis of a direct digital control system that can be used with a continuous process; the term *direct digital control* is used when the digital controller, basically a microprocessor, is in control of the closed-loop control system. The controller receives inputs from sensors, executes control programs and provides the output to the correction elements. Such controllers require inputs which are digital, process the information in digital form and give an output in digital form. Since many control systems have analogue measurements an analogue-to-digital converter (ADC) is used for the inputs. A clock supplies a pulse at regular time intervals and dictates when samples of the controlled variable are taken by the ADC. These samples are then converted to digital signals which are compared by the microprocessor with the set point value to give the error signal. The microprocessor can then initiate a control mode to process the error signal and give a digital output. The control mode used by the microprocessor is determined by the program of instructions used by the microprocessor for processing the digital signals, i.e. the *software*. The digital output, generally after processing by a digital-to-analogue converter since correcting elements generally require analogue signals, can be used to initiate the correcting action.

A digital controller basically operates the following cycle of events:

1   Samples the measured value.
2   Compares it with the set value and establishes the error.
3   Carries out calculations based on the error value and stored values of previous inputs and outputs to obtain the output signal.
4   Sends the output signal to the DAC.
5   Waits until the next sample time before repeating the cycle.

Microprocessors as controllers have the advantage over analogue controllers that the form of the controlling action, e.g. proportional or three mode, can be altered by purely a change in the computer software. No change in hardware or electrical wiring is required. Indeed the control strategy can be altered by the computer program during the control action in response to the developing situation.

They also have other advantages. With analogue control, separate controllers are required for each process being controlled. With a microprocessor many separate processes can be controlled by sampling processes with a multiplexer (see Chapter 3). Digital control gives better accuracy than analogue control because the amplifiers and other components used with analogue systems change their characteristics with time and temperature and so show drift, while digital control, because it operates on signals in only the on-off mode, does not suffer from drift in the same way.

13.8.1 Implementing control modes

In order to produce a digital controller which will give a particular mode of control it is necessary to produce a suitable program for the controller. The program has to indicate how the digital error signal at a particular instant is to be processed in order to arrive at the required output for the following correction element. The processing can involve the present input together with previous inputs and previous outputs. The program is thus asking the controller to carry out a difference equation (see Section 3.10).

The transfer function for a PID analogue controller is

$$\text{Transfer function} = K_\text{P} + \frac{1}{s}K_\text{I} + sK_\text{D}$$

Multiplication by s is equivalent to differentiation. We can, however, consider the gradient of the time response for the error signal at the present instant of time as being (latest sample of the error e_n minus the last sample of the error e_{n-1})/(sampling interval T_s) (Fig. 13.20). Division by s is equivalent to integration. We can, however, consider the integral of the error at the end of a sampling period as being the area under the error–time graph during the last sampling interval plus the sum of the areas under the graph for all previous samples (Int_prev). If the sampling period is short relative to the times involved then the area during the last sampling interval is approximately $\frac{1}{2}(e_n + e_{n-1})/T_s$ (see Section 3.10 for another approximation known as Tustin's approximation). Thus we can write for the controller output x_n at a particular instant the equivalent of the transfer function as:

Fig. 13.20　Error signals

$$x_n = K_\text{P}e_n + K_\text{I}\left(\frac{(e_n + e_{n-1})T_s}{2} + \text{Int}_\text{prev}\right) + K_\text{D}\frac{e_n - e_{n-1}}{T_s}$$

We can rearrange this equation to give:

$$x_n = Ae_n + Be_{n-1} + C(\text{Int}_{\text{prev}})$$

where $A = K_P + 0.5K_IT_s + K_D/T_s$, $B = 0.5K_IT_s - K_D/T_s$ and $C = K_I$.
The program for PID control thus becomes:

1 Set the values of K_P, K_I and K_D.
2 Set the initial values of e_{n-1}, Int_{prev} and the sample time T_s.
3 Reset the sample interval timer.
4 Input the error e_n.
5 Calculate y_n using the above equation.
6 Update, ready for the next calculation, the value of the previous area to $\text{Int}_{\text{prev}} + 0.5(e_n + e_{n-1})T_s$.
7 Update, ready for the next calculation, the value of the error by setting e_{n-1} equal to e_n.
8 Wait for the sampling interval to elapse.
9 Go to 3 and repeat the loop.

13.8.2 A computer control system

Typically a computer control system consists of the elements shown in Figure 13.19 with set points and control parameters being entered from a keyboard. The software for use with the system will provide the program of instructions needed, for example, for the computer to implement the PID control mode, provide the operator display, recognise and process the instructions inputted by the operator, provide information about the system, provide start-up and shut-down instructions, and supply clock/calendar information. An operator display is likely to show such information as the set point value, the actual measured value, the sampling interval, the error, the controller settings and the state of the correction element. The display is likely to be updated every few seconds.

For a more detailed discussion of computer control systems the reader is referred to specialist texts, e.g. *Real-time Computer Control* by S. Bennett (Prentice-Hall 1994).

13.9 Control system performance

The transfer function of a control system is affected by the mode chosen for the controller. Hence the response of the system to, say, a step input is affected. Consider the simple system shown in Figure 13.21. With proportional control the transfer function of the forward path is $K_PG(s)$ and so the transfer function of the feedback system $G(s)$ is

$$G(s) = \frac{K_PG_p(s)}{1 + K_PG_p(s)}$$

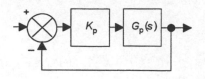

Fig. 13.21   System with proportional control

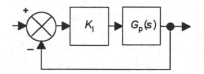

Fig. 13.22   System with integral control

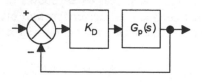

Fig. 13.23   System with derivative control

Suppose we have a process which is first order with a transfer function of $1/(\tau s + 1)$ where τ is the time constant. With proportional control, and unit feedback, the transfer function of the control system becomes

$$G(s) = \frac{K_P/(\tau s + 1)}{1 + K_P/(\tau s + 1)} = \frac{K_P}{\tau s + 1 + K_P}$$

The control system remains a first-order system. The proportional control has had the effect of just changing the form of the first-order response of the process. Without the controller the response to a unit step input was (see Section 11.2)

$$y = 1 - e^{-t/\tau}$$

Now it is

$$y = K_P(1 - e^{-t/(\tau/1 + K_P)})$$

The effect of the proportional control has to been to reduce the time constant from τ to $\tau/(1 + K_P)$.

With integral control (Fig. 13.22) we have a forward-path transfer function of $K_I G_p(s)/s$ and so the system transfer function is

$$G(s) = \frac{K_I G_p(s)}{s + K_I G_p(s)}$$

Thus, now if we have a process which is first order with a transfer function of $1/(\tau s + 1)$, with proportional control and unit feedback the transfer function of the control system becomes

$$G(s) = \frac{K_I/(\tau s + 1)}{s + K_I/(\tau s + 1)}$$

$$= \frac{K_I}{s(\tau s + 1) + K_I} = \frac{K_I}{\tau s^2 + s + K_I}$$

The control system is now a second-order system. Now, with a step input, the system will give a second-order response instead of a first-order response.

With a system having derivative control (Fig. 13.23) the forward path transfer function is $sK_D G(s)$ and so, with unity feedback, the system transfer function is

$$G(s) = \frac{sK_D G_p(s)}{1 + K_D G_p(s)}$$

With a process which is first order with a transfer function of $1/(\tau s + 1)$, derivative control gives an overall transfer function of

$$G(s) = \frac{sK_D/(\tau s + 1)}{1 + sK_D/(\tau s + 1)} = \frac{sK_D}{\tau s + 1 + sK_D}$$

13.10 Controller tuning

The term *tuning* is used to describe the process of selecting the best controller settings. With a proportional controller this means selecting the value of K_P; with a PID controller the three constants K_P, K_I and K_D have to be selected. There are a number of methods of doing this; here just two methods will be discussed, both by Ziegler and Nichols. They assumed that when the controlled system is open loop a reasonable approximation to its behaviour is a first-order system with a built-in time delay. Based on this, they then derived parameters for optimum performance.

13.10.1 Process reaction method

The process control loop is opened, generally between the controller and the correction unit, so that no control action occurs. A test input signal is then applied to the correction unit and the response of the controlled variable determined. The test signal should be as small as possible. Figure 13.24 shows the form of test signal and a typical response. The test signal is a step signal with a step size expressed as the percentage change P in the correction unit. The graph of the measured variable plotted against time is called the *process reaction curve*. The measured variable is expressed as the percentage of the full-scale range.

A tangent is drawn to give the maximum gradient of the graph. For Figure 13.24 the maximum gradient R is M/T. The time between the start of the test signal and the point at which this tangent intersects the graph time axis is termed the lag L. Table 13.1 gives the criteria recommended by Ziegler and Nichols for control settings based on the values of P, R and L.

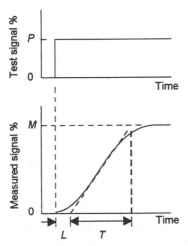

Fig. 13.24 Process reaction curve

Table 13.1 Process reaction curve criteria

Control mode	K_P	T_I	T_D
P	P/RL		
PI	$0.9P/RL$	$3.33L$	
PID	$1.2P/RL$	$2L$	$0.5L$

Consider the following example. Determine the settings required for a three-mode controller which gave the process reaction curve shown in Figure 13.25 when the test signal was a 6% change in the control valve position. Drawing a tangent to the maximum gradient part of the graph gives a lag L of 150 s and a gradient R of 5/300 = 0.017 /s. Hence

$$K_P = \frac{1.2P}{RL} = \frac{1.2 \times 6}{0.017 \times 150} = 2.82$$

Fig. 13.25   Process curve example

$$T_I = 2L = 300 \text{ s}$$

$$T_D = 0.5L = 0.5 \times 150 = 75 \text{ s}$$

13.10.2 Ultimate cycle method

With this method, integral and derivative actions are first reduced to their minimum values. The proportional constant K_P is set low and then gradually increased. This is the same as saying that the proportional band is gradually made narrower. While doing this small disturbances are applied to the system. This is continued until continuous oscillations occur. The critical value of the proportional constant K_{Pc} at which this occurs is noted and the periodic time of the oscillations T_c measured. Table 13.2 shows how the Ziegler and Nichols recommended criteria for controller settings are related to this value of K_{Pc}. The critical proportional band is $100/K_{Pc}$.

Table 13.2   Ultimate cycle criteria

Control mode	K_P	T_I	T_D
P	$0.5K_{Pc}$		
PI	$0.45K_{Pc}$	$T_c/1.2$	
PID	$0.6K_{Pc}$	$T_c/2.0$	$T_c/8$

Consider the following example. When tuning a three-mode control system by the ultimate cycle method it was found that oscillations begin when the proportional band is decreased to 30%. The oscillations have a periodic time of 500 s. What are the suitable settings for the controller? The critical value of K_{Pc} is 100/critical proportional band and is therefore 100/30 = 3.33. Then, using the criteria given in Table 13.2,

$$K_P = 0.6K_{Pc} = 0.6 \times 3.33 = 2.0$$

$$T_1 = T_c/2.0 = 500/2 = 2.5 \text{ s}$$

$$T_D = T_c/8 = 500/8 = 62.5 \text{ s}$$

13.11 Velocity control

Consider the problem of controlling the movement of a load by means of a motor. Because the motor system is likely to be second-order, proportional control will lead to the system output taking time to reach the required displacement when there is, say, a step input to the system and may oscillate for a while about the required value. Time will thus be taken for the system to respond to an input signal. A higher speed of response, with fewer oscillations, can be obtained by using PD rather than just P control. There is, however, an alternative of achieving the same effect and this is by the use of a second feedback loop which gives a measurement related to the rate at which the displacement is changing. This is termed *velocity feedback*. Figure 13.26 shows such a system; the velocity feedback might involve the use of a tachogenerator giving a signal proportional to the rotational speed of the motor shaft and hence the rate at which the displacement is changing and the displacement might be monitored using a rotary potentiometer.

13.12 Adaptive control

There are many control situations where the parameters of the plant change with time or, perhaps, load, e.g. a robot manipulator being used to move loads when the load is changed. If the transfer function of the plant changes then retuning of the system is desirable for the optimum values to be determined for proportional, derivative and integral constants. For the control systems so far considered, it has been assumed that the system once tuned retains its values of proportional, derivative and integral constants until the operator decides to retune. The alternative to this is an *adaptive control system* which 'adapts' to changes and changes its parameters to fit the circumstances prevailing.

The adaptive control system is based on the use of a microprocessor as the controller. Such a device enables the control mode and the control parameters used to be adapted to fit the circumstances, modifying them as the circumstances change.

An adaptive control system can be considered to have three stages of operation:

1 Starts to operate with controller conditions set on the basis of an assumed condition.
2 The desired performance is continuously compared with the actual system performance.
3 The control system mode and parameters are automatically and continuously adjusted in order to minimise the difference between the desired and actual system performance.

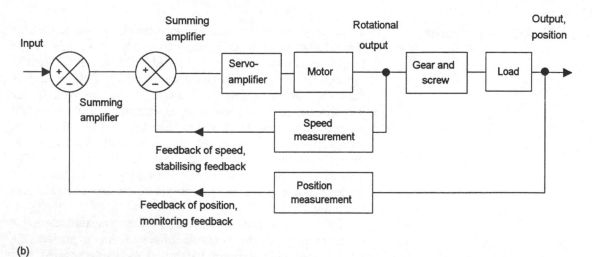

Fig. 13.26 System with velocity feedback: (a) descriptive diagram of the system, (b) block diagram of the system

For example, with a control system operating in the proportional mode, the proportional constant K_P may be automatically adjusted to fit the circumstances, changing as they do.

Adaptive control systems can take a number of forms. Three commonly used forms are:

1 Gain-scheduled control.
2 Self-tuning.
3 Model-reference adaptive systems.

13.12.1 Gain-scheduled control

With *gain-scheduled control* or, as it is sometimes referred to, *pre-programmed adaptive control*, preset changes in the parameters of the controller are made on the basis of some auxiliary measurement of some process variable. Figure 13.27 illustrates this method. The term *gain-scheduled control* was used because the only parameter originally adjusted was the gain, i.e. the proportionality constant K_P.

Fig. 13.27　　Gain-scheduled control

For example, for a control system used to control the positioning of some load, the system parameters could be worked out for a number of different load values and a table of values loaded into the memory of the controller. A load cell might then be used to measure the actual load and give a signal to the controller indicating a mass value which is then used by the controller to select the appropriate parameters.

A disadvantage of this system is that the control parameters have to be determined for many operating conditions so that the controller can select the one to fit the prevailing conditions. An advantage, however, is that the changes in the parameters can be made quickly when the conditions change.

13.12.2 Self-tuning

With *self-tuning control* the system continuously tunes its own parameters based on monitoring the variable that the system is controlling and the output from the controller. Figure 13.28 illustrates the features of this system.

Fig. 13.28   Self-tuning

Self-tuning is often found in commercial PID controllers, it generally then being referred to as *auto-tuning*. When the operator presses a button, the controller injects a small disturbance into the system and measures the response. This response is compared to the desired response and the control parameters adjusted, by a modified Ziegler–Nichols rule, to bring the actual response closer to the desired response.

13.12.3 Model-reference adaptive systems

With the *model-reference* system an accurate model of the system is developed. The set value is then used as an input to both the actual and the model systems and the difference between the actual output and the output from the model compared. The difference in these signals is then used to adjust the parameters of the controller to minimise the difference. Figure 13.29 illustrates the features of the system.

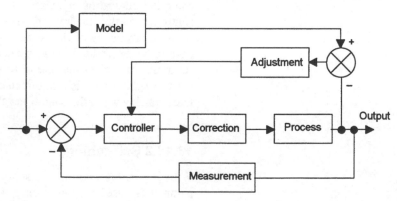

Fig. 13.29   Model-referenced control

For more details on adaptive control, the reader is referred to specialist texts such as *Adaptive Control* by K.J. Åstrom and B. Wittenmark (Addison-Wesley 1989, 1995).

Problems

1 What are the limitations of two-step (on–off) control and in what situation is such a control system commonly used?

2 A two-position mode controller switches on a room heater when the temperature falls to 20°C and off when it reaches 24°C. When the heater is on, the air in the room increases in temperature at the rate of 0.5°C per minute; when the heater is off it cools at 0.2°C per minute. If the time lags in the control system are negligible, what will be the times taken for (a) the heater switching on to off, (b) the heater switching off to on?

3 A two-position mode controller is used to control the water level in a tank by opening or closing a valve which in the open position allows water at the rate of 0.4 m³/s to enter the tank. The tank has a cross-sectional area of 12 m² and water leaves it at the constant rate of 0.2 m³/s. The valve opens when the water level reaches 4.0 m and closes at 4.4 m. What will be the times taken for (a) the valve opening to closing, (b) the valve closing to opening?

4 A proportional controller is used to control the height of water in a tank where the water level can vary from zero to 4.0 m. The required height of water is 3.5 m and the controller is to fully close a valve when the water rises to 3.9 m and fully open it when the water falls to 3.1 m. What proportional band and transfer function will be required?

5 A proportional controller has K_P of 20 and a set point of 50% output. Its output is to have a valve which at the set point allows a flow of 2.0 m³/s. The valve changes its output in direct proportion to the controller output. What will be the controller output and the offset error when the flow has to be changed to 2.5 m³/s?

6 A derivative controller has a set point of 50% and derivative constant K_D of 0.5 s. The error starts at zero and then changes at 2%/s for 3 s before becoming constant for 2 s, after which it decreases at 1%/s to zero. What will be the controller output at (a) 0 s, (b) 1 s, (c) 4 s, (d) 6 s?

7 An integral controller has a set point of 50% and a value of K_I of 0.10 /s. The error starts at zero and changes at 4%/s for 2 s before becoming constant for 3 s. What will be the output after times of (a) 1 s, (b) 3 s?

8 A three-mode controller has K_P as 2, K_I as 0.1 /s, K_D as 1.0 s, and a set point output of 50%. The error starts at zero and changes at 5%/s for 2 s before becoming constant for 3 s. It then decreases at 2%/s to zero and remains at zero. What will be the controller output at (a) 0 s, (b) 3 s, (c) 7 s?

9 Describe and compare the characteristics of (a) proportional control, (b) proportional plus integral control, (c) proportional plus integral plus derivative control.

10 Determine the settings of K_P, T_I and T_D required for a three-mode controller which gave a process reaction curve with a

lag L of 200 s and a gradient R of 0.010%/s when the test signal was a 5% change in the control valve position.

11 When tuning a three-mode control system by the ultimate cycle method it was found that oscillations began when the proportional band was decreased to 20%. The oscillations had a periodic time of 200 s. What are the suitable values of K_P, T_I and T_D?

12 Explain the basis on which the following forms of adaptive control systems function: (a) gain-scheduled, (b) self-tuning, (c) model-reference.

13 A d.c. motor behaves like a first-order system with a transfer function of relating output position to which it has rotated a load to input signal of $1/s(1 + s\tau)$. If the time constant τ is 1 s and the motor is to be used in a closed-loop control system with unity feedback and a proportional controller, determine the value of the proportionality constant which will give a closed-loop response with a 25% overshoot.

14 The small ultrasonic motor used to move the lens for automatic focusing with a camera (see Section 22.3.3) drives the ring with so little inertia that the transfer function relating angular position with input signal is represented by $1/cs$, where c is the constant of proportionality relating the frictional torque to angular velocity. If the motor is to be controlled by a closed-loop system with unity feedback, what type of behaviour can be expected if proportional control is used?

14   Digital logic

14.1 Digital logic

Many control systems are concerned with setting events in motion or stopping them when certain conditions are met. For example, with the domestic washing machine, the heater is only switched on when there is water in the drum and it is to the prescribed level. Such control involves *digital* signals where there are only two possible signal levels. Digital circuitry is the basis of digital computers and microprocessor controlled systems.

This circuitry evolved from the transistor circuits being able to output at one of two voltage levels depending on the levels at its inputs. The two levels, usually 5 V and 0 V are the high and low signals and represented by 1 and 0. The *binary numbering system* involves just the numbers 0 and 1 and is thus widely used with such digital circuitry. These two levels of 0 and 1 may represent levels of on or off, open or closed, yes or no, true or false, +5 V or 0 V, etc.

With *digital control* we might, for example, have the water input to the domestic washing machine switched on if we have both the door to the machine closed and a particular time in the operating cycle has been reached. There are two input signals which are either yes or no signals and an output signal which is a yes or no signal. The controller is here programmed to only give a yes output if both the input signals are yes, i.e. if input A and input B are both 1 then there is an output of 1. Such an operation is said to be controlled by a *logic gate*, in this example an AND gate. There are many machines and processes which are controlled in this way.

The term *combinational logic* is used for the combining of two or more basic logic gates to form a required function. For example, a requirement might be that a buzzer sounds in a car if the key is in the ignition and a door is opened or if the headlights are on and the driver's door is opened.

In addition to a discussion of combinational logic, this chapter also includes a discussion of *sequential logic*. Such digital circuitry is used to exercise control in a specific sequence dictated by a control clock or enable–disable control signals. These are combinational logic circuits with memory.

14.2 Number systems

The *decimal system* is based on the use of 10 symbols or digits: 0, 1, 2, 3, 4, 5, 6, 7, 8, 9. When a number is represented by this system, the digit position in the number indicates that the weight attached to each digit increases by a factor of 10 as we proceed from right to left.

...	10^3	10^2	10^1	10^0
	thousands	hundreds	tens	units

The *binary system* is based on just two symbols or states: 0 and 1. These are termed *binary digits* or *bits*. When a number is represented by this system, the digit position in the number indicates that the weight attached to each digit increases by a factor of 2 as we proceed from right to left.

...	2^3	2^2	2^1	2^0
	bit 3	bit 2	bit 1	bit 0

For example, the decimal number 15 in the binary system is 1111. In a binary number the bit 0 is termed the *least significant bit* (LSB) and the highest bit the *most significant bit* (MSB).

The *octal system* is based on eight digits: 0, 1, 2, 3, 4, 5, 6, 7. When a number is represented by this system, the digit position in the number indicates that the weight attached to each digit increases by a factor of 8 as we proceed from right to left.

...	8^3	8^2	8^1	8^0

For example, the decimal number 15 in the octal system is 17.

The *hexadecimal system* is based on 16 digits/symbols: 0, 1, 2, 3, 4, 5, 6, 7, 8, 9, A, B, C, D, E, F. When a number is represented by this system, the digit position in the number indicates that the weight attached to each digit increases by a factor of 16 as we proceed from right to left.

...	16^3	16^2	16^1	16^0

For example, the decimal number 15 is F in the hexadecimal system. This system is generally used in the writing of programs for microprocessor-based systems since it represents a very compact method of entering data.

The *binary coded decimal system* (BCD) is a widely used system with computers. Each decimal digit is coded separately in binary. For example, the decimal number 15 in BCD is 0001 0101. This code is useful for outputs from microprocessor-based systems where the output has to drive decimal displays, each decimal digit in the display being supplied by the microprocessor with its own binary code.

Table 14.1 gives examples of numbers in the decimal, binary, BCD, octal and hexadecimal systems.

Table 14.1 Number systems

Decimal	Binary	BCD	Octal	Hexadecimal
0	0000	0000 0000	0	0
1	0001	0000 0001	1	1
2	0010	0000 0010	2	2
3	0011	0000 0011	3	3
4	0100	0000 0100	4	4
5	0101	0000 0101	5	5
6	0110	0000 0110	6	6
7	0111	0000 0111	7	7
8	1000	0000 1000	10	8
9	1001	0000 1001	11	9
10	1010	0001 0000	12	A
11	1011	0001 0001	13	B
12	1100	0001 0010	14	C
13	1101	0001 0011	15	D
14	1110	0001 0100	16	E
15	1111	0001 0101	17	F

14.2.1 Binary mathematics

Addition of binary numbers follows the following rules:

$0 + 0 = 0$

$0 + 1 = 1 + 0 = 1$

$1 + 1 = 10$ i.e. $0 + $ carry 1

$1 + 1 + 1 = 11$ i.e. $1 + $ carry 1

In decimal numbers the addition of 14 and 19 gives 33. In binary numbers this addition becomes

Augend	01110
Addend	10111
Sum	100001

For bit 0, 0 + 1 = 1. For bit 1, 1 + 1 = 10 and so we have 0 with 1 carried to the next column. For bit 3, 1 + 0 + carried 1 = 10. For bit 4, 1 + 0 + carried 1 = 10. We continue this through the various bits and end up with the sum plus a carry 1. The final number is thus 100001. When adding binary numbers A and B to give C, i.e. $A + B = C$, then A is termed the *augend*, B the *addend* and C the *sum*.

Subtraction of binary numbers follows the following rules:

$$0 - 0 = 0$$

$$1 - 0 = 1$$

$$1 - 1 = 0$$

$$0 - 1 = 10 - 1 + \text{borrow} = 1 + \text{borrow}$$

When evaluating 0 − 1, a 1 is borrowed from the next column on the left containing a 1. The following example illustrates this. In decimal numbers the subtraction of 14 from 27 gives 13.

Minuend	11011
Subtrahend	01110
Difference	01101

For bit 0 we have 1 − 0 = 1. For bit 1 we have 1 − 1 = 0. For bit 2 we have 0 − 1. We thus borrow 1 from the next column and so have 10 − 1 = 1. For bit 3 we have 0 − 1; remember we borrowed the 1. Again borrowing 1 from the next column, we then have 10 − 1 = 1. For bit 4 we have 0 − 0 = 0; remember we borrowed the 1. When subtracting binary numbers A and B to give C, i.e. we have $A - B = C$, then A is termed the *minuend*, B the *subtrahend* and C the *difference*.

The subtraction of binary numbers is more easily carried out electronically when an alternative method of subtraction is used. The subtraction example above can be considered to be the addition of a positive number and a negative number. The following techniques indicate how we can specify negative numbers and so turn subtraction into addition. It also enables us to deal with negative numbers in any circumstances.

The numbers used so far are referred to as *unsigned*. This is because the number itself contains no indication whether it is negative or positive. A number is said to be *signed* when the most significant bit is used to indicate the sign of the number, a 0 being used if the number is positive and a 1 if it is negative. When we have a positive number then we write it in the normal way with a 0 preceding it. Thus a positive binary number of 10010 would be written as 010010. A negative number of 10010 would be written as 110010. However, this is not the most useful way of representing negative numbers for ease of manipulation by computers.

A more useful way of representing negative numbers is to use the twos complement method. A binary number has two complements, known as the *ones complement* and the *twos complement*. The ones complement of a binary number is obtained by changing all the 1s in the unsigned number into 0s and the 0s into 1s. The twos complement is then obtained by adding 1 to the ones complement. When we have a negative number then we obtain the twos complement and then sign it with a 1, the positive number being signed by a 0. Consider the representation of the decimal number −3 as a signed twos complement number. We first write the binary number for the unsigned 3 as 0011, then obtain the ones complement of 1100, add 1 to give the unsigned twos complement of 1101, and finally sign it with a 1 to indicate it is negative. The result is thus 11101. The following is another example, the signed twos complement being obtained as an 8-bit number for −6.

Table 14.2 Signed numbers

Denary number	Signed number	
+127	0111 1111	Just the
etc.		binary
+6	0000 0110	number
+5	0000 0101	signed with
+4	0000 0101	a 0
+3	0000 0011	
+2	0000 0010	
+1	0000 0001	
+0	0000 0000	
−1	1111 1111	The twos
−2	1111 1110	complement
−3	1111 1101	signed with
−4	1111 1100	a 1
−5	1111 1011	
−6	1111 1010	
etc.		
−127	1000 0000	

Unsigned binary number	000 0110
Ones complement	111 1001
Add 1	1
Unsigned twos complement	111 1010
Signed twos complement	1111 1010

When we have a positive number then we write it in the normal way with a 0 preceding it. Thus a positive binary number of 100 1001 would be written as 01001001. Table 14.2 shows some examples of numbers on this system.

Subtraction of a positive number from a positive number involves obtaining the signed twos complement of the subtrahend and then adding it to the signed minuend. Hence, for the subtraction of the decimal number 6 from the decimal number 4 we have

Signed minuend	0000 0100
Subtrahend, signed twos complement	1111 1010
Sum	1111 1110

The most significant bit of the outcome is 1 and so the result is negative. This is the signed twos complement for −2.

Consider another example, the subtraction of 43 from 57. The signed positive number of 57 is 0011 1001. The signed twos complement for −43 is given by:

Unsigned binary number for 43	010 1011
Ones complement	101 0100
Add 1	1
Unsigned twos complement	101 0101
Signed twos complement	1101 0101

Thus we obtain by the addition of the signed positive number and the signed twos complement number:

Signed minuend	0011 1001
Subtrahend, signed twos complement	1101 0101
Sum	0000 1110 + carry 1

The carry 1 is ignored. The result is thus 0000 1110 and since the most significant bit is 0 the result is positive. The result is the decimal number 14.

If we wanted to add two negative numbers then we would obtain the signed twos complement for each number and then add them. Whenever a number is negative we use the signed twos complement, when positive just the signed number.

14.2.2 Floating numbers

In the decimal number system, large numbers such as 120 000 are often written in *scientific notation* as 1.2×10^5 or perhaps 120×10^3 and small numbers such as 0.000 120 as 1.2×10^{-4} rather than as a number with a fixed location for the decimal point. Numbers in this form of notation are written in terms of 10 raised to some power. Likewise we can use such notation for binary numbers but with them written in terms of 2 raised to some power. For example, we might have 1010 written as 1.010×2^3 or perhaps 10.10×2^2. Because the binary point can be moved to different locations by a choice of the power to which the 2 is raised, this notation is termed *floating point*.

A floating point number is in the form:

$$a \times r^e$$

where a is termed the *mantissa*, r the *radix* or *base* and e the *exponent* or *power*. With binary numbers the base is understood to be 2, i.e. we have $a \times 2^e$.

The advantage of using floating-point numbers is that, compared with fixed-point representation, a much wider range of numbers can be represented by a given number of digits.

Because with floating point numbers it is possible to store a number in a number of different ways, e.g. 0.1×10^2 and $0.01 \times$

10^3, with computing systems such numbers are *normalised*, i.e. they are all put in the form $0.1 \times r^e$. Hence, with binary numbers we have 0.1×2^e and so if we had 0.00001001 it would become 0.1001×2^{-4}. In order to take account of the sign of a binary number we then add a sign bit of 0 for a positive number and 1 for a negative number. Thus the number 0.1001×2^{-4} becomes 1.1001×2^{-4} if negative and 0.1001×2^{-4} if positive.

If we want to add 2.01×10^3 and 10.2×10^2 we have to make the power (the term *exponent* is generally used) the same for each. Thus we can write $2.01 \times 10^3 + 1.02 \times 10^3$. We can then add them digit by digit, taking account of any carry, to give 2.03×10^3. We adopt a similar procedure for binary floating-point numbers. Thus if we want to add 0.101100×2^4 and 0.111100×2^2 we first adjust them to have the same exponents, e.g. 0.101100×2^4 and 0.001111×2^4, and then add them digit by digit to give 0.111011×2^4.

Likewise for subtraction, digit-by-digit subtraction of floating-point numbers can only occur between two numbers when they have the same exponent. Thus 0.1101100×2^{-4} minus 0.1010100×2^{-5} can be written as $0.01010100 \times 2^{-4} - 0.101010 \times 2^{-4}$ and the result given as 0.1000010×2^{-4}.

14.2.3 Gray code

Consider two successive numbers in binary code 0001 and 0010 (denary 2 and 3); two bits have changed in the code group in going from one number to the next. Thus if we had, say, an absolute encoder (see Section 2.3.7) and assigned successive positions to successive binary numbers then two changes have to be made in this case. This can present problems in that both changes must be made at exactly the same instant; if one occurs fractionally before the other then there can momentarily be another number indicated. Thus in going from 0001 to 0010 we might momentarily have 0011 or 0000. Thus an alternative method of coding is likely to be used.

The *Gray code* is such a code, only one bit in the code group changes in going from one number to the next. The Gray code is unweighted in that the bit positions in the code group do not have any specific weight assigned to them. It is thus not suited to arithmetic operations but is widely used for input–output devices such as absolute encoders. Table 14.3 lists decimal numbers and their values in the binary code and in Gray code.

14.2.4 Parity method for error detection

The movement of digital data from one location to another can result in transmission errors, the receiver not receiving the same signal as transmitted by the transmitter as a result of electrical noise in the transmission process. Sometimes a noise pulse may be large enough at some point to alter the logic level of the signal.

Table 14.3 Gray code

Decimal number	Binary code	Gray code
0	0000	0000
1	0001	0001
2	0010	0011
3	0011	0010
4	0100	0110
5	0101	0111
6	0110	0101
7	0111	0100
8	1000	1100
9	1001	1101
10	1010	1111
11	1011	1110
12	1100	1010
13	1101	1011
14	1110	1001
15	1111	1000

For example, the sequence 1001 may be transmitted and be received as though 1101. In order to detect such errors a *parity bit* is often used. A parity bit is an extra 0 or 1 bit attached to a code group at transmission. In the *even-parity* method the value of the bit is chosen so that the total number of 1s in the code group, including the parity bit, is an even number. For example, in transmitting 1001 the parity bit used would be 0 to give 01001 and so an even number of 1s. In transmitting 1101 the parity bit used would be 1 to give 11101 and so an even number of 1s. With *odd parity* the parity bit is chosen so that the total number of 1s, including the parity bit, is odd. Thus if at the receiver the number of 1s in a code group does not give the required parity, the receiver will know that there is an error and can request the code group be retransmitted.

An extension of the parity check is the *sum check* in which blocks of code may be checked by sending a series of bits representing their binary sum. Parity and sum checks can only detect single errors in blocks of code, double errors go undetected. Also the error is not located so that correction by the receiver can be made. Multiple-error detection techniques and methods to pinpoint errors have been devised (see Section 21.3) and texts such as *Audio, Video and Data Telecommunications* by D. Peterson (McGraw-Hill 1992) explain these in more detail.

14.3 Logic gates

Logic gates are the basic building blocks for digital electronic circuits.

14.3.1 AND gate

Suppose we have a gate giving a high output only when both input A and input B are high, for all other conditions it gives a low output. This is an AND logic gate. We can visualise the AND gate as an electrical circuit involving two switches in series (Fig. 14.1). Only when switch A and switch B are closed is there a current.

Fig. 14.1   AND gate representation

An example of an AND gate is an interlock control system for a machine tool such that if the safety guard is in place and gives a 1 signal and the power is on, giving a 1 signal, then there can be an output, a 1 signal, and the machine operates. Another example is a burglar alarm in which it gives an output, the alarm sounding, when the alarm is switched on and when a door is opened to activate a sensor.

The relationships between inputs to a logic gate and the outputs can be tabulated in a form known as a *truth table*. This specifies the relationships between the inputs and outputs. Thus for an AND gate with inputs A and B and a single output Q, we will have a 1 output when, and only when, $A = 1$ and $B = 1$. All other combinations of A and B will generate a 0 output. We can thus write the truth table as:

Fig. 14.2 AND gate

Inputs		Output
A	B	Q
0	0	0
0	1	0
1	0	0
1	1	1

Consider what happens when we have two digital inputs which are functions of time, as in Fig. 14.2. There will only be an output from the AND gate when each of the inputs is high and thus the output is as shown in the figure.

We can express the relationship between the inputs and the outputs of an AND gate in the form of an equation, termed a *Boolean equation*. The Boolean equation for the AND gate is written as:

$$A \cdot B = Q$$

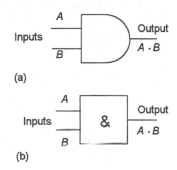

Fig. 14.3 Standard symbols for AND gates

Different sets of standard circuit symbols for logic gates have been used with the main form being that originated in the United States; an international standard form (IEEE/ANSI), however, has now been developed; this removes the distinctive shape and uses a rectangle with the logic function written inside it. Figure 14.3(a) shows the United States form of symbol used for an AND gate and (b) shows the new standardised form, the & symbol indicating AND. Both forms will be used in this book.

14.3.2 OR gate

Fig. 14.4 OR gate representation

An OR gate with inputs A and B gives an output of a 1 when A or B is 1. We can visualise such a gate as an electrical circuit involving two switches in parallel (Fig. 14.4). When switch A or B is closed then there is a current. OR gates can also have more than two inputs. The truth table for the gate is:

Inputs		Output
A	B	Q
0	0	0
0	1	1
1	0	1
1	1	1

We can write the Boolean equation for an OR gate as:

Fig. 14.5   Symbols for OR gate

Fig. 14.6   NOT gate

(a)

(b)

Fig. 14.7   Symbols for NOT gate

$$A + B = Q$$

The symbols used for an OR gate are shown in Figure 14.5; the use of a greater than or equal to 1 sign to depict OR arises from the OR function being true if at least more than one input is true.

14.3.3 NOT gate

A NOT gate has just one input and one output, giving a 1 output when the input is 0 and a 0 output when the input is 1. The NOT gate gives an output which is the inversion of the input and is called an *inverter*. Thus if we have a digital input which varies with time, as in Figure 14.6, the out variation with time is the inverse.

The following is the truth table for the NOT gate:

Input A	Output Q
0	1
1	0

The Boolean equation describing the NOT gate is:

$$\overline{A} = Q$$

A bar over a symbol is used to indicate that the inverse, or complement, is being taken; thus the bar over the A indicates that the output Q is the inverse value of A. Figure 14.7 shows the symbols used for a NOT gate. The 1 representing NOT actually symbolises logic identity, i.e. no operation, and the inversion is depicted by the circle on the output.

14.3.4 NAND gate

The NAND gate can be considered as a combination of an AND gate followed by a NOT gate (Fig. 14.8(a)). Thus when input A is 1 and input B is 1 there is an output of 0, all other inputs giving an output of 1.

The NAND gate is just the AND gate truth table with the outputs inverted. An alternative way of considering the gate is as an AND gate with a NOT gate applied to invert both the inputs before they reach the AND gate. Figure 14.8(b) shows the symbols used for the NAND gate, being the AND symbol followed by the circle to indicate inversion. The following is the truth table:

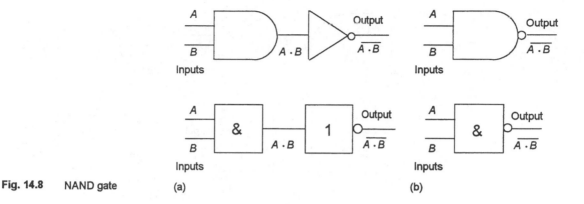

Fig. 14.8   NAND gate   (a)   (b)

| Inputs | | Output |
A	B	Q
0	0	1
0	1	1
1	0	1
1	1	0

Fig. 14.9   NAND gate

The Boolean equation describing the NAND gate is:

$$\overline{A \cdot B} = Q$$

Figure 14.9 shows the output that occurs for a NAND gate when its two inputs are digital signals which vary with time. There is only a low output when both the inputs are high.

14.3.5 NOR gate

The NOR gate can be considered as a combination of an OR gate followed by a NOT gate (Fig. 14.7(a)). Thus when input A or input B is 1 there is an output of 0. It is just the OR gate with the outputs inverted. An alternative way of considering the gate is as an OR gate with a NOT gate applied to invert both the inputs before they reach the OR gate. Figure 14.10(b) shows the symbols used for the NOR gate; it is the OR symbol followed by the circle to indicate inversion. The Boolean equation for the NOR gate is:

$$\overline{A + B} = Q$$

The following is the truth table for the NOR gate:

Fig. 14.10   NOR gate

(a)   (b)

| Inputs | | Output |
A	B	Q
0	0	1
0	1	0
1	0	0
1	1	0

14.3.6 XOR gate

The EXCLUSIVE-OR gate (XOR) can be considered to be an OR gate with a NOT gate applied to one of the inputs to invert it before the inputs reach the OR gate (Fig. 14.11(a)). Alternatively it can be considered as an AND gate with a NOT gate applied to one of the inputs to invert it before the inputs reach the AND gate. The symbols are shown in Figure 14.11(b); the =1 depicts that the output is true if only one input is true. The following is the truth table:

(a)

(b)

Fig. 14.11   XOR gate

Inputs		Output
A	B	Q
0	0	0
0	1	1
1	0	1
1	1	0

14.3.7 Combining gates

It might seem that to make logic systems we require a range of gates. However, as the following shows, we can make up all the gates from just one. Consider the combination of three NOR gates shown in Figure 14.12. The truth table, with the intermediate and final outputs, is as follows:

A	B	C	D	Q
0	0	1	1	0
0	1	1	0	0
1	0	0	1	0
1	1	0	0	1

The result is the same as an AND gate. If we followed this assembly of gates by a NOT gate then we would obtain a truth table the same as a NAND gate.

A combination of three NAND gates is shown in Figure 14.13. The truth table, with the intermediate and final outputs, is as follows:

A	B	C	D	Q
0	0	1	1	0
0	1	1	0	1
1	0	0	1	1
1	1	0	0	1

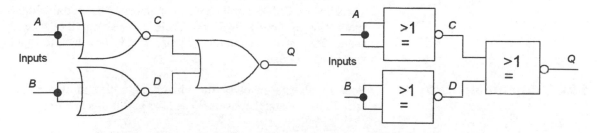

Fig. 14.12 Three NOR gates

Fig. 14.13 Three NAND gates

(a)

(b)

Fig. 14.14   Integrated circuits:
(a) 7408, (b) 7402

The result is the same as an OR gate. If we followed this assembly of gates by a NOT gate then we would obtain a truth table the same as a NOR gate.

The above two illustrations of gate combinations show how one type of gate, a NOR or a NAND, can be used to substitute for other gates, provided we use more than one gate. Gates can also be combined to make complex gating circuits and sequential circuits.

Logic gates are available as integrated circuits. The different manufacturers have standardised their numbering schemes so that the basic part numbers are the same regardless of the manufacturer. For example, Figure 14.14(a) shows the gate systems available in integrated circuit 7408; it has four two-input AND gates and is supplied in a 14-pin package. Power supply connections are made to pins 7 and 14, these supplying the operating voltage for all the four AND gates. In order to indicate at which end of the package pin 1 starts, a notch is cut between pins 1 and 14. Integrated circuit 7411 has three AND gates with each having three inputs; integrated circuit 7421 has two AND gates with each having four inputs. Figure 14.14(b) shows the gate systems available in integrated circuit 7402. This has four two-input NOR gates in a 14-pin package, power connections being to pins 7 and 14. Integrated circuit 7427 has three gates with each having three inputs; integrated circuit 7425 has two gates with each having four inputs.

The above integrated circuits, with their 74xx numbers are *transistor-transistor logic (TTL) circuits* and are based on the use of transistors and basically operate between the 0 and 5 V levels. The standard CMOS family have the numbers 40xx and the high-speed CMOS family 74HCxx. Table 14.4 shows the general characteristics of the families; Figure 14.15 showing the input and output voltages.

14.4 Boolean algebra

Boolean algebra involves the binary digits 1 and 0 and the operations $\cdot$, $+$ and the inverse. The laws of this algebra are:

1   Anything ORed with itself is equal to itself: $A + A = A$.

Table 14.4 General characteristics of TTL and CMOS families

		TTL		CMOS	
Supply voltage		4.75–5.25 V		5–15 V	
Max. supply current		−100 mA		−0.02 mA	
		Input	Output	Input	Output
0 State	Voltage	0.8 V	0.5 V	1.5 V	0.05 V
	Current	−0.4 mA	8 mA	−0.0001 mA	0.5 mA
1 State	Voltage	2.0 V	2.7 V	3.5 V	4.95 V
	Current	0.02 mA	−0.4 mA	0.0001 mA	−0.2 mA
Max. operating frequency		33 MHz		10 MHz	
Active power consumption		8 mW		0.1 mW	

Input voltage (V)   Output voltage (V)

(a)

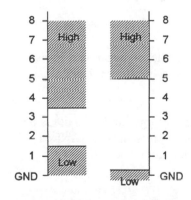

Input voltage (V)   Output voltage (V)

(b)

Fig. 14.15   Defining high and low:
(a) TTL, (b) CMOS

2   Anything ANDed with itself is equal to itself: $A \cdot A = A$.

3   It does not matter in which order we consider inputs for OR and AND gates, e.g.

$$A + B = B + A \quad \text{and} \quad A \cdot B = B \cdot A$$

4   As the following truth table indicates:

$$A + (B \cdot C) = (A + B) \cdot (A + C)$$

A	B	C	$B \cdot C$	$A + B \cdot C$	$A + B$	$A + C$	$(A + B) \cdot (A + C)$
0	0	0	0	0	0	0	0
0	0	1	0	0	0	1	0
0	1	0	0	0	1	0	0
0	1	1	1	1	1	1	1
1	0	0	0	1	1	1	1
1	0	1	0	1	1	1	1
1	1	0	0	1	1	1	1
1	1	1	1	1	1	1	1

5   Likewise we can use a truth table to show that we can treat bracketed terms in the same way as in ordinary algebra, e.g.

$$A \cdot (B + C) = A \cdot B + A \cdot C$$

6   Anything ORed with its own inverse equals 1:

$$A + \overline{A} = 1$$

7   Anything ANDed with its own inverse equals 0:

$$A \cdot \overline{A} = 0$$

8   Anything ORed with a 0 is equal to itself; anything ORed with a 1 is equal to 1. Thus $A + 0 = A$ and $A + 1 = 1$.

9   Anything ANDed with a 0 is equal to 0; anything ANDed with a 1 is equal to itself. Thus $A \cdot 0 = 0$ and $A \cdot 1 = A$.

As an illustration of the use of the above to simplify Boolean expressions, consider simplifying:

$$(A + B) \cdot \overline{C} + A \cdot C$$

Using item 5 for the first term gives:

$$A \cdot \overline{C} + B \cdot \overline{C} + A \cdot C$$

We can regroup this and use item 6 to give:

$$A \cdot (\overline{C} + C) + B \cdot \overline{C} = A \cdot 1 + B \cdot \overline{C}$$

Hence, using item 9 the simplified expression becomes:

$$A + B \cdot \overline{C}$$

14.4.1 De Morgan laws

As illustrated above, the laws of Boolean algebra can be used to simplify Boolean expressions. In addition we have what are known as the *De Morgan* laws:

1   The inverse of the outcome of ORing A and B, is the same as when the inverses of A and B are separately ANDed. The following truth table shows the validity of this:

$$\overline{A + B} = \overline{A} \cdot \overline{B}$$

A	B	$A + B$	$\overline{A+B}$	$\overline{A}$	$\overline{B}$	$\overline{A} \cdot \overline{B}$
0	0	0	1	1	1	1
0	1	1	0	1	0	0
1	0	1	0	0	1	0
1	1	1	0	0	0	0

2   The inverse of the outcome of ANDing A and B is the same as when the inverses of A and B are separately ORed. The following truth table shows the validity of this:

$$\overline{A \cdot B} = \overline{A} + \overline{B}$$

A	B	$A \cdot B$	$\overline{A \cdot B}$	$\overline{A}$	$\overline{B}$	$\overline{A} + \overline{B}$
0	0	0	1	1	1	1
0	1	0	1	1	0	1
1	0	0	1	0	1	1
1	1	1	0	0	0	0

As an illustration of the use of a De Morgan law, consider the simplification of the logic circuit shown in Figure 14.16.

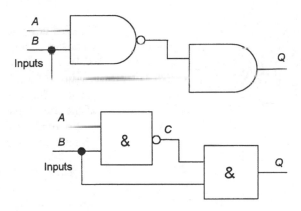

Fig. 14.16   Circuit simplification

The Boolean equation for the output in terms of the input is:

$$Q = \overline{A \cdot B} \cdot B$$

Applying the second law from above gives:

$$Q = (\overline{A} + \overline{B}) \cdot B$$

We can write this as:

$$Q = \overline{A} \cdot B + \overline{B} \cdot B = \overline{A} \cdot B + 0 = \overline{A} \cdot B$$

Hence the simplified circuit is as shown in Figure 14.17.

Fig. 14.17   Circuit simplification

(a)

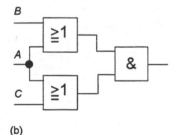

(b)

Fig. 14.18 (a) Sum of products,
(b) product of sums

14.4.2 Boolean function generation from truth tables

Given a situation where the requirements of a system can be specified in terms of a truth table, how can a logic gate system using the minimum number of gates be devised to give that truth table?

Boolean algebra can be used to manipulate switching functions into many equivalent forms, some of which take many more logic gates than others; the form, however, to which most are minimised is AND gates driving a single OR gate or vice versa. Two AND gates driving a single OR gate (Fig. 14.18(a)) give:

$$A \cdot B + A \cdot C$$

This is termed the *sum of products* form. For two OR gates driving a single AND gate (Fig. 14.18(b)), we have:

$$(A + B) \cdot (A + C)$$

This is known as the *product of sums* form.

Thus in considering what minimum form might fit a given truth table, the usual procedure is to find the sum of products or the product of sums form that fits the data. Generally the sum of products form is used. The procedure used is to consider each row of the truth table in turn and find the product that would fit a row. The overall result is then the sum of all these products.

Suppose we have a row in a truth table of:

$$A = 1, B = 0 \text{ and output } Q = 1$$

When A is 1 and B is not 1 then the output is 1, thus the product which fits this is:

$$Q = A \cdot \overline{B}$$

We can repeat this operation for each row of a truth table, as the following table indicates.

A	B	Output	Products
0	0	0	$\overline{A} \cdot \overline{B}$
0	1	0	$\overline{A} \cdot B$
1	0	1	$A \cdot \overline{B}$
1	1	0	$A \cdot B$

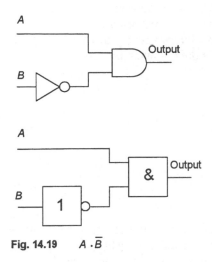

Fig. 14.19 $A \cdot \overline{B}$

However, only the row of the truth table that has an output of 1 need be considered, since the rows with 0 output do not contribute to the final expression; the result is thus:

$$Q = A \cdot \overline{B}$$

The logic gate system that will give this truth table is thus that shown in Figure 4.19.

As a further example, consider the following truth table, only the products terms giving a 1 output being included:

A	B	C	Output	Products
0	0	0	1	$\overline{A} \cdot \overline{B} \cdot \overline{C}$
0	0	1	0	
0	1	0	1	$\overline{A} \cdot B \cdot \overline{C}$
0	1	1	0	
1	0	0	0	
1	0	1	0	
1	1	0	0	
1	1	1	0	

Thus the sum of products which fits this table is:

$$Q = \overline{A} \cdot \overline{B} \cdot \overline{C} + \overline{A} \cdot B \cdot \overline{C}$$

This can be simplified to give:

$$Q = \overline{A} \cdot \overline{C} \cdot (\overline{B} + B) = \overline{A} \cdot \overline{C}$$

The truth table can thus be generated by just a NAND gate.

14.5 Karnaugh maps

The *Karnaugh map* is a graphical method that can be used to produce simplified Boolean expressions from sums of products obtained from truth tables. The truth table has a row for the value of the output for each combination of input values. With two input variables there are four lines in the truth table, with three input variables six lines and with four input variables 16 lines. Thus with two input variables there are four product terms, with three input variables there are six and with four input variables 16. The Karnaugh map is drawn as a rectangular array of cells, with each cell corresponding to a particular product value. Thus with two input variables there are four cells, with three input variables six cells and with four input variables 16 cells. The output values for the rows are placed in their cells in the

Fig. 14.20   Two input variable map

Fig. 14.21   Two input variable map

Karnaugh map, though it is usual to only indicate the 1 output values and leave the cells having 0 output as empty.

Figure 14.20 shows the map for two input variables. The cells are given the output values for the following products:

the upper left cell $\overline{A} \cdot \overline{B}$,

the lower left cell $A \cdot \overline{B}$,

the upper right cell $\overline{A} \cdot B$,

the lower right cell $A \cdot B$

The arrangement of the map squares is such that horizontally adjacent squares differ only in one variable and, likewise, vertically adjacent squares differ in only one variable. Thus horizontally with our two-variable map the variables differ only in A and vertically they differ only in B.

For the following truth table, if we put the values given for the products in the Karnaugh map, only indicating where a cell has a 1 value and leaving blank those with a 0 value, then the map shown in Figure 14.21 is obtained.

A	B	Output	Products
0	0	0	$\overline{A} \cdot \overline{B}$
0	1	0	$\overline{A} \cdot B$
1	0	1	$A \cdot \overline{B}$
1	1	0	$A \cdot B$

Because the only 1 entry is in the lower right square, the truth table can be represented by the Boolean expression:

output $= A \cdot \overline{B}$

As a further example, consider the following truth table:

A	B	Output	Products
0	0	0	$\overline{A} \cdot \overline{B}$
0	1	0	$\overline{A} \cdot B$
1	0	1	$A \cdot \overline{B}$
1	1	1	$A \cdot B$

Fig. 14.22   Two input variable map

Fig. 14.23   Three input variable map

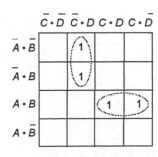

Fig. 14.24   Four-variable map

It gives the Karnaugh map shown in Figure 14.22. This has an output given by:

$$\text{output} = A \cdot \overline{B} + A \cdot B$$

We can simplify this to:

$$A \cdot \overline{B} + A \cdot B = A \cdot (\overline{B} + B) = A$$

When two cells containing a 1 have a common vertical edge we can simplify the Boolean expression to just the common variable. We can do this by inspection of a map, indicating which cell entries can be simplified by drawing loops round them, as in Figure 14.22.

Figure 14.23 shows the Karnaugh map for the following truth table, it having three input variables. As before we can use looping to simplify the resulting Boolean expression to just the common variable. The result is:

$$\text{output} = \overline{A} \cdot \overline{C}$$

A	B	C	Output	Products
0	0	0	1	$\overline{A} \cdot \overline{B} \cdot \overline{C}$
0	0	1	0	$\overline{A} \cdot \overline{B} \cdot C$
0	1	0	1	$\overline{A} \cdot B \cdot \overline{C}$
0	1	1	0	$\overline{A} \cdot B \cdot C$
1	0	0	0	$A \cdot \overline{B} \cdot \overline{C}$
1	0	1	0	$A \cdot \overline{B} \cdot C$
1	1	0	0	$A \cdot B \cdot \overline{C}$
1	1	1	0	$A \cdot B \cdot C$

Figure 14.24 shows the Karnaugh map for the following truth table, it having four input variables. Looping simplifies the resulting Boolean expression to give:

$$\text{output} = \overline{A} \cdot \overline{C} \cdot D + A \cdot B \cdot C$$

The above represents just some simple examples of Karnaugh maps and the use of looping. Note that in looping, adjacent cells can be considered to be those in the top and bottom rows of the left- and right-hand columns. Think of opposite edges of the map being joined together. Looping a pair of adjacent 1s in a map eliminates the variable that appears in complemented and uncomplemented form. Looping a quad of adjacent 1s eliminates the two variables that appear in both complemented and uncomplemented form. Looping an octet of adjacent 1s eliminates the three variables that appear in both complemented and uncomplemented form.

A	B	C	D	Output	Products
0	0	0	0	0	
0	0	0	1	1	$\overline{A} \cdot \overline{B} \cdot \overline{C} \cdot D$
0	0	1	0	0	
0	0	1	1	0	
0	1	0	0	0	
0	1	0	1	1	$\overline{A} \cdot B \cdot \overline{C} \cdot D$
0	1	1	0	0	
0	1	1	1	0	
1	0	0	0	0	
1	0	0	1	0	
1	0	1	0	0	
1	0	1	1	0	
1	1	0	0	0	
1	1	0	1	0	
1	1	1	0	1	$A \cdot B \cdot C \cdot \overline{D}$
1	1	1	1	1	$A \cdot B \cdot C \cdot D$

As a further illustration, consider an automated machine that will only start when two of three sensors A, B and C give signals. The following truth table fits this requirement and Figure 14.25 shows the resulting three-variable Karnaugh diagram.

A	B	C	Output	Products
0	0	0	0	
0	0	1	0	
0	1	0	0	
0	1	1	1	$\overline{A} \cdot B \cdot C$
1	0	0	0	
1	0	1	1	$A \cdot \overline{B} \cdot C$
1	1	0	1	$A \cdot B \cdot \overline{C}$
1	1	1	1	$A \cdot B \cdot C$

Fig. 14.25 Map for machine

The Boolean expression which fits the map and thus describes the outcome from the machine is:

$$\text{Outcome} = A \cdot B + B \cdot C + A \cdot C$$

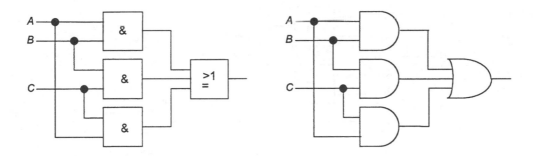

Fig. 14.26 Automated machine

Figure 14.26 shows the logic gates that could be used to generate this Boolean expression. $A \cdot B$ describes an AND gate for the inputs A and B. Likewise $B \cdot C$ and $A \cdot C$ are two more AND gates. The $+$ signs indicate that the outputs from the three AND gates are then the inputs to an OR gate.

In some logic systems there are some input variable combinations for which outputs are not specified. They are termed 'don't care states' When entering these on a Karnaugh map, the cells can be set to either 1 or 0 in such a way that the output equations can be simplified.

14.6 Applications of logic gates

The following are some examples of the uses of logic gates for a number of simple applications.

14.6.1 Parity generators

In Section 14.2.4 the use of parity bits as an error detection method was discussed. A single bit is added to each code block to force the number of 1s in the block, including the parity bit, to be an odd number if odd parity is being used or an even number if even parity is being used.

Figure 14.27 shows a logic gate circuit that could be used to determine and add the appropriate parity bit. The system employs XOR gates; with an XOR gate if all the inputs are 0 or all are 1 the output is 0, and if the inputs are not equal the output is a 1. Pairs of bits are checked and an output of 1 given if they are not equal. If odd parity is required the bias bit is 0, if even parity it is 1. The appropriate bias bit can then be added to the signal for transmission. The same circuit can be used to check the parity at the receiver, with the final output being a 1 when there is an error. Such circuits are available as integrated circuits.

Fig. 14.27 Parity bit generator

14.6.2 Digital comparator

A digital comparator is used to compare two digital words to determine if they are exactly equal. The two words are compared bit by bit and a 1 output given if the words are equal. To compare the equality of two bits an XOR gate can be used; if the bits are both 0 or both 1 the output is 0, and if they are not equal the output is a 1. To obtain a 1 output when the bits are the same we need to add a NOT gate, this combination of XOR and NOT being termed an XNOR gate. To compare each of the pairs of bits in two words we need an XNOR gate for each pair. If the pairs are made up of the same bits then the output from each XNOR gate is a 1. We can then use an AND gate to give a 1 output when all the XNOR outputs are 1s. Figure 14.28 shows the system.

Digital comparators are available as integrated circuits and can generally not only determine if two words are equal but which one is greater than the other. For example, the 7485 4-bit magnitude comparator compares two 4-bit words A and B, giving a 1 output from pin 5 if A is greater than B, a 1 output from pin 6 if A equals B and a 1 output from pin 7 if A is less than B.

Fig. 14.28   Comparator

Fig. 14.29   Decoder

14.6.3 Code converter

In many applications there is a need to change data from one type of code to another. For example, the output from a microprocessor system might be BCD and need to be transformed into a suitable code to drive a seven-segment display. The term *data decoding* is used for the process of converting some code group, e.g. BCD, binary, hex, into an individual active output representing that group. A decoder has n binary input lines for the coded input of an n-bit word and gives m output lines such that only one line is activated for one possible combination of inputs, i.e. only one output line gives an output for a particular word input code. For example, a BCD-to-decimal decoder has a 4-bit input code and 10 output lines so that a particular BCD input will give rise to just one of the output lines being activated and so indicating a particular decimal number with each output line corresponding to a decimal number (Fig. 14.29).

Thus, in general, a *decoder* is a logic circuit that looks at its inputs, determines which number is there, and activates the one output that corresponds to that number. Decoders are widely used in microprocessor circuits.

Decoders can have the active output high and the inactive ones low or the active output low and the inactive ones high. For active-high output a decoder can be assembled from AND gates while for active-low output NAND gates can be used. Figure 14.30 shows how a BCD-to-decimal decoder for active-low output can be assembled and the resulting truth table. Such a decoder is readily available as an integrated circuit, e.g. 74LS145.

A decoder that is widely used is BCD-to-seven, e.g. 74LS244, for taking a 4-bit BCD input and giving an output to drive the seven segments of a display.

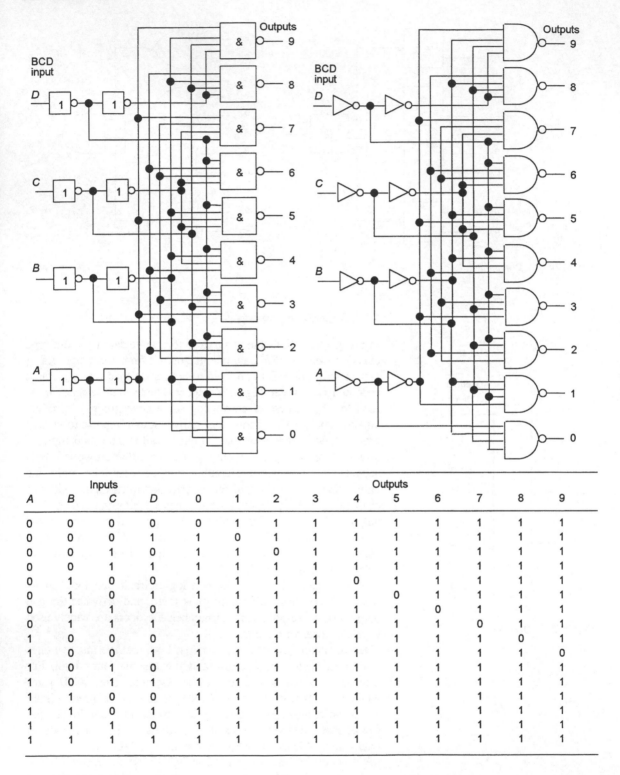

	Inputs			Outputs									
A	B	C	D	0	1	2	3	4	5	6	7	8	9
0	0	0	0	0	1	1	1	1	1	1	1	1	1
0	0	0	1	1	0	1	1	1	1	1	1	1	1
0	0	1	0	1	1	0	1	1	1	1	1	1	1
0	0	1	1	1	1	1	1	1	1	1	1	1	1
0	1	0	0	1	1	1	1	0	1	1	1	1	1
0	1	0	1	1	1	1	1	1	0	1	1	1	1
0	1	1	0	1	1	1	1	1	1	0	1	1	1
0	1	1	1	1	1	1	1	1	1	1	0	1	1
1	0	0	0	1	1	1	1	1	1	1	1	0	1
1	0	0	1	1	1	1	1	1	1	1	1	1	0
1	0	1	0	1	1	1	1	1	1	1	1	1	1
1	0	1	1	1	1	1	1	1	1	1	1	1	1
1	1	0	0	1	1	1	1	1	1	1	1	1	1
1	1	0	1	1	1	1	1	1	1	1	1	1	1
1	1	1	0	1	1	1	1	1	1	1	1	1	1
1	1	1	1	1	1	1	1	1	1	1	1	1	1

Fig. 14.30 BCD-to-decimal decoder: 1 = HIGH, 0 = LOW

The term *3-line-to-8-line decoder* is used where a decoder has three input lines and eight output lines. It takes a 3-bit binary number and activates the one of the eight outputs corresponding to that number. Figure 14.31 shows how such a decoder can be realised from logic gates and its truth table.

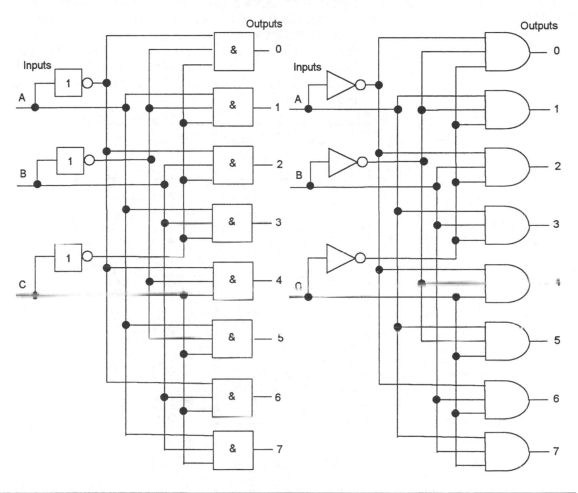

	Inputs					Outputs				
C	B	A	0	1	2	3	4	5	6	7
0	0	0	1	0	0	0	0	0	0	0
0	0	1	0	1	0	0	0	0	0	0
0	1	0	0	0	1	0	0	0	0	0
0	1	1	0	0	0	1	0	0	0	0
1	0	0	0	0	0	0	1	0	0	0
1	0	1	0	0	0	0	0	1	0	0
1	1	0	0	0	0	0	0	0	1	0
1	1	1	0	0	0	0	0	0	0	1

Fig. 14.31 3-line-to-8-line decoder

Some decoders have one or more ENABLE inputs that are used to control the operation of the decoder. Thus with the ENABLE line HIGH the decoder will function in its normal way and the inputs will determine which output is HIGH; with the ENABLE line LOW all the outputs are held LOW regardless of the inputs. Figure 14.32 shows a commonly used 3-line-to-8-line decoder with this facility, the 74LS138. Note that the outputs are active-LOW rather than the active-HIGH of Figure 14.31, and that it has three ENABLE lines with the requirement for normal functioning that E1 and E3 are LOW and E3 is HIGH. All other variations result in the decoder being disabled and just a HIGH output.

	Enable			Inputs					Outputs				
E1	E2	E3	C	B	A	0	1	2	3	4	5	6	7
1	X	X	X	X	X	1	1	1	1	1	1	1	1
X	1	X	X	X	X	1	1	1	1	1	1	1	1
X	X	0	X	X	X	1	1	1	1	1	1	1	1
0	0	1	0	0	0	0	1	1	1	1	1	1	1
0	0	1	0	0	1	1	0	1	1	1	1	1	1
0	0	1	0	1	0	1	1	0	1	1	1	1	1
0	0	1	0	1	1	1	1	1	0	1	1	1	1
0	0	1	1	0	0	1	1	1	1	0	1	1	1
0	0	1	1	0	1	1	1	1	1	1	0	1	1
0	0	1	1	1	0	1	1	1	1	1	1	0	1
0	0	1	1	1	1	1	1	1	1	1	1	1	0

Fig. 14.32 74LS138: 1 = HIGH, 0 = LOW, X = does not matter

Figure 14.33 illustrates the type of response we can get from a 74LS138 decoder for different inputs.

A 74LS138 decoder might be used with a microprocessor with the ENABLE used to switch on the decoder and then depending on the the output from three output lines from the microprocessor so one of the 8 decoder outputs receives the LOW output with all the others remaining HIGH. Thus, we can consider each output device to have an address, i.e. a unique binary output number, so that when a microprocessor sends an address to the decoder it activates the device which has been allocated that address. The 74LS138 can then be referred to as an address decoder.

Fig. 14.33 74LS138

14.7 Sequential logic

The logic circuits considered in earlier sections of this chapter are all examples of combinational logic systems. With such systems the output is determined by the combination of the input variables at a particular instant of time. For example, if input A and input B occur at the same time then an AND gate gives an output. The output does not depend on what the inputs previously were. Where a system requires an output which depends on earlier values of the inputs, a *sequential logic* system is required. The main difference between a combinational logic system and a sequential logic system is that the sequential logic system must have some form of memory.

Figure 14.34 shows the basic form of a sequential logic system. The combinational part of the system accepts logic signals from external inputs and from outputs from the memory. The combinational system then operates on these inputs to produce its outputs. The outputs are thus a function of both its external inputs and the information stored in its memory.

Fig. 14.34 Sequential logic system

Fig. 14.35 SR flip-flop

Fig. 14.36 Timing diagram

Fig. 14.37 SR flip-flop

14.7.1 The flip-flop

The *flip-flop* is a basic memory element which is made up of an assembly of logic gates. There are a number of forms of flip-flops. Figure 14.35 shows one form, the SR (set-reset) flip-flop, involving NOR gates. If initially we have both outputs 0 and $S = 0$ and $R = 0$, then when we set and have S change from 0 to 1, the output from NOR gate 2 will become 0. This will then result in both the inputs to NOR gate 1 becoming 0 and so its output becomes 1. This feedback acts as an input to NOR gate 2 which then has both its inputs at 1 and results in no further change.

Now if S changes from 1 to 0, the output from NOR gate 1 remains at 1 and the output from NOR gate 2 remains at 0. There is no change in the outputs when the input S changes from 1 to 0. It will remain in this state indefinitely if the only changes are to S. It 'remembers' the state it was set to. Figure 14.36 illustrates this with a timing diagram in which a rectangular pulse is used as the input S.

If we change R from 0 to 1 when S is 0, the output from NOR gate 1 changes to 0 and hence the output from NOR gate 2 changes to 1. The flip-flop has been reset. A change then of R to 0 will have no effect on these outputs.

Thus when S is set to 1 and R made 0, the output Q will change to 1 if it was previously 0, remaining at 1 it was previously 1. This is the set condition and it will remain in this condition even when S changes to 0. When S is 0 and R is made 1 the output Q is reset to 0 if it was previously 1, remaining at 0 if it was previously 0. This is the rest condition. The output Q that occurs at a particular time will depend on the inputs S and R and also the last value of the output. The following state table illustrates this:

S	R	$Q_t \rightarrow Q_{t+1}$	$\overline{Q}_t \rightarrow \overline{Q}_{t+1}$
0	0	$0 \rightarrow 0$	$1 \rightarrow 1$
0	0	$1 \rightarrow 1$	$0 \rightarrow 0$
0	1	$0 \rightarrow 0$	$1 \rightarrow 1$
0	1	$1 \rightarrow 0$	$0 \rightarrow 0$
1	0	$0 \rightarrow 1$	$1 \rightarrow 0$
1	0	$1 \rightarrow 1$	$0 \rightarrow 0$
1	1	Not allowed	
1	1	Not allowed	

Note that if S and R are simultaneously made equal to 1, no stable state can occur and so this input condition is not allowed. Figure 14.37 shows the simplified block symbol used for the SR flip-flop.

As a simple illustration of the use of a flip-flop consider a simple alarm system in which an alarm is to sound when a beam

Fig. 14.38   Alarm circuit

Fig. 14.39 Clocked SR flip-flop

Fig. 14.40   Timing diagram

Fig. 14.41   JK flip-flop

of light is interrupted and remain sounding even when the beam is no longer interrupted. Figure 14.38 shows a possible system. A photo transistor might be used as the sensor and so connected that when it is illuminated it gives a virtually 0 V input to S but when the illumination ceases it gives about 5 V input to S. When the light beam is interrupted S becomes 1 and the output from the flip-flop becomes 1 and the alarm sounds. The output will remain as 1 even when S changes to 0. The alarm can only be stopped if the reset switch is momentarily opened to produce a 5 V input to R.

14.7.2 Synchronous systems

It is often necessary for set and reset operations to occur at particular times. With an unclocked or *asynchronous system* the outputs of logic gates can change state any time any one or more of the inputs change. With a clocked or *synchronous system* the exact times at which any output can change state is determined by a signal termed the clock signal. This is generally a rectangular pulse train and when the same clock signal is used for all parts of the system, outputs are synchronised. Figure 14.39 shows the principle of a gated SR flip-flop. The set and clock signal are supplied through an AND gate to the S input of the flip-flop. Thus the set signal only arrives at the flip-flop when both it and the clock signal are 1. Likewise the reset signal is supplied with the clock signal to the R input via another AND gate. As a consequence, setting and resetting can only occur at the time determined by the clock. Figure 14.40 shows the timing diagram.

14.7.3 JK flip-flop

For many applications the indeterminate state that occurs with the SR flip-flop when $S = 1$ and $R = 1$ is not acceptable and another form of flip-flop is used, the *JK flip-flop* (Fig. 14.41). This has become a very widely used flip-flop device. The following is the truth table for this flip-flop; note the only changes from the state table for the SR flip-flop are the entries when both inputs are 1.

Fig. 14.42   D flip-flop

Fig. 14.43   Output from D
flip-flop

Fig. 14.44   Symbol for D flip-flop

J	K	$Q_t \to Q_{t+1}$	$\overline{Q}_t \to \overline{Q}_{t+1}$
0	0	$0 \to 0$	$1 \to 1$
0	0	$1 \to 1$	$0 \to 0$
0	1	$0 \to 0$	$1 \to 1$
0	1	$1 \to 0$	$0 \to 0$
1	0	$0 \to 1$	$1 \to 0$
1	0	$1 \to 1$	$0 \to 0$
1	1	$0 \to 1$	$1 \to 0$
1	1	$1 \to 0$	$0 \to 1$

As an illustration of the use of such a flip-flop, consider the requirement for a high output when input A goes high and then some time later B goes high. An AND gate can be used to determine whether two inputs are both high but its output will be high regardless of which input goes high first. However, if the inputs A and B are used with a JK flip-flop, then A must be high first in order for the output to go high when B subsequently goes high.

14.7.4 D flip-flop

The Data or *D flip-flop* is basically a clocked SR flip-flop or a JK flip-flop with the D input being connected directly to the S or J inputs and via a NOT gate to the R or K inputs (Fig. 14.42); in the symbol for the D flip-flop this joined R and K input is labelled D. This arrangement means that a 0 or a 1 input will then switch the outputs to follow the input when the clock pulse is 1 (Fig. 14.43). A particular use of the D flip-flop is to ensure that the output will only take on the value of the D input at precisely defined times. Figure 14.44 shows the symbol used for a D flip-flop.

With the above form of D flip-flop, when the clock or enable input goes high, the output follows the data presented at input D. The flip-flop is said to be transparent. When there is a high-to-low transition at the enable input, output Q is held at the data level just prior to the transition. The data at transition is said to be *latched*. D flip-flops are available as integrated circuits. The 7475 is an example, it containing four transparent D latches.

The 7474 D flip-flop differs from the 7475 in being an edge-triggered device; there are two such flip-flops in the package. With an edge-triggered D flip-flop, transitions in Q only occur at the edge of the input clock pulse and with the 7474 it is the positive edge, i.e. low-to-high transition. Figure 14.45 illustrates this. The basic symbol for an edge-triggered D flip-flop differs from that of a D flip-flop by a small triangle being included on the CK input (Fig. 14.46). There are also two other inputs called preset and clear. A low on the preset sets the output Q to 1 while a low on clear clears the output, setting Q to 0.

CK

D

Q

Fig. 14.45   Positive edge-triggered

Fig. 14.46   Symbol for edge-triggered D flip-flop

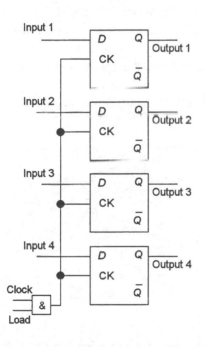

Fig. 14.47 Alarm system

As an illustration of a simple application for such a flip-flop, Figure 14.47 shows a system that could be used to show a green light when the sensor input is low and a red light when it goes high and sound an alarm. The red light is to remain on as long as the sensor input is high but the alarm can be switched off. This might be a monitoring system for the temperature in some process, the sensor and signal conditioning giving a low signal when the temperature is below the safe level and a high signal when it is above. The flip-flop has a high input. When a low input is applied to the CK input and the sensor input is low, the green light is on. When the sensor input changes to high, the green light goes out, the red light on and the alarm sounds. The alarm can be cancelled by applying a high signal to the CK input, but the red light remains on as long as the sensor input is high. Such a system could be constructed using a 7474 and an integrated circuit or circuits giving three NAND gates.

14.7.5 Registers

A *register* is a set of memory elements and is used to hold information until it is needed. It can be implemented by a set of flip-flops. Each flip-flop stores a binary signal, i.e. a 0 or a 1. Figure 14.48 shows the form a 4-bit register can take when using D flip-flops. When the load signal is 0, no clock input occurs to the D flip-flops and so no change occurs to the states of the flip-flops. When the load signal is 1 then the inputs can change the states of the flip-flops. As long as the load signal is 0 the flip-flops will hold their old state values.

Fig. 14.48 Register

Problems

1 What is the largest decimal number that can be represented by the use of an 8-bit binary number?
2 Convert the following binary numbers to decimal numbers: (a) 1011, (b) 10 0001 0001.
3 Convert the decimal numbers (a) 423, (b) 529 to hex.
4 Convert the BCD numbers (a) 0111 1000 0001, (b) 0001 0101 0111 to decimal.

5 What are the twos complement representations of the decimal numbers (a) –90, (b) –35?

6 What even-parity bits should be attached to (a) 100 1000, (b) 100 1111?

7 Subtract the following decimal numbers using twos complements (a) 21 – 13, (b) 15 – 3.

8 Explain what logic gates might be used to control the following situations:

   (a) The issue of tickets at an automatic ticket machine at a railway station.

   (b) A safety lock system for the operation of a machine tool.

9 State the Boolean functions that can be used to described the following situations:

   (a) There is an output when switch A is closed and either switch B or switch C is closed.

   (b) There is an output when either switch A or switch B is closed and either switch C or switch D is closed.

   (c) There is an output when either switch A is opened or switch B is closed.

   (d) There is an output when switch A is opened and switch B is closed.

10 State the Boolean functions for each of the logic circuits shown in Figure 14.49.

Fig. 14.49
Problem 10

(a)   (b)

11 Construct a truth table for the Boolean equation $Q = (A \cdot C + B \cdot C) \cdot (A + C)$.

12 Simplify the following Boolean equations:

   (a) $Q = A \cdot C + A \cdot C \cdot D + C \cdot D$

   (b) $Q = A \cdot \overline{B} \cdot D + A \cdot \overline{B} \cdot \overline{D}$

   (c) $Q = A \cdot B \cdot C + C \cdot D + C \cdot D \cdot E$

13 Use De Morgan's laws to show that a NOR gate with inverted inputs is equivalent to an AND gate.

14 Draw the Karnaugh maps for the following truth tables and hence determine the simplified Boolean equation for the outputs:

(a)

A	B	Q
0	0	1
0	1	1
1	0	1
1	1	1

(b)

A	B	C	Q
0	0	0	0
0	0	1	1
0	1	0	1
0	1	1	1
1	0	0	0
1	0	1	1
1	1	0	0
1	1	1	1

Fig. 14.50   Problem 17

Fig. 14.51   Problem 18

15 Simplify the following Boolean equations by the use of Karnaugh maps;

(a) $Q = \overline{A} \cdot B \cdot C + \overline{A} \cdot \overline{B} \cdot \overline{C} + \overline{A} \cdot B \cdot \overline{C}$

(b) $Q = \overline{A} \cdot B \cdot \overline{C} \cdot D + A \cdot \overline{B} \cdot \overline{C} \cdot D + \overline{A} \cdot \overline{B} \cdot \overline{C} \cdot D$
$+ A \cdot B \cdot \overline{C} \cdot D + A \cdot B \cdot \overline{C} \cdot \overline{D} + A \cdot B \cdot C \cdot D$

16 Devise a system which will allow a door to be opened only when the correct combination of four push-buttons is pressed, any incorrect combination sounding an alarm.

17 Figure 14.50 shows the timing diagram for the S and R inputs for a SR flip-flop. Complete the diagram by adding the Q output.

18 Explain how the arrangement of gates shown in Figure 14.51 gives a SR flip-flop.

15 Microprocessors

15.1 Control

If we take a simple control problem, e.g. the sequencing of the red, yellow, green lights at a traffic crossing, then it would be possible to solve it by an electronic control system involving combinational and sequential logic integrated circuits. However, with a more complex situation there might be many more variables to control in a more complex control sequence. The simplest solution now becomes not one of constructing a system based on hard-wired connection of combinational and sequential logic integrated circuits but of using a microprocessor and using software to make the 'interconnections'.

The microprocessor systems that we are concerned with in this book are for use as control systems and are termed *embedded microprocessors*. This is because such a microprocessor is dedicated to controlling a specific function and is self-starting requiring no human intervention and completely self-contained with its own operating program. For the human, it is not apparent that the system is a microprocessor one. Thus, a modern washing machine contains a microprocessor but all that the operator has to do to operate it is to select the type of wash required by pressing the appropriate button or rotating a switch and then press the button to start.

This chapter is an overview of the structure of microprocessors and microcontrollers with the next two chapters discussing programming and Chapter 17 interfacing. For more details the reader is referred to the publications issued by the microprocessor manufacturers, or texts on microprocessors such as *Microprocessor Systems* by W. Bolton (Longman 2000) and on specific microcontrollers such as *Software and Hardware Engineering, Motorola M68HC11* by F.M. Cady (Oxford University Press 1997), *Microcontroller Technology: The 68HC11* by P. Spasov (Prentice-Hall 1992, 1996), *The 8051 Family of Microcontrollers* by R.H. Barnett (Prentice-Hall 1995), *PIC: Your Personal Introductory Course* by J. Morton (Newnes

2001) or *PIC Basic: Programming and Projects* by I. Ibrahim (Newnes 2001).

15.2 Microprocessor systems

Microprocessor systems have three parts:

1   A *central processing unit* (CPU) to recognise and carry out program instructions. This is the part which uses the microprocessor.

2   *Input and output interfaces* to handle communications between the computer and the outside world; the term *port* is used for the interface.

3   *Memory* to hold the program instructions and data.

Microprocessors which have memory and various input/output arrangements all on the same chip are called *microcontrollers*.

15.2.1 Buses

Digital signals move from one section to another along paths called *buses*. A bus, in the physical sense, is just a number of conductors along which electrical signals can be carried. It might be tracks on a printed circuit board or wires in a ribbon cable. Figure 15.1 illustrates the general arrangement of a microprocessor system and its buses. There are three forms of bus in a microprocessor system:

Fig. 15.1 General form of a microprocessor system

1   *Data bus*

The data associated with the processing function of the CPU is carried by the *data bus*. Thus, it is used to transport a word to or from the CPU and the memory or the input/output interfaces. Word lengths used may be 4, 8, 16, 32 or 64. Each wire in the bus carries a binary signal, i.e. a 0 or a 1. Thus with a four-wire bus we might have the word 1010 being carried, each bit being carried by a separate wire in the bus, as

Word	Bus wire
0 (least significant bit)	First data bus wire
1	Second data bus wire
0	Third data bus wire
1 (most significant bit)	Fourth data bus wire

The more wires the data bus has the longer the word length that can be used. The range of values which a single item of data can have is restricted to that which can be represented by the word length. Thus with a word of length 4 bits the number of values is $2^4 = 16$. Thus if the data is to represent, say, a temperature, then the range of possible temperatures must be divided into 16 segments if we are to represent that range by a 4-bit word. The earliest microprocessors were 4-bit (word length) devices and such 4-bit microprocessors are still widely used in such devices as toys, washing machines and domestic central heating controllers. They were followed by 8-bit microprocessors, e.g. the Motorola 6800, the Intel 8085A and the Zilog Z80. Now, 16-bit, 32-bit and 64-bit microprocessors are available; however, 8-bit microprocessors are still widely used for controllers

2   *Address bus*

The *address bus* carries signals which indicate where data is to be found and so the selection of certain memory locations or input or output ports. Each storage location within a memory device has a unique identification, termed its address, so that the system is able to select a particular instruction or data item in the memory. Each input/output interface also has an address. When a particular address is selected by its address being placed on the address bus, only that location is open to the communications from the CPU. The CPU is thus able to communicate with just one location at a time. A computer with an 8-bit data bus has typically a 16-bit wide address bus, i.e. 16 wires. This size of address bus enables 2^{16} locations to be addressed. 2^{16} is 65 536 locations and is usually written as 64 K, where K is equal to 1024. The more memory that can be addressed the greater the volume of data that can be stored and the larger and more sophisticated the programs that can be used.

3 *Control bus*

The signals relating to control actions are carried by the *control bus*. For example, it is necessary for the microprocessor to inform memory devices whether they are to read data from an input device or write data to an output device. The term READ is used for receiving a signal and WRITE for sending a signal. The control bus is also used to carry the system clock signals; these are to synchronise all the actions of the microprocessor system. The clock is a crystal-controlled oscillator and produces pulses are regular intervals.

15.2.2 The microprocessor

The microprocessor is generally referred to as the *central processing unit* (CPU). It is that part of the processor system which processes the data, fetching instructions from memory, decoding them and executing them. The internal structure, the term *architecture* is used, of a microprocessor depends on the microprocessor concerned. Figure 15.2 indicates, in a simplified manner, the general architecture of a microprocessor.

The following are the functions of the constituent parts of a microprocessor:

1 *Arithmetic and logic unit* (ALU)

The arithmetic and logic unit is responsible for performing the data manipulation.

Fig. 15.2   General internal architecture of a microprocessor

2 *Registers*

Internal data that the CPU is currently using is temporarily held in a group of *registers* while instructions are being executed. These are memory locations within the microprocessor and are used to store information involved in program execution. A microprocessor will contain a group of registers, each type of register having a different function.

3 *Control unit*

The *control unit* determines the timing and sequence of operations. It generates the timing signals used to fetch a program instruction from memory and execute it. The Motorola 6800 uses a clock with a maximum frequency of 1 MHz, i.e. a clock period of 1 μs, and instructions require between two and twelve clock cycles. Operations involving the microprocessor are reckoned in terms of the number of cycles they take.

There are a number of types of register, the number, size and types of registers varying from one microprocessor to another. The following are common types of registers.

1 *Accumulator register*

The accumulator register (A) is where data for an input to the arithmetic and logic unit is temporarily stored. In order for the CPU to be able to access, i.e. read, instructions or data in the memory it has to supply the address of the required memory word using the address bus. When this has been done, the required instructions or data can be read into the CPU using the data bus. Since only one memory location can be addressed at once, temporary storage has to be used when, for example, numbers are combined. For example, in the addition of two numbers, one of the numbers is fetched from one address and placed in the accumulator register while the CPU fetches the other number from the other memory address. Then the two numbers can be processed by the arithmetic and logic section of the CPU. The result is then transferred back into the accumulator register. The accumulator register is thus a temporary holding register for data to be operated on by the arithmetic and logic unit and also, after the operation, the register for holding the results. It is thus involved in all data transfers associated with the execution of arithmetic and logic operations.

2 *Status register*, or *condition code register* or *flag register*

This contains information concerning the result of the latest process carried out in the arithmetic and logic unit. It contains individual bits with each bit having special significance. The bits are called *flags*. The status of the latest operation is indicated by each flag with each flag being set or reset to indicate a specific status. For example, they can be used to indicate whether the last operation resulted in a negative result, a zero result, a carry output occurs (e.g. the sum of two binary numbers such as 1010 and 1100 is (1)0110 which might be bigger than the microprocessor's word size and thus there is a carry of a 1), an overflow occurs or the program is to be allowed to be interrupted to allow an external event to occur. The following are common flags.

Flag	Set, i.e. 1	Reset, i.e. 0
Z	result is zero	result is not zero
N	result is negative	result is not negative
C	carry is generated	carry is not generated
V	overflow occurs	overflow does not occur
I	interrupt is ignored	interrupt is processed normally

As an illustration, consider the state of the Z, N, C and V flags for the operation of adding the hex numbers 02 and 06. The result is 08. Since it is not zero then Z is 0. The result is positive so N is 0. There is no carry so C is 0. The unsigned result is within the range −128 to +127 and so there is no overflow and V is 0. Now consider the flags when the hex numbers added are F9 and 08. the result is (1)01. The result is not zero so Z is 0. Since it is positive then N is 0. The unsigned result has a carry and so C is 1. The unsigned result is within the range −128 to +127 and so V is 0.

3   *Program counter register* (PC) or *instruction pointer* (IP)
This is the register used to allow the CPU to keep track of its position in a program. This register contains the address of the memory location that contains the next program instruction. As each instruction is executed the program counter register is updated so that it contains the address of the memory location where the next instruction to be executed is stored. The program counter is incremented each time so that the CPU executes instructions sequentially unless an instruction, such as JUMP or BRANCH, changes the program counter out of that sequence.

4   *Memory address register* (MAR)
This contains the address of data. Thus, for example, in the summing of two numbers the memory address register is loaded with the address of the first number. The data at the address is then moved to the accumulator. The memory address of the second number is then loaded into the memory address register. The data at this address is then added to the data in the accumulator. The result is then stored in a memory location addressed by the memory address register.

5   *Instruction register* (IR)
    This stores an instruction. After fetching an instruction from the memory via the data bus, the CPU stores it in the instruction register. After each such fetch, the microprocessor increments the program counter by one with the result that the program counter points to the next instruction waiting to be fetched. The instruction can then be decoded and used to execute an operation. This sequence is known as the *fetch-execute cycle*.

6   *General-purpose registers*
    These may serve as temporary storage for data or addresses and used in operations involving transfers between other registers.

7   *Stack pointer register* (SP)
    The contents of this register form an address which defines the top of the stack in RAM. The *stack* is a special area of the memory in which program counter values can be stored when a subroutine part of a program is being used.

The number and form of the registers depends on the microprocessor concerned. For example, the Motorola 6800 microprocessor (Fig. 15.3) has two accumulator registers, a status register, an index register, a stack pointer register and a program counter register. The status register has flag bits to show negative, zero, carry, overflow, half-carry and interrupt. The Motorola 6802 is similar but includes a small amount of RAM and a built-in clock generator.

The Intel 8085A microprocessor is a development of the earlier 8080 processor; the 8080 required an external clock generator whereas the 8085A has an in-built clock generator. Programs written for the 8080 can be run on the 8085A. The 8085A has six general-purpose registers B, C, D, E, H and L, a stack pointer, a program counter, a flag register and two temporary registers. The general-purpose registers may be used as six 8 bit registers or in pairs BC, DE and HL as three 16 bit registers. Figure 15.4 shows a block diagram representation of the architecture.

As will be apparent from Figure 15.3 and Figure 15.4, microprocessors have a range of timing and control inputs and outputs. These provide outputs when a microprocessor is carrying out certain operations and inputs to influence control operations. In addition there are inputs related to interrupt controls. These are designed to allow program operation to be interrupted as a result of some external event.

Buffers to generate sufficient current for external devices and prevent unwanted interactions between such devices and the microprocessor

When HALT input occurs, all activity ceases.
An input to TSC causes the address lines and the READ/WRITE lines to go into high impedance state
VMA gives an output that there is a valid address on the address bus.
Pins 4 and 6 are used to request an interruption in the program.
RESET is used to set the progam counter to zero.
DBE is used to enable other devices to take control of the data bus.
BA indicates that the address bus is available

Fig. 15.3 Motorola 6800 architecture

15.2.3 Memory

The memory unit in a microprocessor system stores binary data and takes the form of one or more integrated circuits. The data may be program instruction codes or numbers being operated on.

The size of the memory is determined by the number of wires in the address bus. The memory elements in a unit consist essentially of large numbers of storage cells with each cell capable of storing either a 0 or a 1 bit. The storage cells are grouped in locations with each location capable of storing one word. In order to access the stored word, each location is identified by a unique address. Thus with a 4-bit address bus we can have 16 different addresses with each, perhaps, capable of storing one byte, i.e. a group of 8 bits.

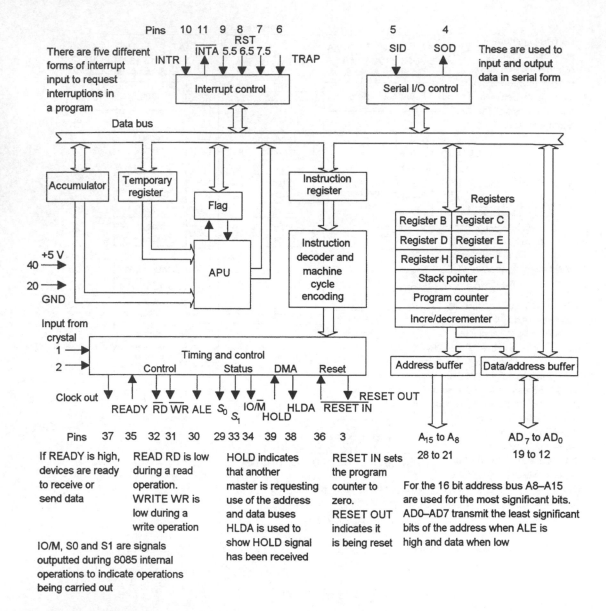

Fig. 15.4 Intel 8085A architecture

As an illustration of the above discussion of data storage related to the size of the address bus, Figure 15.5 shows the possibilities with a 4-bit address bus .

The size of a memory unit is specified in terms of the number of storage locations available; 1 K is $2^{10} = 1024$ locations and thus a 4 K memory has 4096 locations.

There are a number of forms of memory unit:

Fig. 15.5   Address bus size

Fig. 15.6   ROM chip

1   *ROM*
For data that is stored permanently a memory device called a *read-only memory* (ROM) is used. ROMs are programmed with the required contents during the manufacture of the integrated circuit. No data can then be written into this memory while the memory chip is in the computer. The data can only be read and is used for fixed programs such as computer operating systems and programs for dedicated microprocessor applications. They do not lose their memory when power is removed. Figure 15.6 shows the pin connections of a typical ROM chip which is capable of storing $1 K \times 8$ bits.

2   *PROM*
The term *programmable ROM* (PROM) is used for ROM chips that can be programmed by the user. Initially every memory cell has a fusible link which keeps its memory at 0. The 0 is permanently changed to 1 by sending a current through the fuse to permanently open it. One the fusible link has been opened the data is permanently stored in the memory and cannot be further changed.

3   *EPROM*
The term *erasable and programmable ROM* (EPROM) is used for ROMs that can be programmed and their contents altered. A typical EPROM chip contains a series of small electronic circuits, cells, which can store charge. The program is stored by applying voltages to the integrated circuit connection pins and producing a pattern of charged and uncharged cells. The pattern remains permanently in the chip until erased by shining ultraviolet light through a quartz window on the top of the device. This causes all the cells to become discharged. The chip can then be reprogrammed. The Intel 2716 EPROM has 11 address pins and a single chip enable pin which is active when taken low.

4   *EEPROM*
Electrically erasable PROM (EEPROM) is similar to EPROM. Erasure is by applying a relatively high voltage rather than using ultraviolet light.

5   *RAM*
Temporary data, i.e. data currently being operated on, is stored in a read/write memory referred to as a *random-access memory* (RAM). Such a memory can be read or written to. Figure 15.7 shows the typical pin connections for a $1 K \times 8$-bit RAM chip. The Motorola 6810 RAM chip has seven address pins and six chip select pins of which four are active when low and two active when high and all must be made simultaneously active to enable the RAM.

Chip select

Address lines

Data lines

Read

Write

Fig. 15.7   RAM chip

When ROM is used for program storage, then the program is available and ready for use when the system is switched on. Programs stored in ROM are termed *firmware*. Some firmware must always be present. When RAM is used for program storage then such programs are referred to as *software*. When the system is switched on, software may be loaded into RAM from some other peripheral equipment such as a keyboard or hard disc or floppy disc.

15.2.4 Input/output

The input/output operation is defined as the transfer of data between the microprocessor and the external world. The term *peripheral devices* is used for pieces of equipment that exchange data with a microprocessor system. Because the speeds and characteristics of peripheral devices can differ significantly from those of the microprocessor, they are connected via interface chips. A major function of an interface chip is thus to synchronise data transfers between the microprocessor and peripheral device. In input operations the input device places the data in the data register of the interface chip; this holds the data until it is read by the microprocessor. In output operations the microprocessor places the data in the register until it is read by the peripheral.

For the microprocessor to input valid data from an input device, it needs to be certain that the interface chip has correctly latched the input data. It can do this by *polling* or an *interrupt*. With polling, the interface chip uses a status bit set to 1 to indicate when it has valid data. The microprocessor keeps on checking the interface chip until it detects the status bit as 1. The problem with this method is that the microprocessor is having to wait for this status bit to show. With the interrupt method, the interface chip sends an interrupt signal to the microprocessor when it has valid data; the microprocessor then suspends execution of its main program and executes the routine associated with the interrupt in order to read the data.

15.2.5 Examples of systems

Figure 15.8 shows an example of a microprocessor system using the Intel 8085A microprocessor. It has an address latch 74LS373, a 3 line to 8 line address decoder 74LS138, two 1K × 4 RAM chips 2114, a 2K × 8 EPROM chip 2716 and input and output interface chips 74LS244 and 74LS374.

1. *Address latch*

   The output ALE (address latch enable) provides an output to the external hardware to indicate when the lines AD0–AD7 contain addresses and when they contain data. When ALE is made high it activates the latch and the lines A0–A7 pass the lower part of the address to it where it becomes latched. Thus when ALE changes and goes back to being low, so that data can be outputted from the microprocessor, this part of the address remains latched in the 74LS373. The higher part of the address is sent through A8–A15 and is always valid, and the full address is given by the lower part from the latch and the upper part from the microprocessor address bus.

Fig. 15.8 Intel 8085A system

2 *Address decoder*

The 74LS138 is a 3 line to 8 line decoder and provides an active low at one of its eight outputs; the output selected depends on the signals on its three input lines A, B and C (see Fig. 14.32). Before it can make such a selection it has to be enabled by enable 1 and enable 2 being low and enable 3 high.

3 *EPROM*

Address bits A11, A12, A13 and A14 are used to select which device is to be addressed. This leaves the address bits on A0–A10 for addresses. Hence the EPROM can have 2^{11} = 2048 addresses. This is the size of the Intel 2716 EPROM. The EPROM is selected whenever the microprocessor reads an address from 0000 to 07FF and outputs its 8 bit contents to the data bus via O0–O7. The output enable line OE is connected to the read output of the microprocessor to ensure that the EPROM is only written to.

4 *RAM*

Two RAM chips are shown as being used, each being 1K × 4. Together they provide a memory for an 8-bit wide signal. Both the chips use the same address bits A0–A9 for memory selection with one chip providing data D0–D3 and the other D4–D7. With 10 address bits we have 2^{10} = 1024 different addresses 2000 to 23FF. The write enable WE input is used by the RAM to determine whether the RAM is being written to or read. If it is low then the selected RAM address is being written to and if high it is being read.

5. *Input buffer*

The input buffer 74LS244 is set to pass the binary value of the inputs over the data bus whenever OE1 and OE2 are low. It is accessed at any address from 2800 to 2FFF, thus we might use the address 2800. The buffer is to ensure that the inputs impose very little loading on the microprocessor.

6 *Output latch*

The 74LS374 is an output latch. It latches the microprocessor output so that the output devices have time to read it while the microprocessor can get on with other instructions in its program. The output latch is given a range of addresses from 3800 to 3FFF and thus might be addressed by using 3800.

Figure 15.9 shows an example of a system based on the use of a Motorola 6800 microprocessor and having just one RAM chip, one ROM chip and a programmable input/output. No address decoding is necessary with this system because of the small number of devices involved. For parallel inputs/outputs a peripheral interface adapter (PIA) (see Section 18.4) is used and for serial inputs/outputs an asynchronous interface adapter

(ACIA) (see Section 18.5) is used. These can be programmed to deal with both inputs and outputs and give the required buffering.

Fig. 15.9 M6800 system

1 *RAM*

Address lines A14 and A15 are connected to the enable inputs of the RAM chip. When both these lines are low then the RAM chip will be conversing with the microprocessor.

2 *ROM*

Address lines A14 and A15 are connected to the enable inputs of the ROM chip and when the signals on both these lines are high then the ROM chip is addressed.

3 *Inputs/outputs*

Address lines A14 and A15 are connected to the enable inputs of the PIA and ACIA. When the signal on line 15 is low and the signal on A14 high then the input/output interfaces are addressed. In order to indicate which of the devices is being enabled, address line A2 is taken high for the PIA and address line A3 is taken high for the ACIA.

15.3 Microcontrollers

For a microprocessor to give a system which can be used for control, additional chips are necessary, e.g. memory devices for program and data storage and input/output ports to allow it to communicate with the external world and receive signals from it. The *microcontroller* is the integration of a microprocessor with memory and input/output interfaces, and other peripherals such as timers, on a single chip. Figure 15.10 shows the general block diagram of a microcontroller.

Fig. 15.10 Block diagram of a microcontroller

The general microcontroller has pins for external connections of inputs and outputs, power, clock and control signals. The pins for the inputs and outputs are grouped into units called input/output ports. Usually such ports have eight lines in order to be able to transfer an 8-bit word of data. Two ports may be used for a 16-bit word, one to transmit the lower 8 bits and the other the upper 8 bits. The ports can be input only, output only or programmable to be either input or output.

The Motorola 68HC11, the Intel 8051 and the PIC16C6x/7x are examples of 8-bit microcontrollers in that the data path is 8 bits wide. The Motorola 68HC16 is an example of a 16-bit microcontroller and the Motorola 68300 a 32-bit microcontroller. Microcontrollers have limited amounts of ROM and RAM and are widely used for embedded control systems. A microprocessor system with separate memory and input/output chips is more suited to processing information in a computer system.

15.3.1 Motorola M68HC11

Motorola offer two basic 8-bit families of microcontrollers, the 68HC05 being the inexpensive core and the 68HC11 the higher performance core. The Motorola M68HC11 family (Fig. 15.11), this being based on the Motorola 6800 microprocessor, is very widely used for control systems.

Fig. 15.11   Block diagram of M68HC11

There are a number of versions of this family, the differences being due to differences in the RAM, ROM, EPROM, EEPROM and configuration register features. For example, one version (68HC11A8) has 8 K ROM, 512 bytes EEPROM, 256 bytes RAM, a 16-bit timer system, a synchronous serial peripheral interface, an asynchronous non-return-to-zero serial communication interface, an 8-channel 8-bit analogue-to-digital converter for analogue inputs, and five ports A, B, C, D and E.

1   *Port A*
Port A has three input lines only, four output lines only and one line that can serve as either input or output. The port A data register is at address $1000 (Fig. 15.12) with a pulse accumulator control register (Fig. 15.13) at address $1026 being used to control the function of each bit in port A. This port also provides access to the internal timer of the microcontroller, the PAMOD, PEDGE, RTR1 and RTRO bits controlling the pulse accumulator and clock.

Port A data register $1000

Bit

7	6	5	4	3	2	1	0

Fig. 15.12   Port A register

Fig. 15.13   Pulse accummulator control register

2   *Port B*
Port B is output only and has eight output lines (Fig. 15.14). Input data cannot be put on port B pins. Its data register is at the address $1004 and to output data it has to be written to this memory location.

Port B data register $1004

Bit

7	6	5	4	3	2	1	0

Fig. 15.14   Port B register

3   *Port C*
Port C can be either input or output, data being written or read from its data register at address $1003 (Fig. 15.15). Its direction is controlled by the port data direction register at address $1007. The eight bits in this register correspond to the individual bits in port C and determine whether the lines are inputs or outputs; when the data direction register bit is set to 0 it is an input and when set to 1 it is an output. The lines STRA and STRB (when operating in single-chip mode) are associated with ports B and C and are used for handshake signals with those ports. Such lines control the timing of data transfers. The parallel I/O control register PIOC at address $1002 contains bits to control the handshaking mode and the polarity and active edges of the handshaking signals.

Port C data register $1003

Bit

7	6	5	4	3	2	1	0

Port C data direction register $1007

Bit

7	6	5	4	3	2	1	0

When a bit is set to 0 the corresponding bit in the port is an input, when set to 1 an output

Fig. 15.15   Port C registers

Port D data register $1008
Bit

Port D data direction register $1009
Bit

When a bit is set to 0 the corresponding
bit in the port is an input, when set to 1
an output

Fig. 15.16 Port D registers

Port E data register $100A
Bit

7	6	5	4	3	2	1	0

Fig. 15.17 Port E register

4 *Port D*

Port D contains just six lines; these can be either input or output and have a data register at address $1008 (Fig. 15.16); the directions are controlled by a port data direction register at address $1009 with the corresponding bit being set to 0 for an input and 1 for an output. Port D also serves as the connection to the two serial subsystems of the microcontroller. The serial communication interface is an asynchronous system that provides serial communication that is compatible with modems and terminals. The serial peripheral interface is a high-speed synchronous system which is designed to communicate between the microcontroller and peripheral components that can access at such rates.

5 *Port E*

Port E is an 8-bit input only port (Fig. 15.17) which can be used as a general-purpose input port or for analogue inputs to the internal analogue-to-digital converter. The two inputs V_{RH} and V_{RL} provide reference voltages to the ADC. The port E data register is at address $1002.

The 68HC11 has an internal analogue-to-digital (A/D) converter; port E bits 0, 1, 2, 3, 4, 5, 6 and 7 are the analogue input pins. Two lines V_{RH} and V_{LH} provide the reference voltages used by the ADC; the high reference voltage V_{RH} should not be lower than V_{DD}, i.e. 5 V, and the low reference voltage V_{LH} should not be lower than V_{SS}, i.e. 0 V. The ADC must be enabled before it can be used. This is done by setting the A/D power-up (ADPU) control bit in the OPTION register (Fig. 15.18), this being bit 7. Bit 6 selects the clock source for the ADC. A delay of at least 100 μs is required after powering up to allow the system to stabilise.

OPTION register $1039

Fig. 15.18 OPTION register

The analogue-to-digital conversion is initiated by writing to the A/D control/status register (ADCTL) after powering up and the stabilisation delay (Fig. 15.19). This involves selecting the channels and operation modes. The conversion will then start one clock cycle later. For example, if single-channel mode is selected by setting MULT = 0 then four successive A/D conversions will occur of the channel selected by the CD–CA bits. The results of the conversion are placed in the A/D result registers ADR1–ADR4.

ADCTL register $1030

MULT = 0

CD	CC	CB	CA	Channel converted
0	0	0	0	PE0
0	0	0	1	PE1
0	0	1	0	PE2
0	0	1	1	PE3
0	1	0	0	PE4
0	1	0	1	PE5
0	1	1	0	PE6
0	1	1	1	PE7

MULT = 1

				A/D result registers			
CD	CC	CB	CA	ADR1	ADR2	ADR3	ADR4
0	0	X	X	PE0	PE1	PE2	PE3
0	1	X	X	PE4	PE5	PE6	PE7

Fig. 15.19 ADCTL register

6 *Modes*

MODA and MODB are two pins that can be used to force the microcontroller into one of four modes at power-up, these modes being special bootstrap, special test, single chip and expanded.

MODB	MODA	Mode
0	1	Special bootstrap
0	1	Special test
1	0	Single chip
1	1	Expanded

In the single-chip mode the microcontroller is completely self-contained with the exception of an external clock source and a reset circuit. With such a mode the microcontroller may not have enough resources, e.g. memory, for some applications and so the expanded mode can be used so that the number of addresses can be increased. Ports B and C then provide address, data and control buses. Port B functions as the upper eight address pins and port C as the multiplexed data and low address pins. The bootstrap mode allows a manufacturer to load special programs in a special ROM for an M68HC11 customer. When the microcontroller is set in this mode, the special program is loaded. The special test mode is primarily used during Motorola's internal production testing.

After the mode has been selected, the MODA pin becomes a pin which can be used to determine that an instruction is starting to execute. The MODB pin has the other function of giving a means by which the internal RAM of the chip can be powered when the regular power is removed.

7 *Oscillator pins*
The oscillator system pins XTAL and EXTAL are the connections to access the internal oscillator. Figure 15.20 shows an external circuit that might be used. E is the bus clock and runs at one-quarter of the oscillator frequency and can be used to synchronise external events.

8 *Interrupt controller*
The interrupt controller is to enable the microcontroller to interrupt a program (see Chapter 18). The two lines IRQ and XIRQ are for the inputs of the interrupt signals. RESET is for resetting. An interrupt is an event that requires the CPU to stop normal program execution and perform some service related to the event. The two lines IRQ and XIRQ are for external interrupt sources.

9 *Timer*
The M68HC11 contains a timer system. This has a free-running counter, five output compare function, the ability to capture the time when an external event occurs, a real-time periodic interrupt and a counter, called the pulse accumulator, for external events. The free-running counter, called TCNT, is a 16-bit counter which starts counting at 0000 when the CPU is reset and continues thereafter continuously running and cannot be reset by the program. Its value can be read at any time. The source for the counter is the system bus clock and its output can be prescaled by setting the PR0 and PR1 bits at bits 0 and 1 in the TMSK2 register at address $1024 (Fig. 15.21).

Fig. 15.20 Oscillator input

Timer interrupt register 2 at address $1024
Bit

7	6	5	4	3	2	1	0

PR1 PR0

Fig. 15.21   TMSK2 register

Prescale factors

			One count	
		Prescale	Bus frequency	
PR1	PR0	factor	2 MHz	1 MHz
0	0	1	0.5 ms	1 ms
0	1	4	2 ms	4 ms
1	0	8	4 ms	8 ms
1	1	16	8 ms	16 ms

The output compare functions allow times, i.e. timer counts, to be specified at which an output will occur when the preset count is reached. The input capture system captures the value of the counter when an input occurs and so the exact time at which the input occurs is captured. The pulse accumulator can be configured to operate as an event counter and count external clock pulses or as a gated time accumulator and store the number of pulses occurring in a particular time interval as a result of the counter being enabled and then, at some later time, disabled. The pulse accumulator control register PACTL (see Fig. 15.12) at address $1026 is used to select the mode of operation. The bit PAEN is set to 0 to disable the pulse accumulator and to 1 to enable it, the bit PAMOD to 0 for the event counter mode and 1 for the gated time mode and the bit PEDGE to 0 for the pulse accumulator to respond to a falling edge when in the event counter mode and 1 to respond to a rising edge in that mode. In gated mode, bit PEDGE set to 0 causes the count to be disabled when port A, bit 7 is 0 and to accumulate when 1 and when bit PEDGE is set to 1 in this mode, the count is disabled when port A, bit 7 is 1 and enabled when it is 0.

10   *COP*
Another timer function is COP, the *computer operating properly* function. This is a timer which times out and resets the system if an operation is not concluded in what has been deemed to be a reasonable time (see Section 21.2). This is often termed a *watchdog timer*.

11   *PWM*
Pulse width modulation (PWM) is used for the control of the speed of d.c. motors (see Section 7.5.5) by using a square wave signal and, by varying the amount of time for which the signal is on, changing the average value of the signal. A square wave can be generated by a microcontroller by arranging for an output to come on every half-period. However, some versions of M68HC11 have a pulse width modulation module and so, after the PWM module has been

RESET	EA
XTAL2	ALE
XTAL1	PSEN
P0.0	P2.0
P0.1	P2.1
P0.2	P2.2
P0.3	P2.3
P0.4	P2.4
P0.5	P2.5
P0.6	P2.6
P0.7	P2.7
P1.0	P3.0 — RXD
P1.2	P3.1 — TXD
P1.2	P3.2 — INT0
P1.3	P3.3 — INT1
P1.4	P3.4 — T0
P1.5	P3.5 — T1
P1.6	P3.6 — WR
P1.7	P3.7 — RD

Fig. 15.22   Intel 8051

initialised and enabled, the PWM waveforms can be automatically outputted.

As will be apparent from the above, before a microcontroller can be used it is necessary to initialise it, i.e. set the bits in appropriate registers so that it will perform as required.

The above is just a brief indication of the input/output connections to the M68HC11 microcontroller. For more details the reader is referred to the manuals published by the manufacturer or to texts such as *Software and Hardware Engineering, Motorola M68HC11* by F.M. Cady (Oxford University Press 1997) or *Microcontroller Technology, The 68HC11* by P. Spasov (Prentice-Hall 1996, 1992).

15.3.2 Intel 8051

Another common family of microcontrollers is the Intel 8051, Figure 15.22 showing the pin connections and Figure 15.23 the architecture. The 8051 has four parallel input/output ports, ports 0, 1, 2 and 3. Ports 0, 2 and 3 also have alternative functions. The 8051AH version has 4 K ROM bytes, 128 bytes RAM, two timers and interrupt control for five interrupt sources.

Fig. 15.23   Intel 8051

1   *Input/output ports*
    Port 0 is at address 80H, port 1 at address 90H, port 2 at address A0H and port 3 at address B0H (note the use of H, or h, after the address with Intel to indicate that it is in hex). When a port is to be used as an output port, the data is put into the corresponding special function register. When a port is to be used as an input port, the value FFH must first be written to it. All the ports are bit addressable. Thus we might, for example, use just bit 6 in port 0 to switch a motor on or off and perhaps bit 7 to switch a pump on or off.

    Port 0 can be used as an input port or an output port. Alternatively it can be used as a multiplexed address and data bus to access external memory. Port 1 can be used as an input port or an output port. Port 2 can be used as an input port or an output port. Alternatively it can be used for the high address bus to access external memory. Port 3 can be used as an input port or an output port. Alternatively, it can be used as a special-purpose input/output port. The alternative functions of port 3 include interrupt and timer outputs, serial port input and output and control signals for interfacing with external memory. RXD is the serial input port, TXD is the serial output port, INT0 is the external interrupt 0, INT1 is the external interrupt 1, T0 is the timer/counter 0 external input, T1 is the timer/counter 1 external input, WR is the external memory write strobe and RD is the external memory read strobe. The term *strobe* describes a connection used to enable or disable a particular function. Port 0 can be used as either an input or output port. Alternatively it can be used to access external memory.

2   *ALE*
    The *address latch enable* (ALE) pin provides an output pulse for latching the low-order byte of the address during access to external memory. This allows 16-bit addresses to be used. Figure 15.24 illustrates this.

3   *PSEN*
    The *program store enable* (PSEN) pin is the read signal pin for external program memory and is active when low. It is connected to the output enable pin of external ROM or EPROM.

4   *EA*
    The *external access* (EA) pin is taken low for the microprocessor to access only external program code; when high it automatically accesses internal or external code depending on the address. Thus when the 8051 is first reset, the program counter starts at $0000 and points to the first program instruction in the internal code memory unless EA is tied low. Then the CPU issues a low on PSEN to enable the

Fig. 15.24   Use of ALE

30 pF

30 pF

XTL1

XTL2

12 MHz crystal

Fig. 15.25 Crystal

external code memory to be used. This pin is also used on a microcontroller with EPROM to receive the programming supply voltage for programming the EPROM.

5 *XTAL1, XTAL2*
These are the connecting pins for a crystal or external oscillator. Figure 15.25 illustrates how they are used with a crystal. The most commonly used crystal frequency is 12 MHz.

6 *RESET*
A high signal on this pin resets the microcontroller.

7 *Serial input/output*
Writing to the serial data buffer SBUF at address 99H loads data for transmission; reading SBUF accesses received data. The bit addressable serial port control register SCON at address 98H is used to control the various modes of operation. See Chapter 10 for a discussion of such interfaces.

8 *Timing*
The timer mode register TMOD at address 89H is used to set the operating mode for timer 0 and timer 1 (Fig. 15.26). It is loaded at an entity and is not individually bit addressable. The timer control register TCON (Fig. 15.27) contains status and control bits for timer 0 and timer 1. The upper four bits are used to turn the timers on and off or to signal a timer overflow. The lower four bits have nothing to do with timers but are used to detect and initiate external interrupts.

7	6	5	4	3	2	1	0
GATE	C/T	M1	M0	GATE	C/T	M1	M0

89H

Timer 1 · Timer 0

Gate: 0 = timer runs whenever TR0/TR1 set
1 = timer runs only when INT0/INT1 is high along with TR0/TR
C/T: counter/timer select
0 = input from system clock, 1 = input from TX0/TX1
M0 and M1 set the mode

M1	M0	Mode	
0	0	0	13 bit counter, lower 5 bits of TL0 and all 8 bits of TH
0	1	1	16 bit counter
1	0	2	8 bit auto-reload timer/counter
1	1	3	TL0 is an 8 bit timer/counter controlled by timer 0 control bits. TH0 is an 8 bit timer controlled by timer 1 control bits. Timer 1 is off.

Fig. 15.26 TMOD register

	7	6	5	4	3	2	1	0
88H	TF1	TR1	TF0	TR0	IE1	IT1	IE0	IT0

TF0, TF1	Timer overflow flag; set by hardware when time overflows and cleared by hardware when the processor calls the interrupt routine
TR0, TR1	Timer run control bits: 1 = timer on, 0 = timer off
IE0, IE1	Interrupt edge flag set by hardware when external interrupt edge or low level detected and cleared when interrupt processed
IT0, IT1	Interrupt type set by software: 1 = falling edge triggered interrupt, 0 = low level triggered interrupt

Fig. 15.27   TCON register

The source of the bits counted by each timer is set by the C/T bit; if the bit is low the source is the system clock divided by 12 otherwise if high it is set to count an input from an external source. The timers can be started by setting TR0 or TR1 to 1 and stopping by making it 0. Another method of controlling a timer is by setting the GATE to 1 and so allowing a timer to be controlled by the INT0 or INT1 pin on the microcontroller going to 1. In this way an external device connected to one of these pins can control the counter on/off.

9   *Interrupts*
Interrupts force the program to call a subroutine located at a specified address in memory; they are enabled by writing to the interrupt enable register IE at address A8H (Fig. 15.28).

Fig. 15.28   IE register

8D	TH1
8C	TH0
8B	TL1
8A	TL0
89	TMOD
88	TCON
87	PCON
83	DPH
82	DPL
81	SP
80	P0

F0	B
E0	ACC
D0	PSW
B8	IP
B0	P3
A8	IE
A0	P2
99	SBUF
98	SCON
90	P1

Fig. 15.29   Registers

The term *special function registers* is used for the input/output control registers (Fig. 15.29), like IE above, and these are located at addresses 80 to FF. Accumulator A (ACC) is the major register used for data operations; the B register is used for multiplication and division. P0, P1, P2 and P3 are the latch registers for ports 0, 1, 2 and 3.

For more details, the reader is referred to the manuals published by the manufacturer or to texts such as *Programming and Interfacing the 8051 Microcontroller* by S. Yeralan and A. Ahluwalia (Addison-Wesley 1993), *The 8051 Microcontroller* by I. Scott MacKenzie (Prentice Hall 1999, 1995, 1992) or *The 8051 Family of Microcontrollers* by R.H. Barnett (Prentice Hall 1995).

15.3.3 Microchip microcontrollers

Another widely used family of 8-bit microcontrollers is that provided by Microchip as PIC16C6x/7x. The term PIC, Peripheral Interface Controller, is used for its single-chip microcontrollers. These use a form of architecture termed *Harvard architecture*. With this architecture, instructions are fetched from program memory using buses that are distinct from the buses used for accessing variables (Fig. 15.30). In the other microcontrollers discussed in this chapter, separate buses are not used and thus program data fetches have to wait for variable read/write and input/output operations to be completed before the next instruction can be received from memory. With Harvard architecture, instructions can be fetched every cycle without waiting, each instruction being executed during the cycle following its fetch. Harvard architecture enables faster execution speeds to be achieved for a given clock frequency. Figure 15.31 shows the pin connections for one version of the PIC16C74A microcontroller, and Figure 15.32 the architecture.

Fig. 15.30 Harvard architecture

Fig. 15.31 PIC16C74A

Fig. 15.32 PIC 16C74/74A

The basic features of the microcontroller are:

1 *Input/output ports*

   Pins 2, 3, 4, 5, 6 and 7 are for the bi-directional input/output port A. As with the other bi-directional ports, signals are read from and written to via port registers. The direction of the signals is controlled by the TRIS direction registers; there is a TRIS register for each port. TRIS is set as 1 for read and 0 for write (Fig. 15.33)

Port bits

TRIS for 5 outputs and 3 inputs

Fig. 15.33 Port direction

ADCON0 bits			
5	4	3	For analogue input on
0	0	0	Port A, bit 0
0	0	1	Port A, bit 1
0	1	0	Port A, bit 2
0	1	1	Port A, bit 3
1	0	0	Port A, bit 5
1	0	1	Port E, bit 0
1	1	0	Port E, bit 1
1	1	1	Port E, bit 2

Pins 2, 3, 4 and 5 can also be used for analogue inputs, pin 6 for a clock input to timer 0; pin 7 can also be the slave select for the synchronous serial port (see later in this section).

Pins 33, 34, 35, 36, 37, 38, 39 and 40 are for the bi-directional input/output port B; the direction of the signals is controlled by a corresponding TRIS direction register. Pin 33 can also be the external interrupt pin. Pins 37, 38, 39 and 40 can also be the interrupt on change pins. Pin 39 can also be the serial programming clock and pin 40 the serial programming data.

Pins 15, 16, 17, 18, 23, 24, 25 and 26 are for the bi-directional input/output port C; the direction of the signals is controlled by a corresponding TRIS direction register. Pin 15 can also be the timer 1 output or the timer 1 clock input. Pin 16 can also be the timer 1 oscillator input or Capture 2 input/Compare 2 output/PWM2 output.

Pins 19, 20, 21, 22, 27, 28, 29 and 30 are for the bi-directional input/output port D; the direction of the signals is controlled by a corresponding TRIS direction register.

Pins 8, 9 and 10 are for the bi-directional input/output port E; the direction of the signals is controlled by a corresponding TRIS direction register. Pin 8 can also be the read control for the parallel slave port or analogue input 5. The parallel slave port is a feature that facilitates the design of personal computer interface circuitry; when in use the pins of ports D and E are dedicated to this operation.

2 *Analogue inputs*
Pins 2, 3, 4, 5 and 7 of port A and pins 8, 9 and 10 of port E can also be used for analogue inputs, feeding through an internal analogue-to-digital converter. Registers ADCON1 and TRISA for port A (TRISE for port E) must be initialised to select the reference voltage to be used for the conversion and select channels as inputs. Then ADCON0 has to be initialised using the settings indicated opposite.

3 *Timers*
The microcontroller has three timers: timer 0, timer 1 and timer 2. Timer 0 is an 8-bit counter which can be written to or read from and can be used to count external signal transitions, generating an interrupt when the required number of events have occurred. The source of the count can be either the internal bus clock signal or an external digital signal. The choice of count source is made by the TOCS bit in the OPTION register (Fig. 15.34).

Fig. 15.34 OPTION register

If the prescaler is not selected then the count is incremented after every two cycles of the input source. A prescaler can be used so that signals are only passed to the counter after some other fixed number of clock cycles. The following shows the scaling rates possible. WDT gives the scaling factors selected when the watchdog timer is enabled. It is used to time out and reset the system if an operation is not concluded in a reasonable time; the default time is nominally 18 ms.

Prescalar bit values			TMR0 rate	WDT rate
PS2	PS1	PS0		
0	0	0	1 : 2	1 : 1
0	0	1	1 : 4	1 : 2
0	1	0	1 : 8	1 : 4
0	1	1	1 : 16	1 : 8
1	0	0	1 : 32	1 : 16
1	0	1	1 : 64	1 : 32
1	1	0	1 : 128	1 : 64
1	1	1	1 : 256	1 : 128

Timer 1 is the most versatile of the timers and can be used to monitor the time between signal transitions on an input pin or control the precise time of transitions on an output pin. When used with the capture or compare modes, it enables the microcontroller to control the timing of an output on pin 17.

Timer 2 can be used to control the period of a pulse width modulated (PWM) output. Pulse width modulated outputs are supplied at pins 16 and 17.

4 *Serial input/output*

The PIC microcontroller includes a synchronous serial port (SSP) module and a serial communications interface module (SCI). Pin 18 has the alternative functions of the synchronous serial clock input or output for SPI serial peripheral interface mode and I²C mode. The I²C bus provides a two-wire bi-directional interface that can be used with a range of other chips; it can also be used for connecting a master microcontroller to slave microcontrollers. UART, i.e. the universal asynchronous receiver transmitter, can be used to create a serial interface to a personal computer.

5 *Parallel slave port*

The parallel slave port uses ports D and E and enables the microcontroller to provide an interface with a PC.

6 *Crystal input*

Pin 13 is for the oscillator crystal input or external clock source input; pin 14 is for the oscillator crystal output. Figure 15.35(a) shows the arrangement that might be used for accurate frequency control. Figure 15.35(b) that which might be used for a low cost frequency control; for a frequency of 4 MHz we can have $R = 4.7$ kΩ and $C = 33$ pF. The internal clock rate is the oscillator frequency divided by 4.

File addr.	Bank 0	Bank 1	File addr
00h	INDF	INDF	80h
01h	TMR0	OPTION	81h
02h	PCL	PCL	82h
03h	STATUS	STATUS	83h
04h	FSR	FSR	84h
05h	PORTA	TRISA	85h
06h	PORTB	TRISB	86h
07h	PORTC	TRISC	87h
08h	PORTD	TRISD	88h
09h	PORTE	TRISE	89h
0Ah	PCLATH	PCLATH	8Ah
0Bh	INTCON	INTCON	8Bh
0Ch	PIR1	PIE1	8Ch
0Dh	PIR2	PIE2	8Dh
0Eh	TMR1L	PCON	8Eh
0Fh	TMR1H		8Fh
10h	T1CON		90h
11h	TMR2		91h
12h	T2CON	PR2	92h
13h	SSPBUF	SSPADD	93h
14h	SSPCON	SSPSTAT	94h
15h	CCPR1L		95h
16h	CCPR1H		96h
17h	CCP1CON		97h
18h	RCSTA	TXSTA	98h
19h	TXREG	SPBRG	99h
1Ah	RCREG		9Ah
1Bh	CCPR2L		9Bh
1Ch	CCPR2H		9Ch
1Dh	CCPR2CON		9Dh
1Eh	ADRES		9Eh
1Fh	ADCON0	ADCON1	9Fh
20h	General-purpose registers	General-purpose registers	A0h
7Fh			FFh

Fig. 15.36 Special-purpose registers

Fig. 15.35 Frequency control

7 *Master clear*

Pin 1 is the master clear, i.e. reset input, and is taken low to reset the device.

The *special-purpose registers* (Fig. 15.36) are used for input/output control, as illustrated above in relation to a few of these registers. The registers for the PIC16C73/74 are arranged in two banks and before a particular register can be selected, the bank has to be chosen by setting a bit in the status register (Fig. 15.37).

Fig. 15.37   STATUS register

For more details of PIC microcontrollers, see the publications of the manufacturer or a book such as *Design with PIC Microcontrollers* by J.B. Peatman (Prentice-Hall 1998).

15.3.4 Selecting a microcontroller

In selecting a microcontroller the following factors need to be considered:

1   *Number of input/output pins*
How many input/output pins are going to be needed for the task concerned?

2   *Interfaces required*
What interfaces are going to be required? For example, is pulse width modulation required? Many microcontrollers have PWM outputs, e.g. the PIC17C42 has two.

3   *Memory requirements*
What size memory is required for the task?

4   *The number of interrupts required*
How many events will need interrupts?

5   *Processing speed required*
The microprocessor takes time to execute instructions (see Section 16.2.2), this time being determined by the processor clock.

As an illustration of the variation of microcontrollers available, Table 15.1 shows details of members of Intel 8051 family, Table 15.2 the PIC16Cxx family and Table 15.3 the M68HC11 family.

Table 15.1   Intel 8051 family members

	ROM	EPROM	RAM	Timers	I/O ports	Interrupts
8031AH	0	0	128	2	4	5
8051AH	4K	0	128	2	4	5
8052AH	8K	0	256	3	4	6
8751H	0	4K	128	2	4	5

Table 15.2 PIC family members

	I/O	EPROM	RAM	ADC channels	USART	CCP modules
PIC16C62A	22	2K	128	0	0	1
PIC16C63	22	4K	192	0	1	2
PIC16C64A	33	2K	128	0	0	1
PIC16C65A	33	4K	192	0	1	2
PIC16C72	22	2K	128	5	0	1
PIC16C73A	22	4K	192	5	1	2
PIC16C74A	33	4K	192	8	1	3

Table 15.3 M68HC11 family members

	ROM	EEPROM	RAM	ADC	Timer	PWM	I/O	Serial	E-clock MHz
68HC11AO	0	0	256	8 ch, 8-bit	(1)	0	22	SCI, SPI	2
68HC11A1	0	512	256	8 ch, 8-bit	(1)	0	22	SCI, SPI	2
68HC11A7	8 K	0	256	8 ch,8-bit	(1)	0	38	SCI, SPI	3
68HC11A8	8 K	512	256	8 ch, 8-bit	(1)	0	38	SCI, SPI	3
68HC11C0	0	512	256	4 ch, 4-bit	(2)	2 ch, 8-bit	36	SCI, SPI	2
68HC11D0	0	0	192	None	(2)	0	14	SCI, SPI	2

Timer: (1) is 3 input capture, 5 output compare, real-time interrupt, watchdog timer, pulse accumulator, (2) is 3 or 4 input capture, 5 or 4 output compare, real-time interrupt, watchdog timer, pulse accumulator. Serial: SCI is asynchronous serial communication interface, SPI is synchronous serial peripheral interface.

15.4 Applications

The following are two examples of how microcontrollers are used; more case studies are given in Chapter 22.

15.4.1 Temperature measurement system

As a brief indication of how a microcontroller might be used, Figure 15.38 shows the main elements of a temperature measurement system using a MC68HC11. The temperature sensor gives a voltage proportional to the temperature (e.g. a thermo-transistor such as LM35; see Section 2.9.4). The output from the temperature sensor is connected to an ADC input line of the microcontroller. The microcontroller is programmed to convert the temperature into a BCD output which can be used to switch on the elements of a two-digit seven-element display. However, because the temperature may be fluctuating it is necessary to use a storage register which can hold data long enough for the display to be read. The storage register, 74HCT273, is an octal D-type flip-flop which is reset on the next positive-going edge of the clock input from the microcontroller.

Fig. 15.38 Temperature measurement system

15.4.2 Domestic washing machine

Figure 15.39 shows how a microcontroller might be used as the controller for a domestic washing machine. The microcontroller often used is the Motorola M68HC05B6; this is simpler and cheaper than the Motorola M68HC11 microcontroller discussed earlier in this chapter and is widely used for low cost applications.

The inputs from the sensors for water temperature and motor speed are via the analogue-to-digital input port. Port A provides the outputs for the various actuators used to control the machine and also the input for the water level switch. Port B gives outputs to the display. Port C gives outputs to the display and also receives inputs from the keyboard used to input to the machine the various program selections. The PWM section of the timer provides a pulse width modulated signal to control the motor speed. The entire machine program is interrupted and stopped if the door of the washing machine is opened.

Fig. 15.39 Washing machine

15.5 Programming

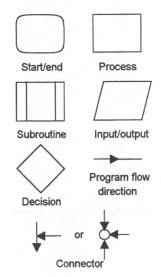

Start/end   Process

Subroutine   Input/output

Decision   Program flow direction

or

Connector

Fig. 15.40   Flow chart symbols

START

Operation A

Is output high?

YES

NO

Operation B

Operation C

Fig. 15.41   Flow chart

A

B

Fig. 15.42   Sequence

A commonly used method for the development of programs follows the following steps:

1. Define the problem, stating quite clearly what function the program is to perform, the inputs and outputs required, any constraints regarding speed of operation, accuracy, memory size, etc.

2. Define the algorithm to be used. An *algorithm* is the sequence of steps which define a method of solving the problem.

3. For systems with fewer than thousands of instructions a useful aid is to represent the algorithm by means of a *flow chart*. Figure 15.40 shows the standard symbols used in the preparation of flow charts. Each step of an algorithm is represented by one or more of these symbols and linked together by lines to represent the program flow. Another useful design tool is *pseudocode*. Pseudocode is a way of describing the steps in an algorithm in an informal way which can later be translated into a program.

4. Translate the flow chart/algorithm into instructions which the microprocessor can execute. This can be done by writing the instructions in some language, e.g. assembly language or perhaps C, and then converting these, either manually or by means of an assembler computer program, into a code which is acceptable to the microprocessor, i.e. machine code.

5. Test and debug the program. Errors in programs are referred to as *bugs* and the process of tracking them down and eliminating them as *debugging*.

Flow charts and pseudocode are aids to the development of systematic, structured, programs. Figure 15.41 shows part of a flow chart where, following the program start, there is operation A, followed by a branch to either operation B or operation C depending on whether the decision to the query is yes or no.

15.5.1 Pseudocode

Pseudocode is rather like drawing a flow chart and involves writing a program as a sequence of functions or operations with the decision element IF-THEN-ELSE and the repetition element WHILE-DO. A sequence (Fig. 15.42) would be written as:

```
BEGIN A
    ...
END A
    ...
```

```
BEGIN B
...
END B
```

and a decision as:

```
IF X
THEN
  BEGIN A
  ...
  END A
ELSE
  BEGIN B
  ...
  END B
EDNIF X
```

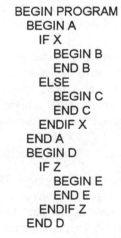

Fig. 15.43   IF-THEN-ELSE

Figure 15.43 shows such a decision in a flow chart. A repetition is written as:

```
WHILE  X
DO
  BEGIN A
  ...
  END A
  BEGIN B
  ...
  END B
ENDO WHILE  X
```

Figure 15.44 shows the WHILE-DO as a flow chart. A program written in this way might then appear as:

```
BEGIN PROGRAM
  BEGIN A
    IF X
      BEGIN B
      END B
    ELSE
      BEGIN C
      END C
    ENDIF X
  END A
  BEGIN D
    IF Z
      BEGIN E
      END E
    ENDIF Z
  END D
```

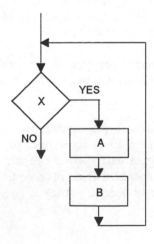

Fig. 15.44   WHILE-DO

Chapter 16 shows how programs can be written in assembly language and Chapter 17 in C.

Problems

1 Explain, for a microprocessor, the roles of (a) accumulator, (b) status, (c) memory address, (d) program counter registers.

2 A microprocessor uses eight address lines for accessing memory. What is the maximum number of memory locations that can be addressed?

3 A memory chip has 8 data lines and 16 address lines. What will be its size?

4 How does a microcontroller differ from a microprocessor?

5 Draw a block diagram of a basic microcontroller and explain the function of each subsystem.

6 Which of the M68HC11 ports is used for (a) the A/D converter, (b) a bi-directional port, (c) serial input/output, (d) as just an 8-bit output-only port?

7 How many bytes of memory does the M68HC11A7 have for data memory?

8 For the Motorola M68HC11, port C is bi-directional. How is it configured to be (a) an input, (b) an output?

9 The Motorola M68HC11 can be operated in single-chip and in extended mode. Why these modes?

10 What is the purpose of the ALE pin connection with the Intel 8051?

11 What input is required to reset an Intel 8051 microcontroller?

12 Write pseudocode to represent the following:
(a) If A is yes then B, else C.
(b) While A is yes do B.

16 Assembly language

16.1 Languages

Software is the term used for the *instructions* that tell a microprocessor or microcontroller what to do. The collection of instructions that a microprocessor will recognise is its *instruction set*. The form of the instruction set depends on the microprocessor concerned. The series of instructions that is needed to carry out a particular task is called a *program*.

Microprocessors work in binary code. Instructions written in binary code are referred to as being in *machine code*. Writing a program in such a code is a skilled and very tedious process. It is prone to errors because the program is just a series of 0s and 1s and the instructions are not easily comprehended by just looking at the pattern. An alternative is to use an easily comprehended form of shorthand code for the patterns of 0s and 1s. For example, the operation of adding data to an accumulator might be represented by just ADDA. Such a shorthand code is referred to as a *mnemonic code*, a mnemonic code being a 'memory aiding' code. The term *assembly language* is used for such a code. Writing a program using mnemonics is easier because they are an abbreviated version of the operation performed by the instruction. Also, because the instructions describe the program operations and can easily be comprehended, they are less likely to be used in error than the binary patterns of machine code programming. The assembler program has still, however, to be converted into machine code since this is all the microprocessor will recognise. This conversion can be done by hand using the manufacturer's data sheets which list the binary code for each mnemonic. However, computer programs are available to do the conversion, such programs being referred to as *assembler programs*.

High-level languages are available which provide a type of programming language which is even closer to describing in easily comprehended language the types of operations required. Examples of such languages are BASIC, C, FORTRAN and PASCAL. Such languages have still, however, to be converted into machine code, by a computer program, for the microprocessor to be able to use. This chapter is an outline of how programs might be written using assembly language, Chapter 17 using C.

16.2 Instruction sets

The following are commonly used instructions that may be given to a microprocessor, the entire list of such instructions being termed the *instruction set*. In general, instructions can be classified as:

1   Data transfer
2   Arithmetic
3   Logical
4   Program control

The instruction set differs from one microprocessor to another. Certain instructions are, however, reasonably common to most microprocessors. These are:

Data transfer

1   *Load*
    This instruction reads the contents of a specified memory location and copies it to a specified register location in the CPU, e.g.

Before instruction	After instruction
Data in memory location 0010	Data in memory location 0010
	Data from 0010 in accumulator

2   *Store*
    This instruction copies the current contents of a specified register into a specified memory location, e.g.

Before instruction	After instruction
Data in accumulator	Data in accumulator
	Data copied to memory location 0011

Arithmetic

3   *Add*
    This instruction adds the contents of a specified memory location to the data in some register, e.g.

Before instruction	After instruction
Accumulator with data 0001	Accumulator with data 0011
Memory location with data 0010	

4 *Decrement*

This instruction subtracts 1 from the contents of a specified location. For example, we might have the accumulator as the specified location and so:

Before instruction	After instruction
Accumulator with data 0011	Accumulator with data 0010

5 *Compare*

This instruction indicates whether the contents of a register are greater than, less than or the same as the contents of a specified memory location. The result appears in the status register as a flag.

Logical

6 *AND*

This instruction carries out the logical AND operation with the contents of a specified memory location and the data in some register. Numbers are ANDed bit by bit, e.g.

Before instruction	After instruction
Accumulator with data 0011	Accumulator with data 0001
Memory location with data 1001	

Only in the least significant bit in the above data have we a 1 in both sets of data and the AND operation only gives a 1 in the least significant bit of the result.

7 *EXCLUSIVE-OR*

This instruction carries out the logical EXCLUSIVE-OR operation with the contents of a specified memory location and the data in some register, bit by bit.

8 *Logical shift (left or right)*

Logical shift instructions involve moving the pattern of bits in the register one place to the left or right by moving a 0 into the end of the number. For example, for logical shift right a 0 is shifted into the most significant bit and the least significant bit is moved to the carry flag in the status register.

Before instruction	After instruction
Accumulator with data 0011	Accumulator with data 0001
	Status register indicates
	Carry 1

9 *Arithmetic shift (left or right)*

Arithmetic shift instructions involve moving the pattern of bits in the register one place to the left or right but preserves the sign bit at the left end of the number, e.g. for an arithmetic shift right:

Before instruction	After instruction
Accumulator with data 1011	Accumulator with data 1001
	Status register indicates
	Carry 1

10 *Rotate (left or right)*

Rotate instructions involve moving the pattern of bits in the register one place to the left or right and the bit that spills out is written back into the other end, e.g. for a rotate right:

Before instruction	After instruction
Accumulator with data 0011	Accumulator with data 1001

Program control

11 *Jump*

This instruction changes the sequence in which the program steps are carried out. Normally the program counter causes the program to be carried out sequentially in strict numerical sequence. However, the jump instruction causes the program counter to jump to some other specified location in the program. For example, a program might require the following sequence of instructions:

Decrement the accumulator
Jump if the accumulator is not zero to instruction ...

12 *Branch*

This is a conditional instruction which might be *branch if zero* or *branch if plus*. This branch instruction is followed if the right conditions occur. For example, a program might require the sequence of instructions shown in the flow chart in Figure 16.1.

Fig. 16.1 Example of a branch

13 *Halt*

This instruction stops all further microprocessor activity.

The types of mnemonic codes used for instructions in assembly language depend on the microprocessor/microcontroller used. Appendix C shows the instruction sets for the microcontroller M68HC11, Intel 8051 and PIC16Cxx. For full details, of this or other instruction sets, see the publications of the microprocessor/microcontroller manufacturer.

Numerical data may be binary, octal, hex or decimal. Generally in the absence of any indicator the assembler assumes the number is decimal. With Motorola, a number is indicated by the prefix #; a binary number is preceded by % or followed by B; an octal number is preceded by @ or followed by O; a hex number is preceded by $ or followed by H; and a decimal number requires no indicating letter or symbol. With Intel, numerical values must be preceded by # to indicate a number and by B for binary, O or Q for octal, H or h for hex and D or nothing for decimal. With PIC microcontrollers the header file has R = DEC for decimal to be the default. Then for binary the number is enclosed in quotation marks and preceded by B and for hex by H.

16.2.1 Addressing

When a mnemonic, such as LDA, is used to specify an instruction it will be followed by additional information to specify the source and destination of the data required by the instruction. The data following the instruction is referred to as the *operand*.

There are several different methods that are used for specifying data locations, i.e. addressing, and hence the way in which the program causes the microprocessor to obtain its instructions or data. Different microprocessors have different addressing modes. The Motorola 68HC11 has the six addressing modes of immediate, direct, extended, indexed, inherent and relative; the Intel 8051 has the five modes of immediate, direct, register, indirect and indexed; the PIC microcontroller has the three modes of immediate, direct and indirect with the indirect mode allowing indexing. The following are commonly used addressing modes:

1 *Immediate addressing*
   The data immediately following the mnemonic is the value to be operated on and is used for the loading of a predetermined value into a register or memory location. For example, with the Motorola code, LDA B #$25 means load the number 25 into accumulator B. The # signifies immediate mode and a number, the $ that the number is in hexadecimal notation. With the Intel code we might have MOV A,#25H to move the number 25 to the accumulator A. The # indicates a number and the H indicates a hex number. With the PIC code we might have movlw H'25' to load the number 25 into the working register w, the H indicating it is a hex number.

2 *Direct, absolute, extended or zero-page addressing*

With this form of addressing the data byte that follows the opcode directly gives an address that defines the location of the data to be used in the instruction. With Motorola the term *direct addressing* is used when the address given is only 8 bits wide; the term *extended addressing* is used when it is 16 bits wide. For example, with Motorola code, LDAA $25 means load the accumulator with the contents of memory location 0025, the 00 is assumed. With Intel code, for the same operation, we can have the direct address instruction MOV A,20H to copy the data at address 20 to the accumulator A. With the PIC code we might have movwf Reg1 to copy the contents of Reg1 into the working register, the address of Reg1 having been previously defined.

3 *Implied addressing, or inherent addressing*

With this mode of addressing, the address is implied in the instruction. For example, with Motorola code and Intel code, CLR A means clear accumulator A. With PIC code clrw means clear the working register.

4 *Register*

With this form of addressing, the operand is specified as the contents of one of the internal registers. For example, with Intel ADD R7,A to add the contents of the accumulator to register R7.

5 *Indirect*

This form of addressing means that that the data is to be found in a memory location whose address is given by the instruction. For example, with the PIC system the INDF and FSR registers are used. The address is first written to the FSR register and then this serves as an address pointer. A subsequent direct access of INDF with the instruction movf INDF,w will load the working register w using the contents of FSR as a pointer to the data location.

6 *Indexed addressing*

Indexed addressing means that the data is in a memory location whose address is held in an index register. The first byte of the instruction contains the opcode and the second byte contains the offset; the offset is added to the contents of the index register to determine the address of the operand. A Motorola instruction might thus appear as LDA A $FF,X; this means load accumulator A with data at the address given by adding the contents of the index register and FF. Another example is STA A $05,X; this means store the contents of accumulator A at the address given by the index register plus 05.

7 *Relative addressing*

This is used with branch instructions. The opcode is followed with a byte called the relative address. This indicates the displacement in address that has to be added to the program counter if the branch occurs. For example, Motorola code BEQ $F1 indicates that if the data is equal to zero then the next address in the program is F1 further on. The relative address of F1 is added to the address of the next instruction.

As an illustration, Table 16.1 shows some instructions with the modes of addressing used in Motorola systems.

Table 16.1 Examples of addressing

Address mode	Instruction	
Immediate	LDA A #$F0	Load accumulator A with data F0
Direct	LDA A $50	Load accumulator A with data at address 0050
Extended	LDA A$0F01	Load accumulator A with data at address 0F01
Indexed	LDA A $CF,X	Load accumulator with data at the address given by the index register plus CF
Inherent	CLR A	Clear accumulator A
Extended	CLR $2020	Clear address 2020, i.e. store all 0s at address 2020
Indexed	CLR $10,X	Clear the address given by the index register plus 10, i.e. store all 0s at that address

16.2.2 Data movement

The following is an example of the type of information that will be found in a manufacturer's (6800) instruction set sheet:

		Addressing modes					
		IMMED			DIRECT		
Operation	Mnemonic	OP	~	#	OP	~	#
Add	ADDA	8B	2	2	9B	3	2

~ is the number of microprocessor cycles required and # is the number of program bytes required.

This means that when using the immediate mode of addressing with this processor the Add operation is represented by the mnemonic ADDA. When the immediate form of addressing is used the machine code for it is 8B and it will take two cycles to be fully expressed. The operation will require two bytes in the program. The *op-code* or *operation code* is the term used for the instruction that the microprocessor will act on and is expressed in hexadecimal form. A byte is a group of eight binary digits recognised by the microprocessor as a word. Thus two words are required. With the direct mode of addressing the machine code is 9B and takes three cycles and two program bytes.

To illustrate how information passes between memory and microprocessor, consider the following tasks. The addresses in RAM where a new program may be placed are just a matter of convenience. For the following examples, the addresses starting at 0010 have been used. For direct addressing to be used, the addresses must be on the zero page, i.e. addresses between 0000 and 00FF. The examples are based on the use of the instruction set for the M6800 microprocessor.

Task: Enter all zeros in accumulator A.

Memory address	Op-code	
0010	8F	CLR A

The next memory address that can be used is 0011 because CLR A only occupies one program byte. This is the inherent mode of addressing.

Task: Add to the contents of accumulator A the data 20.

Memory address	Op-code	
0010	8B 20	ADD A #$20

This uses the immediate form of addressing. The next memory address that can be used is 0012 because, in this form of addressing, ADD A occupies two program bytes.

Task: Load accumulator A with the data contained in memory address 00AF.

Memory address	Op-code	
0010	B6 00AF	LDA A $00AF

This uses the absolute form of addressing. The next memory address that can be used is 0013 because, in this form of addressing, LDA A occupies three program bytes.

Task: Rotate left the data contained in memory location 00AF.

Memory address	Op-code	
0010	79 00AF	ROL $00AF

This uses the absolute form of addressing. The next memory address that can be used is 0013 since ROL, in this mode, occupies three program bytes.

Task: Store the data contained in accumulator A into memory location 0021.

Memory address	Op-code	
0010	D7 21	STA A $21

This uses the direct mode of addressing. The next memory address that can be used is 0012 since STA A, in this mode, occupies two program bytes.

Task: Branch forward four places if the result of the previous instruction is zero.

Memory address	Op-code	
0010	27 04	BEQ $04

This uses the relative mode of addressing. The next memory address if the result is not zero is 0012 since BEQ, in this mode, occupies two program bytes. If the result is zero then the next address is $0012 + 4 = 0016$.

16.3 Assembly language programs

An assembly language program should be considered as a series of instructions to an assembler which will then produce the machine code program. A program written in assembly language consists of a sequence of statements, one statement per line. A statement contains from one to four sections or *fields*, these being:

Label Op-code Operand Comment

A special symbol is used to indicate the beginning or end of a field, the symbols used depending on the microprocessor machine code assembler concerned. With the Motorola 6800 spaces are used. With the Intel 8080 there are a colon after a label, a space after the op-code, commas between entries in the address field and a semicolon before a comment. In general, a semicolon is used to separate comments from the operand.

The *label* is the name by which a particular entry in the memory is referred to. Labels can consist of letters, numbers and some other characters. With the Motorola 6800, labels are restricted to one to six characters, the first of which must be a letter, and cannot consist of a single letter A, B or X since these are reserved for reference to the accumulator or index register. With the Intel 8080, five characters are permitted with the first character a letter, @, or ?. The label must not use any of the names reserved for registers, instruction codes or pseudo-operations (see later in this section). All labels within a program must be unique. If there is no label then a space must be included in the label field. With the Motorola 6800, an asterisk (*) in the label field indicates that the entire statement is a comment, i.e. a comment included to make the purpose of the program clearer. As such, the comment will be ignored by the assembler during the assembly process for the machine code program.

The *op-code* specifies how data is to be manipulated and is specified by its mnemonic, e.g. LDA A. It is the only field that can never be empty.

In addition, the op-code field may contain directives to the assembler. These are termed *pseudo-operations* since they appear in the op-code field but are not translated into instructions in machine code. They may define symbols, assign programs and data to certain areas of memory, generate fixed tables and data, indicate the end of the program, etc. Common assembly directives are:

Set program counter

ORG	This defines the starting memory address of the part of the program that follows. A program may have several origins.

Define symbols

EQU, SET,	Equates/sets/defines a symbol for a numerical
DEF	value, another symbol or an expression.

Reserve memory locations

RMB, RES	Reserves memory bytes/space.

Define constant in memory

FCB	Forms constant byte.
FCC	Forms constant character string.
FDB	Forms double byte constant.
BSW	Block storage of zeros.

The information included in the *operand* field depends on the mnemonic preceding it and the addressing mode used. It gives the address of the data to be operated on by the process specified by the op-code. It is thus often referred to as the *address field*. This field may be empty if the instructions given by the op-code do not need any data or address. Numerical data in this field may be hexadecimal, decimal, octal, or binary. The assembler assumes that numbers are decimal unless otherwise specified. With the Motorola 6800 a hexadecimal number is preceded by $ or followed by H, an octal number preceded by @ or followed by O or Q, a binary number preceded by % or followed by B. With the Intel 8080, a hexadecimal number is followed by H, an octal number by O or Q, a binary number by B. Hexadecimal numbers must begin with a decimal digit, i.e. 0 to 9, to avoid confusion with names. With the Motorola 6800, the immediate mode of address can be indicated by preceding the operand by #, the indexed mode of address involving the operand being followed by X. No special symbols are used for direct or extended addressing modes. If the address is on the zero page, i.e. FF or less, then the assembler automatically assigns the direct mode. If the address is greater than FF, the assembler assigns the extended mode.

The comment field is optional and is there to allow the programmer to include any comments which may make the program more understandable by the reader. The comment field is ignored by the assembler during the production of the machine code program.

16.3.1 Examples of assembly language programs

The following examples illustrate how some simple programs can be developed.

Problem: The addition of the two 8-bit numbers located in different memory addresses and the storage of the result back into memory.

The algorithm is:

1 Start.
2 Load the accumulator with the first number. The accumulator is where the results of arithmetic operations are accumulated. It is the working register, i.e. like a notepad on which the calculations are carried out before the result is transferred elsewhere. Thus we have to copy data to the accumulator before we can carry out the arithmetic. With PIC the term working (w) register is used.
3 Add on the second number.
4 Store the sum in the designated memory location.
5 Stop.

Figure 16.2 shows these steps represented as a flow chart.

The following are the programs written for three different microcontrollers. In each, the first column gives the label, the second the op-code, the third the operand and the fourth the comment. Note that all comments are preceded by a semicolon.

M68HC11 program

```
                ; Addition of two numbers

NUM1    EQU     $00     ; location of number 1
NUM2    EQU     $01     ; location of number 2
SUM     EQU     $02     ; location for the sum
```

Fig. 16.2 Flow chart for the addition of two numbers

```
        ORG    $C000   ; address of start of user RAM
START   LDAA   $NUM1   ; load number 1 into acc. A
        ADDA   $NUM2   ; add number 2 to A
        STAA   SUM     ; save the sum to $02
        END
```

The first line in the program specifies the address of the first of the numbers to be added. The second line specifies the address of the number to be added to the first number. The third line specifies where the sum is to be put. The fourth line specifies the memory address at which the program should start. The use of the labels means that the operand involving that data does not have to specify the addresses but merely the labels.

When translated into machine code the pseudo-operations indicate the addresses for the items. In machine code the program would become:

```
0010  96  70
0012  9B  71
0014  97  72
0016  3F
```

The same program for an Intel 8051 would be:

8051 program

```
                ; Addition of two numbers

NUM1    EQU    20H      ; location of number 1
NUM2    EQU    21H      ; location of number 2
SUM     EQU    22H      ; location for the sum


        ORG    8000H    ; address of start of user RAM
START   MOV    A,NUM1   ; load number 1 into acc. A
        ADD    A,NUM2   ; add number 2 to A
        MOV    SUM,A    ; save the sum to address 22H
        END
```

The same program for a PIC microcontroller would be:

PIC program

```
                ; Addition of two numbers

Num1    equ    H'20'    ; location of number 1
Num2    equ    H'21'    ; location of number 2
Sum     equ    H'22'    ; location for the sum


        org    H'000'   ; address of start of user RAM
Start   movlw  Num1     ; load number 1 into w
        addlw  Num2     ; add number 2 to w
        movwf  Sum      ; save the sum H'22'
        End
```

Fig. 16.3   A loop

In many programs there can be a requirement for a task to be carried out a number of times in succession. In such cases the program can be designed so that the operation passes through the same section a number of times. This is termed *looping*, a *loop* being a section of a program that is repeated a number of times. Figure 16.3 shows a flow diagram of a loop. With such a loop, a certain operation has to be performed a number of times before the program proceeds. Only when the number of such operations is completed can the program proceed. The following problem indicates such a looping.

Problem: the addition of numbers located at 10 different addresses (these might be, for example, the results of inputs from 10 different sensors which have each to be sampled).

The algorithm could be:

1   Start.
2   Set the count as 10.
3   Point to location of bottom address number.
4   Add bottom address number.
5   Decrease the count number by 1.
6   Add 1 to the address location pointer.
7   Is count 0 ? If not branch to 4. If yes proceed.
8   Store sum.
9   Stop.

Figure 16.4 shows the flow chart. The program is:

```
COUNT     EQU      $0010
POINT     EQU      $0020
RESULT    EQU      $0050
          ORG      $0001
          LDA B    COUNT     ; Load count
          LDX      POINT     ; Initialise index register
                             ; at start of numbers
SUM       ADD A    X         ; Add addend
          INX                ; Add 1 to index register
          DEC B              ; Subtract 1 from
                             ; accumulator B
          BNE      SUM       ; Branch to sum
          STA A    RESULT    ; Store
          WAI                ; Stop program
```

The count number of 10 is loaded into accumulator B. The index register gives the initial address of the data being added. The first summation step is to add the contents of the memory location addressed by the index register to the contents of the accumulator, initially assumed zero (a CLR A instruction could be used to clear it first). The instruction INX adds 1 to the index register so that the next address that will be addressed is 0021. DEC B subtracts

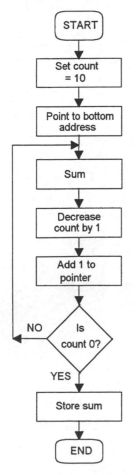

Fig. 16.4   Flow chart for adding 10 numbers

1 from the contents of accumulator B and so indicates that a count of 9 remains. BNE is then the instruction to branch to SUM if not equal to zero, i.e. if the Z flag has a 0 value. The program then loops and repeats the loop until ACC B is zero.

Problem: the determination of the biggest number in a list of numbers (it could be the determination of, say, the largest temperature value inputted from a number of temperature sensors).

The algorithm could be:

1 Clear answer address.
2 List starting address.
3 Load the number from the starting address.
4 Compare the number to the number in answer address.
5 Store answer if bigger.
6 Otherwise save number.
7 Increase starting address by 1.
8 Branch to 3 if the address is not the last address.
9 Stop.

Figure 16.5 shows the flow chart. The program is:

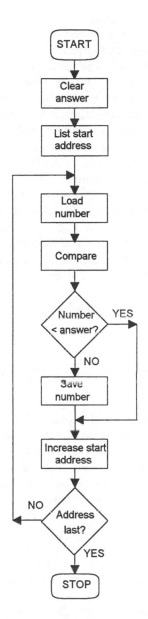

Fig. 16.5 Flow chart for biggest number

```
FIRST    EQU      $0030
LAST     EQU      $0040
ANSW     EQU      $0041
         ORG      $0000
         CLR      ANSW      ; Clear answer
         LDX      FIRST     ; Load first address
NUM      LDA A    $30,X     ; Load number
         CMP A    ANSW      ; Compare with answer
         BLS      NEXT      ; Branch to NEXT if lower or
                            ; same
         STA A    ANSW      ; Store answer
NEXT     INX                ; Increment index register
         CPX      LAST      ; Compare index register
                            ; with LAST
         BNE      NUM       ; Branch if not equal to zero
         WAI                ; Stop program
```

The procedure is that first the answer address is cleared. The first address is then loaded and the number in that address put into accumulator A. LDA A $30,X means load accumulator A with data at address given by the index register plus 30. Compare the number with the answer, keeping the number if it is greater than the number already in the accumulator, otherwise branch to repeat the loop with the next number.

16.4 Subroutines

It is often the case that a block of programming, a subroutine, might be required a number of times in a program. For example, a block of programming might be needed to produce a time delay. It would be possible to duplicate the subroutine program a number of times in the main program. This, however, is an inefficient use of memory. Alternatively we could have a single copy of it in memory and branch or jump to it every time the subroutine was required. This, however, presents the problem of knowing, after completion of the subroutine, the point at which to resume in the main program. What is required is a mechanism for getting back to the main program and continuing with it from the point at which it was left to carry out the subroutine. To do this we need to store the contents of the program counter at the time of branching to the subroutine so that this value can be reloaded into the program counter when the subroutine is complete. The two instructions which are provided with most microprocessors to enable a subroutine to be implemented in this way are:

1   JSR (jump to routine), or CALL, which enables a subroutine to be called.

2   RTS (return from subroutine), or RET (return), which is used as the last instruction in a subroutine and returns it to the correct point in the calling program.

Subroutines may be called from many different points in a program. It is thus necessary to store the program counter contents in such a way that we have a last-in-first-out store (LIFO). Such a register is referred to as a *stack*. It is like a stack of plates in that the last plate is always added to the top of the pile of plates and the first plate that is removed from the stack is always the top plate and hence the last plate that was added to the stack. The stack may be a block of registers within a microprocessor or, more commonly, using a section of RAM. A special register within the microprocessor, called the *stack pointer register*, is then used to point to the next free address in the area of RAM being used for the stack.

In addition to the automatic use of the stack when subroutines are used, a programmer can write a program which involves the use of the stack for the temporary storage of data. The two instructions that are likely to be involved are:

1   PUSH, which causes data in specified registers to be saved to the next free location in the stack.

2   PULL, or POP, which causes data to be retrieved from the last used location in the stack and transferred to a specified register.

For example, prior to some subroutine, data in some registers may have to be saved and then, after the subroutine, the data restored. The program elements might thus be, with the Motorola 6800:

```
SAVE      PSH A         ; Save accumulator A to
                        ; stack
          PSH B         ; Save accumulator B to
                        ; stack
          TPA           ; Transfer status register to
                        ; accumulator A
          PSH A         ; Save status register to stack
; Subroutine
RESTORE   PUL A         ; Restore condition code from
                        ; stack to accumulator A
          TAP           ; Restore condition code from
                        ; A to status register
          PUL B         ; Restore accumulator B from
                        ; stack
          PUL A         ; Restore accumulator A from
                        ; stack
```

16.4.1 Delay subroutine

Delay loops are often required when the microprocessor has an input from a device such as an analogue-to-digital converter. The requirement is often to signal to the converter to begin its conversion and then wait a fixed time before reading the data from the converter. This can be done by providing a loop which makes the microprocessor carry out a number of instructions before proceeding with the rest of the program. A simple delay program might be:

```
DELAY   LDA A   #$05    ; Load accumulator A with 05
LOOP    DEC A           ; Decrement accumulator A by 1
        BNE     LOOP    ; Branch if not equal to zero
        RTS             ; Return from subroutine
```

For each movement through the loop a number of machine cycles are involved. The delay program, when going through the loop five times, thus takes

Instruction	Cycles	Total cycles
LDA A	2	2
DEC A	2	10
BNE	4	20
RTS	1	1

The total delay is thus 33 machine cycles. If each machine cycle takes, say, 1 μs then the total delay is 33 μs. For a longer delay a bigger number can be initially put into accumulator A.

An example of a time delay subroutine for a PIC micro-controller is:

```
            movlw    Value    ; load count value required
            movwf    Count    ; loop counter
Delay       decfsz   Count    ; decrement counter
            goto     Delay    ; loop
```

The decfsz instruction takes one cycle and the goto instruction takes two cycles. This loop will be repeated (count − 1) times. In addition we have the movlw and movwf instructions, each taking one cycle, and when the count equals 1 we have decfsz which gives a further two cycles. Thus the total number of cycles is:

$$\text{number of instruction cycles} = 3(\text{count} - 1) + 4$$

Each instruction cycle takes four clock cycles and so the number of delay cycles introduced by this subroutine is:

$$\text{number of clock cycles} = 4[2(\text{count} - 1) + 4]$$

With a 4 MHz clock each clock cycle takes $1/(4 \times 10^6)$ s.

Often the delay obtained by using just the single loop described above is not enough. One way to obtain a longer delay is to use a nested loop. Figure 16.6 shows the flow chart for a nested loop delay. The inner loop is the same as the single-loop program described earlier. It will decrement register E 255 times before the looping is completed and the zero flag set. The outer loop causes the inner loop routine to be repeatedly executed as register D is decremented down to zero. Thus with register D initially set with a loop count of, say, 140 then the time delay will be 140 × 2.298 = 321.72 ms. The program is thus:

Fig. 16.6   Nested loop delay

```
DELAY     MOV    D,8CH     ; set D to 8CH, i.e. 140
OLOOP     MOV    E,FFH     ; set E to FFH, i.e. 255
ILOOP     DEC    E         ; decrement E, i.e. inner loop
                           ; counter
          JNZ    ILOOP     ; repeat ILOOP 255 times
          DEC    D         ; decrement D, i.e. outer loop
                           ; counter
          JNZ    OLOOP     ; repeat OLOOP 140 times
```

The following are some examples of programs where time delay subroutines are involved.

1   *Problem*: Switch a LED on and off repeatedly.

With this problem a subroutine DELAY is used with loops to provide time delays; the microprocessor takes a finite amount of time to process the instructions in a loop and thus to complete the loop. The structure of the program is:

1 If LED on
        Turn LED off
        While LED off, do subroutine TIME_DELAY
2 ELSE
        Turn LED on
        Do subroutine TIME_DELAY

Subroutine: TIME_DELAY
        Do an instruction, or instructions, or a loop, or a double
        loop depending on the length of time delay required.

Because of the length of time delay required, a double loop
is likely to be used for the time delay. With Intel 8051
programming, it is possible to use the instruction DJNZ,
decrement and jump if the result is not zero. It decrements the
location indicated by the first operand and jumps to the second
operand if the resulting value is not zero. The LED is
connected to bit 0 of port 1 of the microcontroller. The
program with Intel 8051 assembly instructions might thus be:

```
FLAG      EQU     0FH          ; flag set when LED is on
          ORG     8000H

START     JB      FLAG,LED_OFF ; jump if LED_OFF bit set, i.e. LED is on
          SETB    FLAG         ; else set FLAG bit
          CLR     P1.0         ; turn LED on
          LCALL   DELAY        ; call up delay subroutine
          SJMP    START        ; jump to START
LED_OFF   CLR     FLAG         ; clear the LED on flag to indicate the LED is off
          SETB    P1.0         ; turn LED off
          LCALL   DELAY        ; call up delay subroutine
          LJMP    START        ; jump to START

DELAY     MOV     R0,#0FFH     ; outer loop delay value
ILOOP     MOV     R1,#0FFH     ; inner loop delay value
OLOOP     DJNZ    R1,ILOOP     ; wait through inner loop
          DJNZ    R0,OLOOP     ; wait through outer loop
          RET                  ; return from subroutine

          END
```

2 *Problem*: Switch on in sequence eight LEDS.

The rotate instruction can be used to successively turn on
LEDS so that we have initially the bit pattern 0000 0001 which
is then rotated to give 0000 0011, then 0000 0111, and so on.
The following is a program in Motorola 68HC11 assembly
language that can be used, the LEDS being connected to port
B; a short delay is incorporated in the program:

```
COUNT          EQU       8             ; the count gives the number of loops required,
                                       ; i.e. the number of bits to be switched on
FIRST          EQU       %00000001     ; turn on 0 bit
PORTB          EQU       $1004         ; address of port B
               ORG       $C000
               LDAA      #FIRST        ; load initial value
               LDAB      #COUNT        ; load count
LOOP           STAA      PORTB         ; turn on bit 1 and so LED 1
               JSR       DELAY         ; jump to delay subroutine
               SEC                     ; set carry bit to rotate into least significant bit to
                                       ; maintain bit as 1
               ROLA                    ; rotate left
               DECB                    ; decrement count
               BNE       LOOP          ; branch to loop eight times

DELAY          RTS                     ; simple short delay

               END
```

16.5 Look-up tables

Indexed addressing can be used to enable a program to look up values in a table. For example, in determining the squares of integers, a possible method is to look up the value corresponding to a particular integer in a table of squares, instead of doing the arithmetic to determine the square. Look-up tables are particularly useful when the relationship is non-linear and not described by a simple arithmetic equation, e.g. the engine management system described in Section 1.5.2 where ignition timing settings are a function of the angle of the crankshaft and the inlet manifold pressure. Here the microcontroller has to give a timing signal that depends on the input signals from the speed sensor and crankshaft sensors.

To illustrate how look-up tables are used, consider the problem of determining the squares of integers. We can place a table of squares of the integers 0, 1, 2, 3, 4, 5, 6, ... in program memory and have the entries of the squares 0, 1, 4, 9, 16, 25, 36, ... at successive addresses. If the number to be squared is 4 then this becomes the index for the index address of the data in the table, the first entry being index 0. The program adds the index to the base address of the table to find the address of the entry corresponding to the integer. Thus we have:

Index	0	1	2	3	4	5	6
Table entry	0	1	4	9	16	25	36

For example, with the Motorola 68HC11 microcontroller we might have the following look-up program to determine the squares:

```
REGBAS    EQU     $B600           ; base address for the table
          ORG     $E000
          LDAB    $20             ; load acc. B with the
                                  ; integer to be squared
          LDX     #REGBAS         ; point to table
          ABX                     ; add contents of acc. B to
                                  ; index register X
          LDAA    $00,X           ; load acc. A with the
                                  ; indexed value
```

and we could have loaded the table into memory by using the pseudo-operation FDB:

```
          ORG     $B600
          FDB     $00,$01,$04,$09   ; giving values to the
                                    ; reserved memory
                                    ; block
```

With the Intel 8051 microcontroller the instruction MOVC A,@A+DPTR fetches data from the memory location pointed to by the sum of DPTR and the accumulator A and stores it in the accumulator. This instruction can be used to look up data in a table where the data pointer DPTR is initialised to the beginning of the table. As an illustration, suppose we want to use a table for the conversion of temperatures on the Celsius scale to the Fahrenheit scale. The program involves parameter passing of the temperature requiring conversion to a subroutine, so it might include the following instructions:

```
          MOV   A,#NUM                      ; load the
                                            ; value to
                                            ; be
                                            ; converted
          CALL  LOOK_UP                     ; call the
                                            ; LOOK_UP
                                            ; subroutine

LOOK_UP   MOV   DPTR,#TEMP                  ; point to
                                            ; table
          MOVC  A,@A+DPTR                   ; get the
                                            ; value from
                                            ; the table
          RET                               ; return
                                            ; from the
                                            ; subroutine
TEMP      DB    32, 34, 36, 37, 39, 41, 43, 45   ; giving
                                            ; values to
                                            ; the table
```

Another example of the use of a table is to sequence a number of outputs. This might be a sequence to operate traffic lights to give the sequence red, red plus amber, green, amber. The red light is illuminated when there is an output from RD0, the amber is illuminated from RD1, and the green is illuminated from RD2. The data table might then be:

	Red	Red + amber	Green	Amber
Index	0	1	2	3
	0000 0001	0000 0011	0000 0100	0000 0010

16.5.1 Delay with a stepper motor

In operating a stepper motor delays have to be used between each instruction to advance by a step to allow time for that step to occur before the next program instruction. A program to generate a continuous sequence of step pulses could thus have the algorithm:

1　Start.
2　State sequence of outputs needed to obtain the required step sequence.
3　Set to initial step position.
4　Advance a step.
5　Jump to delay routine to give time for the step to be completed.
6　Is this the last step in the step sequence for one complete rotation? If not continue to next step, if yes loop back to 3.
7　Continue until infinity.

The following is a possible program for a stepper in the full-step configuration and controlled by the microcontroller M68HC11 using outputs from PB0, PB1, PB2 and PB3. A 'look-up' table is used for the output code sequence necessary from the outputs to drive the stepper in the step sequence. The following is the table used:

Step	The outputs required from Port B				Code
	PB0	PB1	PB2	PB3	
1	1	0	1	0	A
2	1	0	0	1	9
3	0	1	0	1	5
4	0	1	1	0	6
1	1	0	1	0	4

The code sequence required for full-step stepper operation is A, 9, 5, 6, A and so these values constitute the sequence that the pointer has to look up in the table. FCB is the op-code for 'form constant byte' and is used to initialise data bytes for the table.

```
        BASE    EQU     $1000
        PORTB   EQU     $4          ; Output port
        TFLG1   EQU     $23         ; Timer interrupt flag register 1
        TCNT    EQU     $0E         ; Timer counter register
        TOC2    EQU     $18         ; Output compare 2 register
        TEN_MS  EQU     20000       ; 10 ms on clock

                ORG     $0000
        STTBL   FCB     $A          ; This is the look-up table
                FCB     $9
                FCB     $5
                FCB     $6
        ENDTBL  FCB     $A          ; End of look-up table

                ORG     $C000
                LDX     #BASE
                LDAA    #$80
                STAA    TFLG1,X     ; Clear flag
        START   LDY     #STTBL
        BEG     LDAA    0,Y         ; Start with first position in
                                    ; table
                STAA    PORTB,X
                JSR     DELAY       ; Jump to delay
                INY                 ; Increment in table
                CPY     #ENTBL      ; Is it end of table?
                BNE     BEG         ; If not branch to BEG
                BRA     START       ; If yes, go to start again

        DELAY   LDD     TCNT,X
                ADDD    #TEN_MS     ; Add a 10 ms time delay
                STD     TOC2,X
        HERE    BRCLR   TFLG1, X,   ; Wait till time delay elapsed
                        $80, HERE

                LDAA    #$80
                STAA    TFLG1,X     ; Clear flag
                RTS
```

Note that, in the above program, the label TEN_MS has the space underlined to indicate that both TEN and MS are part of the same label.

The delay in the above program is obtained by using the timer block in the microcontroller. A time delay of 10 ms is considered. For a microcontroller system with a 2 MHz clock a 10 ms delay is 20 000 clock cycles. Thus to obtain such a delay, the current value of the TCNT register is found and then 20 000 cycles added to it and the TOC2 register loaded with this value.

Problems

1 Using the following extract from a manufacturer's instruction set (6800), determine the machine codes required for the operation of adding with carry in (a) the immediate address mode, (b) the direct address mode.

		Addressing modes					
		IMMED			DIRECT		
Operation	Mnemonic	OP	~	#	OP	~	#
Add with carry	ADC A	89	2	2	99	3	2

2 The clear operation with the Motorola 6800 processor instruction set has an entry only in the implied addressing mode column. What is the significance of this?

3 What are mnemonics for, say, the Motorola 6800 for (a) clear register A, (b) store accumulator A, (c) load accumulator A, (d) compare accumulators, (e) load index register?

4 Write a line of assembler program for (a) load the accumulator with 20 (hex), (b) decrement the accumulator A, (c) clear the address $0020, (d) ADD to accumulator A the number at address $0020.

5 Explain the operations specified by the following instructions: (a) STA B $35, (b) LDA A #$F2, (c) CLC, (d) INC A, (e) CMP A #$C5, (f) CLR $2000, (g) JMP 05,X.

6 Write programs in assembly language to:
   (a) Subtract a hexadecimal number in memory address 0050 from a hexadecimal number in memory location 0060 and store the result in location 0070.
   (b) Multiply two 8-bit numbers, located at addresses 0020 and 0021 and store the product, an 8-bit number, in location 0022.
   (c) Store the hexadecimal numbers 0 to 10 in memory locations starting at 0020.
   (d) Move the block of 32 numbers starting at address $2000 to a new start address of $3000.

7 Write, in assembly language, a subroutine that can be used to produce a time delay and which can be set to any value.

8 Write, in assembly language, a routine that can be used so that if the input from a sensor to address 2000 is high the program jumps to one routine starting at address 3000, and if low the program continues.

17 C language

17.1 Why C?

C is a high-level language that is often used in place of assembly language (see Chapter 16) for the programming of microprocessors. It has the advantages when compared with the assembly language of being easier to use and that the same program can be used with different microprocessors; all that is necessary for this is that the appropriate compiler is used to translate the C program into the relevant machine language for the microprocessor concerned. Assembly language is different for the different microprocessors while C language is standardised, the standard being that of the American National Standards Institute (ANSI).

This chapter is intended to give an introduction to the C language and the writing of programs; for a more detailed consideration the reader is referred to texts such as *Teach Yourself C Programming in 21 Days* by P. Aitken and B.L. Jones (Sams Publishing 1995) or *C for Engineers and Scientists* by G. Bronson (West 1993) and, specifically for microcontrollers, *Programming Microcontrollers in C* by T. Van Sickle (High Text 1994), *C Programming for Embedded Systems* by K. Zurell (R & D Books 2000), *C and the 8051*, vols 1 and 2, by T. Schultz (Prentice Hall PTR 1998) or *Embedded C* by M. J. Pont (Addison Wesley 2002).

17.2 Program structure

Figure 17.1 gives an overview of the main elements of a C program. There is a pre-processor command that calls up a standard file, followed by the main function. Within this main function there are other functions which are called up as subroutines. Each function contains a number of statements.

17.2.1 Key features

The following are key features of programs written in C language. Note that in C programs spaces and carriage returns are ignored by the compiler and purely used for the convenience of the programmer to make it easier to read the program.

Fig. 17.1   Structure of a C program

1   *Keywords*
In C certain words are reserved as keywords with specific meanings. For example, *int* is used to indicate that integer values are concerned; *if* is used for when a program can change direction based on whether a decision is true or false. C requires that all keywords are in lower-case letters. Such words should not be used for any other purpose in a C program. The following are the ANSI C standard keywords:

auto	double	int	struct
break	else	long	switch
case	enum	register	typedef
char	extern	return	union
const	float	short	unsigned
continue	for	signed	void
default	goto	sizeof	volatile
do	if	static	while

2   *Statements*
These are the entries which make up a program, every statement being terminated by a semicolon. Statements can be grouped together in blocks by putting them between braces, i.e. { }. Thus for a two statement group we have:

```
{
  statement 1;
  statement 2;
}
```

3   *Functions*
The term *function* is used for a self-contained block of program code which performs a specific set of actions and has a name by which it can be referred (it is like a subroutine in an assembly language program). A function is written as a name followed by parentheses, i.e. name( ). The parentheses

enclose arguments; a function's argument is a value that is passed to the function when the function is called. A function is executed by calling it up by its name in the program statement. For example, we might have the statement

```
printf("Mechatronics");
```

This would mean that the word Mechatronics is passed to the function printf( ), a pre-written function which is called up by the pre-processor command, and, as a result, the word is displayed on screen. In order to indicate that characters form a sequence, e.g. those making up the word Mechatronics, they are enclosed within double quotes.

4   *Return*
A function may return a value to the calling routine. The *return type* appears in front of the function name, this specifying the type of value to be returned to the calling function when execution of the function is completed. For example, int main( ) is used for an integer return from the main function. The return type may be specified as void if the function does not return a value, e.g. void main(void). Often a header file will contain this return information and so it will not need to be specified for functions defined by the header file.

To return a value from a function back to the calling point, the keyword return is used, e.g. to return the result:

```
return result;
```

The return statement terminates the function.

5   *Standard library functions*
C packages are supplied with libraries containing a large number of predefined functions containing C code that have already been written and so saving you the effort of having to write them. These can be called up by naming them. In order to use the contents of any particular library that library has to be specified in a header file. Examples of such library files are:

```
math.h for mathematical functions
stdio.h for input and output functions
time.h for date and time functions
```

For example, the function printf( ) is a function that can be called up from the stdio.h library and is the function for printing to the screen of the monitor. Another function is scanf( ) which can be used to read data from a keyboard.

6 *Pre-processor*

The *pre-processor* is a program that is identified by *pre-processor commands* so that it is executed prior to the compilation. All such commands are identified by having # at the beginning of the line. Thus we might have:

```
# include < >
```

to include the file named between the angle brackets < >. When this command is reached the specified file will be inserted into the program. It is frequently used to add the contents of standard header files, these giving a number of declarations and definitions to enable standard library functions to be used. The entry would then be:

```
# include <stdio.h>
```

As an illustration, consider the simple program:

```
# include <stdio.h>

main( )
{
   printf("Mechatronics");
}
```

Before starting the main program the file stdio.h is added. Thus when the main program starts we are able to access the function printf( ) which results in the word Mechatronics being displayed on the screen.

Another type of pre-processor command is:

```
# define pi 3.14
```

and this can be used to define values that will be inserted whenever a particular symbol is encountered in the program. Thus whenever pi is encountered the value 3.14 will be used.

```
# define square(x) (x)*(x)
```

will replace the term square in the program by (x)*(x).

7 *Main function*

Every C program must have a function called *main( )*. This function is the one that exercises control when the program is executed and is the first function to be called up. Execution starts with its first statement. Other functions may be called up within statements, each one in turn being executed and control returned to the main function. The statement:

```
void main(void)
```

indicates that no result is to be returned to the main program and there is no argument. By convention a return value of 0 from main( ) is used to indicate normal program termination, i.e. the entry:

```
return 0;
```

8 *Comments*
/* and */ are used to enclose comments. Thus we might have an entry such as:

```
/* Main program follows */
```

Comments are ignored by the compiler and are just used to enable a programmer to more easily comprehend the program. Comments can span more than one line, e.g.

```
/* An example of a program used to
illustrate programming */
```

9 *Variables*
A *variable* is a named memory location that can hold various values. Variables that can hold a character are specified using the keyword *char*, such a variable being 8 bits long and generally used to store a single character. Signed integers, i.e. numbers with no fractional parts and which are signed to indicate positive or negative, are specified using the keyword *int*. The keyword *float* is used for floating-point numbers, these being numbers which have a fractional part. The keyword *double* is also used for floating-point numbers but provides about twice the number of significant digits as *float*. To declare a variable the type is inserted before the variable name, e.g.

```
int counter;
```

This declares the variable 'counter' to be of the integer type. As another example we might have:

```
float x, y;
```

This indicates that the variables x and y are both floating-point numbers.

10 *Assignments*
An assignment statement assigns the value of the expression to the right of the = sign to the variable on its left. For example, a = 2 assigns the value 2 to the variable a.

11 *Arithmetic operators*

The arithmetic operators used are: addition +, subtraction –, multiplication *, division /, modulus %, increment + + and decrement – –. Increment operators increase the value of a variable by 1, decrement operators decrease it by 1. The normal rules of arithmetic hold for the precedence of operations. For example, 2*4 + 6/2 gives 11. An example of a program involving arithmetic operators is:

```
/* program to determine area of a circle*/

#include <stdio.h> /*identifies the function library*/

int radius, area /*variables radius and area are integers*/

int main(void ) */starts main program, the int specifies
      that an integer value is returned, the void indicates
      that main( ) has no parameters*/
{
   printf("Enter radius:"); */ "Enter radius" on screen*/
   scanf("%d", &radius); */Reads an integer from
      keyboard and assigns it to the variable radius*/
   area = 3.14 * radius * radius; */Calculates area*/
   printf("\nArea = %d", area); */On new line prints Area
      = and puts in numerical value of the area*/
   return 0; */returns to the calling point*/
}
```

12 *Relational operators*

Relational operators are used to compare expressions, asking questions such as "Is x equal to y?" or "Is x greater than 10?" The relational operators are: is equal to = =, is not equal to !=, less than <, less than or equal to <=, greater than >, greater than or equal to >=. Note that = = has to be used when asking if two variables are the same, = is used for assignment when you are stating that they are the same. For example, we might have the question "Is x equal to 2?" and represent this by (a = = 2).

13 *Logical operators*

The logical operators are:

Operator	Symbol
AND	&&
OR	||
NOT	!

Note that in C the outcome is equal to 1 if true and 0 if false.

14 *Bitwise operations*

The bitwise operators treat their operands as a series of individual bits rather than a numerical value, comparing corresponding bits in each operand, and only work with integer variables. The operators are:

Bitwise operation	Symbol
AND	&
OR	|
EXCLUSIVE-OR	^
NOT	~
Shift right	>>
Shift left	<<

Thus, for example, we might have the statement:

```
portA = portA | 0x0c;
```

The prefix 0x is used to indicate that the 0c is a hex value, being 0000 1100 in binary. The value ORed with port A is thus a binary number that forces bits 2 and 3 on, all the other bits remaining unchanged.

```
portA = portA ^ 1;
```

This statement causes all the bits except for bit 1 of port A to remain unchanged. If bit 0 is 1 in port A the XOR will force it to 0 and if it is 0 it will force it to 1.

15 *String*

A sequence of characters enclosed within double quotes, i.e. " ", is termed a string. As the term implies, the characters within the double quotes are treated as a linked entity. For example, we might have:

```
printf("Sum = %d", x)
```

The argument in ( ) specifies what is passed to the print function. There are two arguments, the two being separated by a comma. The first argument is the string between the double quotes and specifies how the output is to be presented, the %d specifying that the variable is to be displayed as a decimal integer. Other format specifiers are:

%c	character
%d	signed decimal integer
%e	scientific notation
%f	decimal floating point

%o	unsigned octal
%s	string of characters
%u	unsigned decimal integer
%x	unsigned hexadecimal
%%	prints a % sign

The other argument x specifies the value that is to be displayed.

As another example, the statement:

```
scanf("%d", &x);
```

reads a decimal integer from the keyboard and assigns is to the integer variable x. The & symbol in front of x is the 'address of' operator. When placed before the name of a variable, it returns the address of the variable. The command thus scans for data and stores the item using the address given.

16 *Escape sequences*
Escape sequences are characters that 'escape' from the standard interpretation of characters and are used to control the location of output on a display by moving the screen cursor or indicating special treatments. Thus we might have:

```
printf("\nSum = %d", d)
```

with the \n indicating that a new line is to be used when it is printed on the screen. Escape sequences commonly used are:

\a	sound a beep
\b	backspace
\n	new line
\t	horizontal tab
\\	backslash
\?	question mark
\'	singe quotation

17.2.2 Example of a C program

An example of a simple program to illustrate the use of some of the above terms is:

```
/* A simple program in C*/

# include <stdio.h>
void main(void)
{
```

```
int a, b, c, d;  /*a, c, c and d are integers*/
a = 4;  /*a is assigned the value 4*/
b = 3; /*b is assigned the value 3*/
c = 5; /*c is assigned the value 5*/
d = a * b * c;  /*d is assigned the value of  a % b % c*/
printf("a * b * c = %d\n", d);
}
```

The statement int a, b, c, d; declares the variables a, b, c and d to be integer types. The statements a = 4, b = 3, c = 5 assign initial values to the variables, the = sign being used to indicate assignment. The statement d = a * b * c directs that a is to be multiplied by b and then by c and stored as d. The printf in the statement printf("a * b * c = %d\n", d) is the display on screen function. The argument contains %d and this indicates that it is to be converted to a decimal value for display. Thus it will print a * b * c = 60. The character \n at the end of the string is to indicate that a new line is to be inserted at that point.

17.3 Branches and loops

Statements to enable branching and looping in programs include *if*, *if/else*, *for*, *while* and *switch*.

1   *If*

The *if* statement allows branching (Fig. 17.2). For example, if an expression is true then the statement is executed, if not true it is not and the program proceeds to the next statement. Thus we might have statements of the form:

```
if(condition 1= = condition 2);
printf("\nCondition is OK.");
```

An example of a program involving if statements is:

```
#include <studio.h>

int x, y;
main( )
{
    printf("\nInput an integer value for x: ");
    scanf("%d", &x);
    printf("\nInput an integer value for y: ");
    scanf(%d", &y);
    if(x = = y)
       printf("x is equal to y");
    if(x > y)
       printf("x is greater than y");
    if(x < y)
       printf("x is less than y");
    return 0;
}
```

Fig. 17.2   If

Fig. 17.3 If/else

The screen shows Input an integer value for x: and then a value is to be keyed in. The screen then shows Input a value for y: and then a value is to be keyed in. The if sequence then determines whether the keyed in values are equal or which is greater than the other and displays the result on the screen.

2 *If/else*
The *if* statement can be combined with the *else* statement. This allows one statement to be executed if the result is yes and another if it is no (Fig. 17.3). Thus we might have:

```
#include <studio.h>

main( )
{
  int temp;
  if(temp > 50)
    printf("Warning");
  else
    printf("System OK");
}
```

3 *For*
The term *loop* is used for the execution of a sequence of statements until a particular condition reaches the required condition of being true, or false. Figure 17.4 illustrates this. One way of writing statements for a loop is to use the function *for*. The general form of the statement is:

```
for(initialising expression; test expression; increment
    expression)
loop statement;
```

Thus we might have:

```
#include <studio.h>

int count

main( )
{
  for(count = 0; count < 7; count ++)
  printf("\n%d", count);
}
```

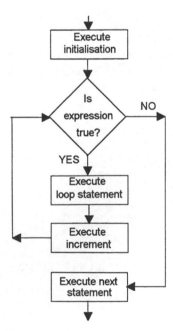

Fig. 17.4 For

Initially the count is 0 and will be incremented by 1 and then looped to repeat the for statement as long as the count is less than 7. The result is that the screen shows 0 1 2 3 4 5 6 with each number being on a separate line.

Fig. 17.5 While

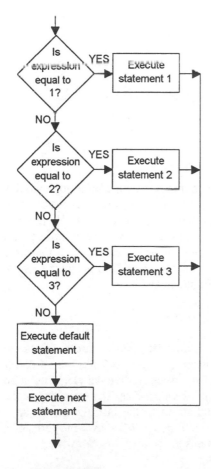

Fig. 17.6 Switch

4 *While*

The *while* statement allows for a loop to be continually repeated as long as the expression is true (Fig. 17.5). When the expression becomes false then the program continues with the statement following the loop. As an illustration we could have the following program where the while statement is used to count as long as the number is less than 7, displaying the results.

```
#include <studio.h>

int count;
int main( );
{
    count = 1;
    while(count < 7)
        {
            printf("\n%d", count);
            count++;
        }
        return 0;
}
```

The display on the screen is 1 2 3 4 5 6 with each number on a separate line.

5 *Switch*

The *switch* statement allows for the selection between several alternatives, the test condition being in parentheses. The possible choices are identified by case labels, these identifying the expected values of the test condition. For example, we might have the situation where if case 1 occurs we execute statement 1, if case 2 occurs we execute statement 2, etc. If the expression is not equal to any of the cases then the default statement is executed. After each case statement there is normally a break statement to transfer execution to the statement after the switch and stop the switch continuing down the list of cases. The sequence is thus (Fig. 17.6):

```
switch(expression)
{
    case 1;
        statement 1;
        break
    case 2;
        statement 2;
        break;
    case 3;
        statement 3;
        break;
    default;
        default statement;
```

```
}
next statement
```

The following is an example of a program which recognises the numbers 1, 2 and 3 and will display whichever one is entered from the keyboard.

```
#include <stdio.h>

int main ( );
{
  int x;

  printf("Enter a number 0, 1, 2 or 3:  ");
  scanf("%d", &x);

  switch (x)
  {
    case 1:
      printf("One");
      break;
    case 2:
      printf("Two");
      break;
    case 3:
      printf("Three");
      break;
    default;
      printf("Not 1, 2 or 3");
  }
  return 0;
}
```

17.4 Arrays

Suppose we want to record the mid-day temperature for each day for a week and then later be able to find the temperature corresponding to any one particular day. This can be done using an array. An *array* is a collection of data storage locations with each having the same data type and referenced though the same name. To declare an array with the name Temperature to store values of type float we use the statement:

```
float Temperature[7];
```

The size of the array is indicated between square brackets [ ] immediately after the array name. In this case 7 has been used for the data for the even days of the week. Individual elements in an array are referenced by an index value. The first element has the number 0, the second 1, and so on to the last element in an n sequence which will be $n - 1$. Figure 17.7 shows the form of a sequential array. To store values in the array we can write:

| array[0] | array[1] | array[2] | array[3] |

Fig. 17.7   A four-element sequential array

```
temperature[0] = 22.1;
```

```
temperature [1] = 20.4;
etc.
```

If you want to use scanf( ) to input a value into an array element, put & in front of the array name, e.g.

```
scanf("%d", &temperature [3]);
```

The following is an example of a simple program to store and display the squares of the numbers 0, 1, 2, 3 and 4:

```
#include <stdio.h>

int main(void)
{
    int sqrs[5];
    int x;

    for(x = 1; x<5; x++)
        sqrs[x – 1] = x * x;
    for(x = 0; x < 4; x++)
        printf("%d", sqrs[x]);

    return 0;
}
```

Arrays can be given initial values when first declared, e.g.

```
int array[7] = {10, 12, 15, 11, 10, 14, 12};
```

If you omit the array size, the compiler will create an array just large enough to hold the initialisation values.

```
int array[ ] = {10, 12, 15, 11, 10, 14, 12};
```

Multidimensional arrays can be used. For example, a table of data is a two-dimensional array (Fig. 17.8), where x represents the row and y the column, and is written as:

```
array[x][y];
```

Fig. 17.8 Two-dimensional array

17.5 Pointers

Each memory location has a unique address and this provides the means by which data stored at a location can be accessed. A *pointer* is a special kind of variable that can store the address of another variable. Thus if a variable called p contains the address of another variable called x, then p is said to *point* to x. Thus if x is at the address 100 in the memory then p would have the value 100. A pointer is a variable and, as with all other variables, has to be declared before it can be used. A pointer declaration takes the form:

type *name;

The * indicates that the name refers to a pointer. Often names used for pointers are written with the prefix p, i.e. in the form pname. Thus we might have:

int *pnumber;

To initialise a pointer and give it an address to point to we can use &, which is the address-of operator, in a statement of the form:

pointer = &variable;

The following short program illustrates the above:

```
#include <stdio.h>

int main(void)

{
  int *p, x;
  x = 12;
  p = &x;   /* assigns p the address of x*/
  printf("%d", *p);  /*displays the value of x using pointer*/

  return 0;
{
```

The program thus displays the number 12 on the screen. Accessing the contents of a variable by using a pointer, as above, is called *indirect access*. The process of accessing the data in the variable pointed to by a pointer is termed *dereferencing* the pointer.

17.5.1 Pointer arithmetic

Pointer variables can have the arithmetic operators +, −, + + and − − applied to them. Incrementing or decrementing a pointer results in it pointing to the next or previous element in an array. Thus to increment a pointer to the next item in an array we can use

pa++; /*using the increment by 1 operator*/

or

pa = pa + 1;   /*adding 1*/

17.5.2 Pointers and arrays

Pointers can be used to access individual elements in an array, the following program showing such access:

```
#include <stdio.h>

int main(void)
{
    int x[5] = (0, 2, 4, 6, 8);
    int *p;
    p = x;   /*assigns to p the address of the start of x*/
    printf("%d %d", x[0], x[2]);

    return 0;
}
```

The statement printf("%d %d", x[0], x[2]); results in pointing to the address given by x and hence the values at the addresses [0] and [2] being displayed, i.e. 0 and 4, on separate lines.

17.6 Program development

In developing programs the aim is to end up with a set of machine language instructions which can be used to operate a microprocessor/microcontroller system. These instructions are called the *executable file*. In order to arrive at such a file the following sequence of events occurs:

1   *Creating the source code*
    This is the writing of the sequence of statements in C that will constitute the program. Many compilers come with an editor and so the programmer can simply type in the source code from the keyboard. Otherwise a program such as Notepad with Microsoft Windows can be used. Using a word processor can present problems in that additional formatting information is included which can prevent compilation unless the file is saved without formatting information.

2   *Compiling the source code*
    Once the source code has been written, the programmer can direct the compiler to translate it into machine code. Before the compilation process starts, all the pre-processor commands are executed. The compiler can detect several different forms of error during the translation and generate messages indicating the errors. Sometimes a single error may result in the cascading sequence of errors all following from that single first error. Errors usually involve going back to the editor stage and re-editing the source code. The compiler then stores the resulting machine code in another file.

3   *Linking to create an executable file*

Then the compiler is used to brink together, i.e. link, the generated code with library functions to give a single executable file. The program is then stored as the executable file.

17.6.1 Header files

Pre-processor commands are used at the beginning of a program to define the functions used in that program; this is so they can be referred to by simple labels. However, to save having to write long lists of standard functions for each program, a pre-processor instruction can be used to indicate that a file should be used which includes the relevant standard functions. All that is necessary is to indicate which file of standard functions should be used by the compiler; this file being is a *header* since it comes at the head of the program. For example, <stdio.h> contains standard input and output functions such as gets (inputs, i.e. reads a line from a device), puts (outputs, i.e. writes a line to a device) and scanf (reads data); <math.h> contains mathematical functions such as cos, sin, tan, exp (exponential), and sqrt (square root).

Header files are also available to define the registers and ports of microcontrollers and save the programmer having to define each register and port by writing pre-processor lines for each. Thus for an Intel 8051 microcontroller we might have the header <reg.51.h>; this defines registers, e.g. the ports P0, P1, P2, and P3, and individual bits in bit-addressable registers, e.g. bits TF1, TR1, TF0, TR0, IE1, IT1, IE0 and IT0 in register TCON. Thus we can write instructions referring to port 0 inputs/outputs by purely using the label P0 or TF1 to refer to the TF1 bit in register TCON. Similarly, for a Motorola M68HC11E9 the header <hc11e9.h.> defines registers, e.g. PORTA, PORTB, PORTC and PORTD, and individual bits in bit-addressable registers, e.g. bits STAF, STAI, CWOM, HNDS, OIN, PLS, EGA and INVB in register PIOC. Thus, for example, we can write instructions referring to port A inputs/outputs by purely using the label PORTA. Libraries might also supply routines to help with the use of hardware peripheral devices such as keypads and liquid crystal displays.

The main program written for perhaps one specific microcontroller can, as a result of changing the header file, be easily adapted for running with another microcontroller. Libraries thus make C programs highly portable.

17.7 Examples of programs

The following are examples of programs written in C for microcontroller systems.

Fig. 17.9   Motor control

17.7.1 Switching a motor on and off

Consider the programming of a microcontroller, M68HC11, to start and stop a d.c. motor. Port C is used for the inputs and port B for the output to the motor, via a suitable driver (Fig. 17.9). The start button is connected to PC0 to switch from a 1 to a 0 input when the motor is to be started. The stop button is connected to PC1 to switch from a 1 to a 0 input when the motor is to be stopped. The port C data direction register DDRC has to be set to 0 so that port C is set for inputs. A program might be:

```
#include <hc11e9.h>  /*includes the header file*/

void main(void)
{
    PORTB.PB0 &=0; /*initially ensures motor off*/
    DDRC = 0; /*sets port C to be input*/
    while (1)  /*repeats while this condition holds*/
    {
        if (PORTC.PC0 = =0)   /*is start button pressed?*/
            PORTB.PB0 |=1; /*start output if pressed*/
        else if(PORTC.PC1 = =0)   /*is stop button pressed?*/
            PORTB.PB0 &=0;  /*stop output if pressed*/
    }
}
```

Note that | is the OR operator and sets a bit in the result to 0 only if the corresponding bits in both operands are 0, otherwise it sets to 1. It is used to turn on, or set, one or more bits in a value. Thus Port B.PB0 |=1 has 1 ORed with the value in PB0 and thus switches the motor on. It is a useful way of switching a number of bits in a port simultaneously. The & in PORTB.PB0 &=0 is used to AND the PB0 bit with 0 and so, since PB0 is already 1, assign to PORTB.PB0 the value 0.

17.7.2 Reading an ADC channel

Consider the task of programming a microcontroller (M68HC11) so that a single channel of its ADC can be read. The M68HC11 contains an eight-channel multiplexed, 8-bit, successive approximations ADC with inputs via port E (Fig. 17.10). The ADC control/status register ADCTL contains the conversion complete flag CCF at bit 7 and other bits to control the multiplexer and the channel scanning. When CCF = 0 the conversion is not complete and when 1 it is complete. The analogue-to-digital conversion is initiated by writing a 1 to the ADPU bit in the OPTION register. However, the ADC must have been turned on for at least 100 μs prior to reading a value.

To convert the analogue input to PE0, the first four bits in the ADCTL register, i.e. CA, CB, CC and CD, have to be all set to 0. When operating to convert just a single channel, bit 5 SCAN might be set to 0 and bit 4 MULT to 0. A simple program to read a particular channel might thus involve, after powering-up the ADC, turning all the bits in the ADCTL register to 0s, putting in the channel number and then reading the input while CCF is 0.

Fig. 17.10 ADC converter

The program might thus be as follows:

```
#include <hc11e9.h>   /*the header file*/

void main(void)
  {
    unsigned int k; /* this enters the channel number*/

    OPTION=0; /*this and following line turns the ADC on*/
    OPTION.ADPU=1;

    ADCTL &=~0x7;  /*clears the bits*/
    ADCTL |=k; /*puts the channel number to be read*/
```

```
        while (ADCTL.CCF= =0);
        return ADR1; /*returns converted value to address 1*/
    }
```

Note that ~ is the complement operator and its action is to reverse every bit in its operand, changing all the 0s to 1s and vice versa. Thus bit 7 is set. | is the OR operator and sets a bit in the result to 0 only if the corresponding bits in both operands are 0, otherwise it sets to 1. It is used to turn on, or set, one or more bits in a value. In this case with k = 1 it just sets CA to 1. A delay subroutine can be included to ensure that after power-up the value is not read too quickly.

Problems

1 The following questions are all concerned with components of programs.
   (a) State what is indicated by int in the statement:

      int counter;

   (b) State what the following statement indicates:

      num = 10

   (c) State what the result of the following statement will be:

      printf("Name");

   (d) State what the result of the following statement will be:

      printf("Number %d", 12);

   (e) State what the effect of the following is:

      #include <stdio.h>

2 For the following program: what are the reasons for including the line (a) #include <stdio.h>, (b) the { and }, (c) the /d, and (d) what will appear on the screen when the program is exccuted?

```
      #include <stdio.h>

      main( )
      {
          printf(/d"problem 3");
      }
```

3 For the following program, what will be displayed on the screen?

```
#include <stdio.h>

int main(void );
{
   int num;
   num = 20;

printf("The number is %d", num);

return 0;
}
```

4   Write a program to compute the area of a rectangle given its length and width at screen prompts for the length and width and then display the answer preceded by the words 'The area is'.

5   Write a program that displays the numbers 1 to 15, each on a separate line.

6   Explain the reasons for the statements in the following program for the division of two numbers:

```
#include <stdio.h>

int main(void);
{
   int num1, num2;

   printf("Enter first number:");
   scanf("%d", &num1);

   printf("Enter second number: ");
   scanf("%d", &num2);

   if(num2 = = 0)
      print f("Cannot divide by zero")
   else
      printf("Answer is: %d", num1/num2);

   return 0;
}
```

18 Input/output systems

18.1 Interfacing

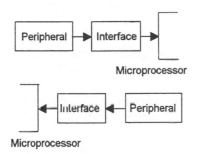

Fig. 18.1   The interfaces

When a microprocessor is used to control some system it has to accept input information, respond to it and produce output signals to implement the required control action. Thus there can be inputs from sensors to feed data in and outputs to such external devices as relays and motors. The term *peripheral* is used for a device, such as a sensor, keyboard, actuator, etc. which is connected to a microprocessor. It is, however, not normally possible to connect directly such peripheral devices to a microprocessor bus system due to a lack of compatibility in signal forms and levels. Because of such incompatibility, a circuit known as an *interface* is used between the peripheral items and the microprocessor. Figure 18.1 illustrates the arrangement. The interface is where this incompatibility is resolved.

This chapter discusses the requirements of such interfaces and the very commonly used Motorola MC6820 Peripheral Interface Adapter and Motorola MC6850 Asynchronous Communications Interface Adapter.

18.2 Input/output addressing

There are two ways microprocessors can select input/output devices. Some microprocessors, e.g. the Zilog Z80, have *isolated input/output* and special input instructions such as IN are used to read from an input device and special output instructions such as OUT are used to output to an output device. For example, with the Z80 we might have:

    IN A,(B2)

to read input device B2 and put the data in the accumulator A. An output instruction might be:

    OUT (C), A

to write the data in accumulator A to port C.

More commonly, microprocessors do not have separate instructions for input and output but use the same instructions as they use for reading from or writing to memory. This is termed *memory-mapped input/output*. With this method, each input/output device has an address, just like a memory location. The Motorola 68HC11, Intel 8051 and PIC microcontrollers have no separate input/output instructions and use memory mapping. Thus, with memory mapping we might use:

LDAA $1003

to read the data input at address $1003 and:

STAA $1004

to write data to the output at address $1004.

Microprocessors use parallel ports to input or output bytes of data. Many peripherals often require several input/output ports. This can be because the data word of the peripheral is longer than that of the CPU. The CPU must then transfer the data in segments. For example, if we require a 16-bit output with an 8-bit CPU the procedure is:

1   The CPU prepares the eight most significant bits of the data.
2   The CPU sends the eight most significant bits of the data to the first port.
3   The CPU prepares the eight least significant bits of the data.
4   The CPU sends the eight least significant bits of the data to the second port.
5   Thus, after some delay, all the 16 bits are available to the peripheral.

18.2.1 Input/output registers

The Motorola 68HC11 microcontroller has five ports A, B, C, D and E (see Section 15.3.1). Ports A, C and D are bi-directional and can be used for either input or output. Ports B is output only and port E input only. Whether a bi-directional port is used for input or output depends on the setting of a bit in its control register. For example, port A at address $1000 is controlled by the pulse accumulator control register PACTL at address $1026. To set port A for input requires bit 7 to be 0; output requires bit 7 to be 1 (see Fig. 15.13). Port C is bi-directional and the eight bits in its register at address $1003 are controlled by the corresponding bits in its port data direction register at address $1007. When the corresponding data direction bit is set to 0 it is an input, when set to 1 it is an output. Port D is bi-directional and contains just six input/output lines at address $1008. It is controlled by a port direction register at address $1009. The direction of each line is controlled by the corresponding bit in the control register; it is set to 0 for an input and 1 for an output. Some of the ports can also be set to carry out other functions by setting other bits in their control registers.

For a fixed-direction port, e.g. port B in the Motorola 68HC11 is output only, the instructions needed to output some value, e.g. $FF, are simply those needed to load the data to that address. The instructions might be:

```
REGBAS  EQU   $1000     ; base address of I/O registers
PORTB   EQU   $04       ; offset of PORTB from REGBAS
        LDX   #REGBAS   ; load index register X
        LDAA  #$FF      ; load $FF into accumulator
        STAA  PORTB,X   ; store value at PORTB address
```

For the fixed-direction port E, which is input only, the instruction to read a byte from it might be:

```
REGBAS  EQU   $1000     ; base address of I/O registers
PORTE   EQU   $0A       ; offset of PORTE from REGBAS
        LDAA  PORTE,X   ; load value at PORTE into
                        ; the accumulator
```

For a bi-directional port such as C, before we can use it for an input we have to configure the port so that it acts as an input. This means setting all the bits to 0. Thus we might have:

```
REGBAS  EQU   $1000     ; base address of I/O registers
PORTC   EQU   $03       ; offset of PORTC from REGBAS
DDRC    EQU   $07       ; offset of data direction
                        ; register from REGBAS
        CLR   DDRC,X    ; set DDRS to all 0
```

For the Intel 8051 microcontroller (see Section 15.3.2) there are four parallel bi-directional input/output ports. When a port bit is to be used for output, the data is just put into the corresponding special function register bit; when it is used for input a 1 must be written to each bit concerned, thus FFH might be written for an entire port to be written to. Consider an example of Intel 8051 instructions to light an LED when a push-button is pressed. The push-button provides an input to P3.1 and an output to P3.0; the push-button pulls the input low when it is pressed.

```
        SETB  P3.1      ; make bit P3.1 a 1 and so an input
LOOP    MOV   C,P3.1    ; read the state of the push-button
                        ; and store it in the carry flag
        CPL   C         ; complement the carry flag
        MOV   P3.0, C   ; copy state of carry flag to output
        SJMP  LOOP      ; keep on repeating the sequence
```

With PIC microcontrollers the direction of the signals at its bi-directional ports is set by the TRIS direction registers (see Section 15.3.3). TRIS is set as 1 for read and 0 for write. The registers for the PIC16C73/74 are arranged in two banks and before a particular register can be selected the bank has to be chosen by setting bit 5 in the STATUS register. This register is in both banks and so we do not have to select the bank in order to use this register. The TRIS registers are in bank 1 and the PORT registers in bank 0. Thus to set port B as output we have to first select bank 1 and then set TRISB to 0. We can then select bank 0 and write the output to PORTB. The bank is selected by setting a bit in the STATUS register. The instructions to select port B as an output are thus:

```
Output   clrf    PORTB          ; clear all the bits in port B
         bsf     STATUS,RP0     ; use status register to
                                ; select bank 1 by setting
                                ; RP0 to 1
         clrf    TRISB          ; clear bits so output
         bcf     STATUS,RP0     ; use status register to
                                ; select bank 0
                                ; port B is now an output
                                ; set to 0
```

18.3 Interface requirements

The following are some of the actions that are often required of an interface circuit:

1   *Electrical buffering/isolation*
    This is needed when the peripheral operates at a different voltage or current to that on the microprocessor bus system or there are different ground references. The term *buffer* is used for a device that provides isolation and current or voltage amplification. For example, if the output of a microprocessor is connected to the base of a transistor, the base current required to switch the transistor is greater than that supplied by the microprocessor and so a buffer is used to step up the current. There also has often to be isolation between the microprocessor and the higher power system.

2   *Timing control*
    Timing control is needed when the data transfer rates of the peripheral and the microprocessor are different, e.g. when interfacing a microprocessor to a slower peripheral. This can be achieved by using special lines between the micro-processor and the peripheral to control the timing of data transfers. Such lines are referred to as *handshake lines* and the process as *handshaking*. The peripheral sends a DATA READY signal to the input/output section. The CPU then determines that the DATA READY signal is active. The

CPU then reads the data from the input/output section and sends an INPUT ACKNOWLEDGED signal to the peripheral. This signal indicates that the transfer has been completed and thus the peripheral can send more data. For an output, the peripheral sends an OUTPUT REQUEST or PERIPHERAL READY signal to the input/output section. The CPU determines that the PERIPHERAL READY signal is active and sends the data to the peripheral. The next PERIPHERAL READY signal may be used to inform the CPU that the transfer has been completed.

3   *Code conversion*
This is needed when the codes used by the peripherals differ from those used by the microprocessor. For example, an LED display might require a decoder to convert the BCD output from a microprocessor into the code required to operate the seven display elements.

4   *Changing the number of lines*
Microprocessors operate on a fixed word length of 4 bits, 8 bits, or 16 bits. This determines the number of lines in the microprocessor data bus. Peripheral equipment may have a different number of lines, perhaps requiring a longer word than that of the microprocessor.

5   *Serial to parallel, and vice versa, data transfer*
Within an 8-bit microprocessor, data is generally manipulated 8 bits at a time. To transfer 8 bits simultaneously to a peripheral thus requires eight data paths. Such a form of transfer is termed *parallel data transfer*. It is, however, not always possible to transfer data in this way. For example, data transfer over the public telephone system can only involve one data path. The data has thus to be transferred sequentially one bit at a time. Such a form of transfer is termed *serial data transfer*. Serial data transfer is a slower method of data transfer than parallel data transfer. Thus, if serial data transfer is used, there will be a need to convert incoming serial data into parallel data for the microprocessor and vice versa for outputs from the microprocessor.

6   *Conversion from analogue to digital and vice versa*
The output from sensors is generally analogue and this requires conversion to digital signals for the microprocessor. The output from a microprocessor is digital and this might require conversion to an analogue signal in order to operate some actuator.

(a)

(b)

Fig. 18.2   Buffers

Fig. 18.3   Use of a buffer

18.3.1 Buffers

Figure 18.2(a) shows the symbol for a buffer that does not change the logic of the input but just changes the current or voltage levels. It is the symbol for an amplifier since that this is the function of the buffer. Figure 18.2(b) shows the symbol for a buffer which not only amplifies but inverts the input.

As an illustration of the use of a buffer, consider the output being used to drive a LED indicator (Fig. 18.3). When the output from the microprocessor is high, the buffer gives a low output. The buffer then provides a low-resistance path to ground for the LED current so that the LED comes on. When the output from the microprocessor is low, the buffer gives a high output and there is no path to ground for the LED current and so it is off.

Often a number of peripheral devices have to share the same data lines from the microprocessor, i.e. they are connected to the data bus, and thus there is a need for the microprocessor to be able to enable just one of the devices at a time with the others disabled. This is achieved by using *three-state buffers*. The term three-state occurs because the output can be low, high or floating. This floating state is high impedance so that the output looks like an open circuit to anything connected to it. Figure 18.4 shows how such buffers can be used and Figure 18.5 their symbols. Such buffers are available as integrated circuits, e.g. the 74125 with four non-inverting, active-low buffers and the 74126 with four non-inverting, active-high buffers.

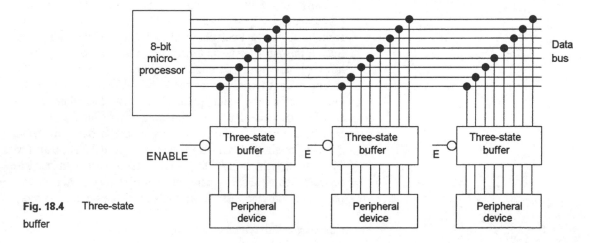

Fig. 18.4   Three-state buffer

18.3.2 Handshaking

Unless two devices can send and receive data at identical rates, handshaking is necessary to exchange data. With handshaking the slower device controls the rate at which the data is transferred. For parallel data transfer *strobe-and-acknowledge* is the commonly used form of handshaking. Readiness to receive is indicated, then data is sent and while it is being received the

Fig. 18.5   Three-state buffers

Fig. 18.6   Handshaking control

Fig. 18.7   PIOC

receiving device indicates not ready for further data, indicating ready again when the transfer has been completed.

With the microcontroller MC68HC11, the basic strobed input/output operates as follows. The handshaking control signals use pins STRA and STRB (Fig. 18.6, also see Fig. 15.9 for the full block model), port C is used for the strobed input and port B for the strobed output. When data is ready to be sent by the microcontroller a pulse is produced at STRA and sent to the peripheral device. When the microcontroller receives either a rising or falling edge to a signal on STRB, then the relevant output port of the microcontroller sends the data to the peripheral. When data is ready to be transmitted to the microcontroller, the peripheral sends a signal to STRA that it is ready and then a rising or falling edge to a signal on STRB is used to indicate readiness to receive. Before handshaking can occur, the parallel input/output register PIOC at address $1002 has to be first configured. Figure 18.7 shows the states required of the relevant bits in that register.

Full handshake input/output involves two signals being sent along STRB, the first being to indicate ready to receive data and the next one that the data has been read. This mode of operation requires that in PIOC the HNDS bit is set to 1 and if PLS is set to 0 the full handshake is said to be pulsed and if to 1 it is interlocked. With pulsed operation a pulse is sent as acknowledgement, with interlocked STRB there is a reset (Fig. 18.8).

18.3.3 Polling and interrupts

Consider the situation where all input/output transfers of data are controlled by the program. When peripherals need attention they signal the microprocessor by changing the voltage level of an input line. The microprocessor can then respond by jumping to a program service routine for the device. On completion of the routine, a return to the main program occurs. Program control of inputs/outputs is thus a loop to continuously read inputs and update outputs, with jumps to service routines as required. This

Fig. 18.8   Full handshaking:
(a) pulsed, (b) interlocked

Fig. 18.9   Interrupt request

process of repeated checking each peripheral device to see if it is ready to send or accept a new byte of data is called *polling*.

An alternative to program control is *interrupt control*. An interrupt involves a peripheral device activating a separate interrupt request line. The reception of an interrupt results in the microprocessor suspending execution of its main program and jumping to the service routine for the peripheral. The interrupt must not lead to a loss of data and an interrupt handling routine has to be incorporated in the software so that the state of processor registers and the last address accessed in the main program are stored in dedicated locations in memory. After the interrupt service routine, the contents of the memory are restored and the microprocessor can continue executing the main program from where it was interrupted. Thus, when an interrupt occurs (Fig. 18.9):

1 The CPU waits until the end of the instruction it is currently executing before dealing with the interrupt.

2 All the CPU registers are pushed onto the stack and a bit set to stop further interrupts occurring during this interruption. The stack is a special area of memory in which program counter values can be stored when a subroutine is to be executed. The program counter gives the address of the next program instruction in a program and thus storing this value enables the program to be resumed at the place where it broke off to execute the interrupt.

3 The CPU then determines the address of the interrupt service routine to be executed. Some microprocessors have dedicated interrupt pins and the pin that is chosen determines which address is to be used. Other microprocessors have only one interrupt pin and the interrupting device must then supply data that tells the microprocessor where the interrupt service routine is located. Some microprocessors have both kinds of interrupt inputs. The starting address of an interrupt service routine is called an *interrupt vector*. The block of memory assigned to store the vectors is known as the *vector table*. Vector addresses are fixed by the chip manufacturer.

4 The CPU branches to the interrupt service routine.

5 After completion of this routine, the CPU registers are returned from the stack and the main program continues from the point it left off.

Unlike a subroutine call, which is located at a specific point in a program, an interrupt can be called from any point in the program. Note that the program does not control when an interrupt occurs; control lies with the interrupting event.

Input/output operations frequently use interrupts since often the hardware cannot wait. For example, a keyboard may generate an interrupt input signal when a key is pressed. The microprocessor then suspends the main program to handle the input from the keyboard; it processes the information and then returns to the main program to continue from where it left off. This ability to code a task as an interrupt service routine and tie it to an external signal simplifies many control tasks, enabling them to be handled without delay. For some interrupts it is possible to program the microprocessor to ignore the interrupt request signal unless an enable bit has been set. Such interrupts are termed *maskable*.

The Motorola 68HC1 has two external interrupt request inputs. XIRQ is a a non-maskable interrupt and will always be executed on completion of the instruction currently being executed. When the XIRQ interrupt occurs, the CPU jumps to the interrupt service routine whose interrupt vector is held at address $FFF4/5 (the low and high bytes of the address). IRQ is a maskable interrupt. When the microcontroller receives a signal at the interrupt request pin IRQ by it going low, the microcontroller jumps to the interrupt service routine indicated by the interrupt vectors $FFF2/3. IRQ can be masked by the instruction set interrupt mask SEI and unmasked by the instruction clear interrupt mask CLI. At the end of an interrupt service routine the instruction RTI is used to return to the main program.

With the Intel 8051, interrupt sources are individually enabled or disabled through the bit-addressable register IE (interrupt enable) at address 0A8H (see Fig. 15.28), a 0 disabling an interrupt and a 1 enabling it. In addition there is a global enable/disable bit in the IE register that is set to enable all external interrupts or cleared to disable all external interrupts. The TCON register (Fig. 15.27) is used to determine the type of interrupt input signal that will initiate an interrupt.

With the PIC microcontrollers, interrupts are controlled by the INTCON register (Fig. 18.10). To use bit 0 of port B as an interrupt, it must be set as an input and the INTCON register must be initialised with a 1 in INTE and a 1 in GIE. If the interrupt is to occur on a rising edge then INTEDG (bit 6) in the OPTION register (see Fig. 15.34) must be set to 1; if on a falling edge it must be set to 0. When an interrupt occurs, INTF is set. It can be cleared by the instruction bcf INTCON,INTF.

Fig. 18.10 INTCON

As an illustration of a program involving external interrupts, consider a simple on/off control program for a central heating

8051

INT0

Temp.
sensors

P3.2

P1.7

P3.3

INT1

Output

to furnace

Fig. 18.11 Central heating system

system involving an Intel 8051 microcontroller (Fig. 18.11). The central heating furnace is controlled by an output from P1.7 and two temperature sensors are used, one to determine when the temperature falls below, say, 20.5°C and the other when it rises above 21.0°C. The sensor for the 21.0°C temperature is connected to interrupt INT0, port 3.2, and the sensor for the 20.5°C temperature is connected to interrupt INT1, port 3.3. By selecting the IT1 bit to be 1 in the TCON register, the external interrupts are edge-triggered, i.e. activated when there is a change from 1 to 0. When the temperature rises to 21.0°C the external interrupt INT0 has an input which changes from 1 to 0 and the interrupt is activated to give the instruction CLR P1.7 for a 0 output to turn the furnace off. When the temperature falls to 20.5°C the external interrupt INT1 has an input which changes from 0 to 1 and the interrupt is activated to give the instruction SETB P1.7 for a 1 output to turn the furnace on. The MAIN program is just a set of instructions to configure and enable the interrupts, establish the initial condition of the furnace to be on if the temperature is less than 21.0° or off if above, and then to wait doing nothing until an interrupt occurs. With the program, a header file has been assumed.

```
        ORG     0
        LJMP    MAIN

        ORG     0003H       ; gives the entry address for ISR0
ISR0    CLR     P1.7        ; interrupt service routine to turn
                            ; the furnace off
        RETI                ; return from interrupt

        ORG     0013H       ; gives the entry address for ISR1
ISR1    SETB    P1.7        ; interrupt service routine to turn
                            ; furnace off
        RETI                ; return from interrupt

        ORG     30H
MAIN    SETB    EX0         ; to enable external interrupt 0
        SETB    EX1         ; to enable external interrupt 1
        SETB    IT0         ; set to trigger when change from
                            ; 1 to 0
        SETB    IT1         ; set to trigger when change from
                            ; 1 to 0
        SETB    P1.7        ; turn the furnace on
        JB      P3.2,HERE   ; if temperature greater than
                            ; 21.0°C jump to HERE and
                            ; leave furnace on
        CLR     P1.7        ; turn the furnace off
HERE    SJMP    HERE        ; just doing nothing until an
                            ; interrupt occurs
        END
```

Microcontrollers, in addition to the interrupt request, have a reset interrupt and a non-maskable interrupt. The *reset interrupt* a special type of interrupt and when this occurs the system resets and thus when this is activated all activity in the system stops, the starting address of the main program is loaded and the start-up routine is executed. The microcontroller M68HC11 has a *computer operating properly* (COP) *watchdog timer*. This is intended to detect software processing errors when the CPU is not executing certain sections of code within an allotted time. When this occurs the COP timer times out and a system reset is initiated.

The *non-maskable interrupt* cannot be masked and so there is no method of preventing the interrupt service routine being executed when it is connected to this line. An interrupt of this type is usually reserved for emergency routines such as those required when there is a power failure, e.g. switching to a back-up power supply.

18.3.4 Serial interfacing

With the parallel transmission of data, one line is used for each bit; serial systems, however, use a single line to transmit data in sequential bits. There are two basic types of serial data transfer: asynchronous and synchronous.

With *asynchronous transmission*, the receiver and the transmitter each use their own clock signals so it is not possible for a receiver to know when a word starts or stops. Thus it is necessary for each transmitted data word to carry its own start and stop bits so that it is possible for the receiver to tell where one word stops and another starts (Fig. 18.12). With such a mode of transmission, the transmitter and receiver are typically remote (see Chapter 20 for details of standard interfaces). With *synchronous transmission*, the transmitter and receiver have a common clock signal and thus transmission and reception can be synchronised.

Fig. 18.12   Asynchronous transmission

The microcontroller MC68HC11 (see Fig. 15.9) has a serial communications interface (SCI) which can be used for asynchronous transmission and thus can be used to communicate with remote peripheral devices. The SCI uses port D pin PD1 as a transmit line and port PD0 as a receive line. These lines can be enabled or disabled by the SCI control register. The microcontroller also has a serial peripheral interface (SPI) for synchronous transmission. This can be used for local serial communication, local meaning essentially inside the machine in which the chip is located.

18.4 Peripheral interface adapters

Interfaces can be specifically designed for particular inputs/outputs; however, programmable input/output interface devices are available which permit various different input and output options to be selected by means of software. Such devices are known as *peripheral interface adapters* (PIAs).

A commonly used PIA parallel interface is the *Motorola MC 6821*. It is part of the MC6800 family and thus can be directly attached to Motorola MC6800 and MC68HC11 buses. The device can be considered to be essentially just two parallel input/output ports, with their control logic, to link up with the host microprocessor. Figure 18.13 shows the basic structure of the MC6821 PIA and the pin connections.

Fig. 18.13 MC6821 PIA

The PIA contains two 8-bit parallel data ports, termed A and B. Each port has:

1   A *peripheral interface register*. An output port has to operate in a different way to an input port because the data must be held for the peripheral. Thus for output a register is used to temporarily store data. The register is said to be *latched*, i.e. connected, when a port is used for output and unlatched when used for input.

2   A *data direction register* that determines whether the input/output lines are inputs or outputs.

3   A *control register* that determines the active logical connections in the peripheral.

4   Two *control lines*, CA1 and CA2 or CB1 and CB2.

Two microprocessor address lines connect the PIA directly through the two register select lines RS0 and RS1. This gives the PIA four addresses for the six registers. When RS1 is low, side A is addressed and when high it is side B. RS0 addresses registers on a particular side, i.e. A or B. When RS0 is high, the control register is addressed, when low the data register or the data direction register. For a particular side, the data register and the data direction register have the same address. Which of them is addressed is determined by bit 2 of the control register (see below).

Each of the bits in the A and B control registers is concerned with some features of the operation of the ports. Thus for the A control register we have the bits shown in Figure 18.14. A similar pattern is used for the B control register.

B7	B6	B5	B4	B3	B2	B1	B0
IRQA1	IRQA2	CA2 control			DDRA access	CA1 control	

Fig. 18.14   Control register

Bits 0 and 1
The first two bits control the way that CA1 or CB1 input control lines operate. Bit 0 determines whether the interrupt output is enabled. B0 = 0 disables the IRQA(B) microprocessor interrupt, B0 = 1 enables the interrupt. CA1 and CB1 are not set by the static level of the input but are edge triggered, i.e. set by a changing signal. Bit 1 determines whether bit 7 is set by a high-to-low transition (a trailing edge) or a low-to-high transition (a leading edge). B1 = 0 sets a high-to-low transition, B1 = 1 sets a low-to-high transition.

Bit 2

Bit 2 determines whether data direction registers or peripheral data registers are addressed. With B2 set to 0 data direction registers are addressed, with B2 set to 1 peripheral data registers are selected.

Bits 3, 4 and 5

These bits allow the PIA to perform a variety of functions. Bit 5 determines whether control line 2 is an input or an output. If bit 5 is set to 0 control line 2 is an input, if set to 1 it is an output. In input mode, both CA2 and CB2 operate in the same way. Bits 3 and 4 determine whether the interrupt output is active and which transitions set bit 6.

With B5 = 0, i.e. CA2(CB2) set as an input: B3 = 0 disables IRQA(B) microprocessor interrupt by CA2(CB2), B3 = 1 enables IRQA(B) microprocessor interrupt by CA2(CB2); B4 = 0 determines that the interrupt flag IRQA(B), bit B6, is set by a high-to-low transition on CA2(CB2), B4 = 1 determines that it is set by a low-to-high transition.

B5 = 1 sets CA2(CB2) as an output. In output mode CA2 and CB2 behave differently. For CA2: with B4 = 0 and B3 = 0, CA2 goes low on the first high-to-low ENABLE (E) transition following a microprocessor read of peripheral data register A and is returned high by the next CA1 transition; B4 = 0 and B3 = 1, CA2 goes low on the first high-to-low ENABLE transition following a microprocessor read of the peripheral data register A and is returned to high by the next high-to-low ENABLE transition. For CB2: with B4 = 0 and B3 = 0, CB2 goes low on the first low-to-high ENABLE transition following a microprocessor write into peripheral data register B and is returned to high by the next CB1 transition; B4 = 0 and B3 = 1, CB2 goes low on the first low-to-high ENABLE transition following a microprocessor write into peripheral data register B and is returned high by the next low-to-high ENABLE transition. With B4 = 1 and B3 = 0, CA2(CB2) goes low as the microprocessor writes B3 = 0 into the control register. With B4 = 0 and B3 = 1, CA2(CB2) goes high as the microprocessor writes B3 = 1 into the control register.

Bit 6

This is the CA2(CB2) interrupt flag, being set by transitions on CA2(CB2). With CA2(CB2) as an input (B5 = 0), it is cleared by a microprocessor read of the data register A(B). With CA2(CB2) as an output (B5 = 1), the flag is 0 and is not affected by CA2(CB2) transitions.

Bit 7

This is the CA1(CB1) interrupt flag, being cleared by a microprocessor read of data register A(B).

The process of selecting which options are to be used is termed *configuring* or *initialising* the PIA. The RESET connection is used to clear all the registers of the PIA. The PIA must then be initialised.

18.4.1 Initialising the PIA

Before the PIA can be used, a program has to be written and used so that the conditions are set for the desired peripheral data flow. The PIA program is placed at the beginning of the main program so that, thereafter, the microprocessor can read peripheral data. The initialisation program is thus only run once.

The initialisation program to set which port is to be input and which is to be output can have the following steps:

1   Clear bit 2 of each control register by a reset, so that data direction registers are addressed. Data direction register A is addressed as XXX0 and data direction register B as XXX2.
2   For A to be an input port, load all 0s into direction register A.
3   For B to be an output port, load all 1s into direction register B.
4   Load 1 into bit 2 of both control registers. Data register A is now addressed as XXX0 and data register B as XXX2.

Thus an initialisation program in assembly language to make side A an input and side B an output could be, following a reset:

```
INIT    LDAA    #$00    Loads 0s
        STAA    $2000   Make side A input port
        LDAA    #$FF    Load 1s
        STAA    $2000   Make side B output port
        LDAA    #$04    Load 1 into bit 2, all other bits 0
        STAA    $2000   Select port A data register
        STAA    $2002   Select port B data register
```

Peripheral data can now be read from input port A with the instruction LDAA 2000 and the microprocessor can write peripheral data to output port B with the instruction STAA 2002.

18.4.2 Connecting interrupt signals via a PIA

The Motorola MC6821 PIA (Fig. 18.15) has two connections IRQA and IRQB through which interrupt signals can be sent to the microprocessor so that an interrupt request from CA1, CA2 or CB1, CB2 can drive the IRQ pin of the microprocessor to the active-low state. When the initialisation program for a PIA was considered in the previous section, only bit 2 of the control register was set as 1, the other bits being 0. These 0s disabled interrupt inputs. In order to use interrupts, the initialisation step which stores $04 into the control register must be modified. The form of the modification will depend on the type of change in the input which is required to initiate the interrupt.

Fig. 18.15   Interfacing with a PIA

Suppose, for example, we want CA1 to enable an interrupt when there is a high-to-low transition, with CA2 and CB1 not used and CB2 enabled and used for a set/reset output. The control register format to meet this specification is, for CA:

B0 is 1 to enable interrupt on CA1.
B1 is 0 so that the interrupt flag IRQA1 is set by a high-to-low transition on CA1.
B2 is 1 to give access to the data register.
B3, B4, B5 are 0 because CA2 is disabled.
B6, B7 are read-only flags and thus a 0 or 1 may be used.

Hence the format for CA1 can be 00000101 which is 05 in hexadecimal notation. The control register format for CB2 is:

B0 is 0 to disable CB1.
B1 may be 0 or 1 since CB1 is disabled.
B2 is 1 to give access to the data register.
B3 is 0, B4 is 1 and B5 is 1, to select the set/reset.
B6, B7 are read-only flags and thus a 0 or 1 may be used.

Hence the format for CA1 can be 00110100 which is 34 in hexadecimal notation. The initialisation program might then read:

```
INIT    LDAA    #$00    Loads 0s
        STAA    $2000   Make side A input port
        LDAA    #$FF    Load 1s
        STAA    $2000   Make side B output port
        LDAA    #$05    Load the required control register
                        format
        STAA    $2000   Select port A data register
        LDAA    #$34    Load the required control register
                        format
        STAA    $2002   Select port B data register
```

18.4.3 An example of interfacing with a PIA

As an example of interfacing with a PIA, Figure 18.16 shows a circuit that can be used with a unipolar stepper motor (see Section 7.7.2). The inductive windings can generate a large back e.m.f. when switched so some way of isolating the windings from the PIA is required. Optoisolators, diodes or resistors might be used. Diodes give a cheap and simple interface, resistors not completely isolating the PIA.

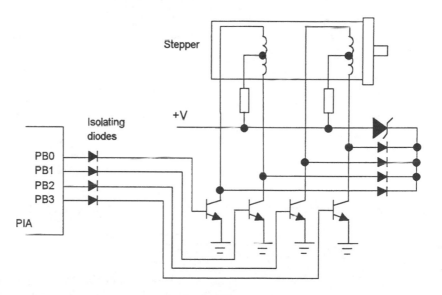

Fig. 18.16   Interfacing a stepper

18.5 Serial communications interface

The *universal asynchronous receiver/transmitter* (UART) is the essential element of a serial communication system, the function being to change serial data to parallel for input and parallel data to serial for output. A common programmable form of a UART is the *asynchronous communications interface adapter* (ACIA) from Motorola MC6850; Figure 18.17 shows a block diagram of the constituent elements.

Data flow between the microprocessor and the ACIA is via eight bi-directional lines D0 to D7. The direction of the data flow is controlled by the microprocessor through the read/write input to the ACIA. The three chip select lines are used for addressing a particular ACIA. The register select line is used to select particular registers within the ACIA; if the register select line is high then the data transmit and data receive registers are selected, if low then the control and status registers are selected. The status register contains information on the status of serial data transfers as they occur and is used to read data carrier detect and clear-to-send lines. The control register is initially used to reset the ACIA and subsequently to define the serial data transfer rate and data format.

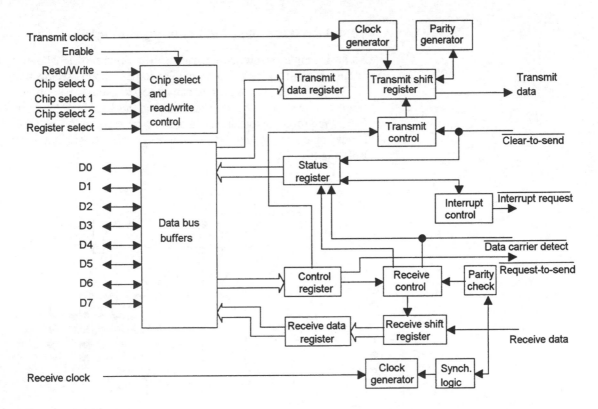

Fig. 18.17 MC6850 ACIA

The peripheral side of the ACIA includes two serial data lines and three control lines. Data is sent by the transmit data line and received by the receive data line. Control signals are provided by clear-to-send, data carrier detect and request-to-send. Figure 18.18 shows the bit formats of the control and Figure 18.19 the status registers.

Defines whether the ACIA interrupt mode to the MPU is enabled for data input
0 disabled, 1 enabled

Defines whether the ACIA interrupt is enabled for data output and provides control over the request-to-send output
0 0 : interupt disabled for data output
0 1 : Interrupt enabled for data output
1 0 : RTS line goes high

Defines serial data format
b2: 0 even, 1 odd parity
b3: 0 two, 1 one stop bit
b4: 0 seven, 1 eight data bits in word

To reset and select the divide ratios between the external receive and transmit clocks and the receive and transmit data rates
0 0 : +1, 0 1: +16, 1 0: ÷64
1 1: master reset

Fig. 18.18   Control register

Interrupt request bit
A copy of the IRQ output to the MPU

Indicates if there is a parity error

Indicates a receiver overrun error

Indicates if there is a framing error

Clear-to-send bit
Provides handshaking input to a modem

Receive data register full flag
0 for register empty,
1 for full

Transmit data register full flag
0 for register empty, 1 for full

Data carrier detect flag
0 for carrier present, 1 for loss of carrier
Use when transferring data down a telephone line using a modem

Fig. 18.19 Status register

Asynchronous serial data transfer is generally used for communications between two computers, with or without a modem, or a computer and a printer (see Chapter 20 for further discussion).

18.5.1 The serial interface of the M68HC11

Many microcontrollers have serial interfaces, i.e. built-in UARTs. For example, the Motorola M68HC11 has a serial peripheral interface (SPI), a synchronous interface, and a serial communication interface (SCI), an asynchronous interface (see Fig. 15.9). The SPI requires the same clock signal to be used by the microcontroller and the externally connected device or devices (Fig. 18.20(a)). The SPI allows several microcontrollers, with this facility, to be interconnected. The SCI is an asynchronous interface and so allows different clock signals to be used by the SCI system and the externally connected device (Fig. 18.20(b)). General-purpose microprocessors do not have a serial communication interface so a UART, e.g. Motorola MC6850, has then to be used to enable serial communication to take place. In some situations more than one serial communication interface is required and thus the microcontroller M68HC11 requires supplementing with a UART.

Fig. 18.20   (a) SPI, (b) SCI

The SPI is initialised by bits in the SPI control register (SPCR) and the port D data direction control register (DDRD). The SPI status register contains status and error bits. The SCI is initialised by using the SCI control register 1, the SCI control register 2 and the baud rate control register. Status flags are in the SCI status register. For details see *Software and Hardware Engineering: Motorola M68HC11* by F.M. Cady (OUP 1997) or *Micro-controller Technology: The 68HC11* by P. Spasov (Prentice-Hall 1992, 1996). For details of a similar system in the Microchip PIC16Cx microcontrollers, see *Proramming and Customizing the PIC Microcontrollers* by M. Predko (McGraw-Hill 1998).

18.6 Examples of interfacing

The following are examples of interfacing.

18.6.1 Interfacing a seven-segment display with a decoder

Consider where the output from a microcontroller is used to drive a seven-segment LED display unit (see Section 4.4). A single LED is an on-off indicator and thus the display number indicated will depend on which LEDs are on. Figure 18.21 shows how we can use a microcontroller to drive a common anode display using a decoder driver (see Section 14.6.3), this being able to take in a BCD input and convert it to the appropriate code for the display.

For the 7447 decoder, pins 7, 1, 2, and 6 are the input pins of the decoder for the BCD input with pins 13, 12, 11, 10, 9, 15, and 14 being the outputs for the segments of the display. Pin 9 of the display is the decimal point. Table 18.1 shows the input and output signals for the decoder.

Fig. 18.21   Driving a display

Table 18.1   7447 BCD decoder for a seven-segment display

Display	Input pins				Output pins						
	6	2	1	7	13	12	11	10	9	15	14
0	L	L	L	L	ON	ON	ON	ON	ON	ON	OFF
1	L	L	L	H	OFF	ON	ON	OFF	OFF	OFF	OFF
2	L	L	H	L	ON	ON	OFF	ON	ON	OFF	ON
3	L	L	H	H	ON	ON	ON	ON	OFF	OFF	ON
4	L	H	L	L	OFF	ON	ON	OFF	OFF	ON	ON
5	L	H	H	L	ON	OFF	ON	ON	OFF	ON	ON
6	L	H	H	L	OFF	OFF	ON	ON	ON	ON	ON
7	L	H	H	H	ON	ON	ON	OFF	OFF	OFF	OFF
8	H	L	H	H	ON	ON	ON	OFF	OFF	OFF	OFF
9	H	L	H	L	ON	ON	ON	OFF	OFF	OFF	OFF

Fig. 18.22   Ripple blanking

Blanking is when none of the segments are lit. This is used to prevent a leading 0 occurring when we have, say, three display units and want to display just 10 rather than 010 and so blank out the leading 0 and prevent it being illuminated. This is achieved by the *ripple blanking input* RBI being set low. When RBI is low and the BCD inputs A, B, C and D are low then the output is blanked. If the input is not zero the ripple blanking output RBO is high regardless of the RBI status. The RBO of the first digit in the display can be connected to the RBI of the second digit and the RBO of the second connected to the RBI of the third digit, thus only allowing the final 0 to be blanked (Fig. 18.22).

With displays having many display elements, rather than use a decoder for each element, multiplexing is used with a single decoder. Figure 18.23 shows the circuit for multiplexing a four-element common cathode type of display. The BCD data is outputted from port A and the decoder presents the decoder output to all the displays. Each display has its common cathode connected to ground through a transistor. The display cannot light up unless the transistor is switched on by an output from port B. Thus by switching between PB0, PB1, PB2 and PB3 the output from port A can be switched to the appropriate display. To maintain a constant display, a display is repeatedly turned on sufficiently often for the display to appear flicker-free. Time division multiplexing (see Section 3.7.2) can be used to enable more than one digit to be displayed at a time.

Fig. 18.23 Multiplexing four displays

18.6.2 Interfacing analogue

Digital-to-analogue conversion is required when the output from a microprocessor or microcontroller is required to provide an analogue signal output. For example, the DAC Analog Devices AD557 can be used for this purpose. It produces an output voltage proportional to its digital input and input latches are available for microprocessor interfacing. If the latches are not required then pins 9 and 10 are connected to ground. Data is latched when there is a positive edge, i.e. a change from low-to-high, on either the input to pin 9 or pin 10. The data is then held until both these pins return to the low level. When this happens, the data is transferred from the latch to the digital-to-analogue converter for conversion to an analogue voltage. Figure 18.24 shows the AD557 with the latch not being used and connected to a Motorola

M68HC11 so that, when the following program is run, it generates a sawtooth voltage output. Other voltage waveforms can readily be generated by changing the program.

```
BASE        EQU $1000        Base address of I/O registers
PORTB       EQU $04          Offset of PORTB from BASE

            ORG  $C000
            LDX  #BASE        Point X to register base
            CLR  PORTB,X      Send 0 to the DAC
AGAIN       INC  PORTB,X      Increment by 1
            BRA  AGAIN        Repeat
            END
```

Fig. 18.24   Waveform generation

Pin 8 is LSB
Pin 1 is MSB

Problems

1 Describe the functions that can be required of an interface.
2 Explain the difference between a parallel and a serial interface.
3 Explain what is meant by a memory-mapped system for inputs/outputs.
4 What is the function of a peripheral interface adapter?
5 Describe the architecture of the Motorola MC6821 PIA.
6 Explain the function of an initialisation program for a PIA.
7 What are the advantages of using external interrupts rather than software polling as a means of communication with peripherals?
8 For a Motorola MC6821 PIA, what value should be stored in the control register if CA1 is to be disabled, CB1 is to be an enabled interrupt input and set by a low-to-high transition, CA2 is to be enabled and used as a set/reset output, and CB2 is to be enabled and go low on first low-to-high E transition following a microprocessor Write into peripheral data register B and return high by the next low-to-high E transition?
9 Write, in assembly language, a program to initialise the Motorola MC6821 PIA to achieve the specification given in problem 8.
10 Write, in assembly language, a program to initialise the Motorola MC6821 PIA to read eight bits of data from port A.

19 Programmable logic controllers

19.1 Programmable logic controllers

Fig. 19.1   Programmable logic controller

A *programmable logic controller* (PLC) is a digital electronic device (Fig. 19.1) that uses a programmable memory to store instructions and to implement functions such as logic, sequencing, timing, counting and arithmetic in order to control machines and processes and has been specifically designed to make programming easy. The term *logic* is used because the programming is primarily concerned with implementing logic and switching operations. Inputs devices, e.g. switches, and output devices, e.g. motors, being controlled are connected to the PLC and then the controller monitors the inputs and outputs according to the program stored in the PLC by the operator and so controls the machine or process. Originally they were designed as a replacement for hard-wired relay (e.g. Fig. 7.2) and timer logic control systems. PLCs have the great advantage that it is possible to modify a control system without having to rewire the connections to the input and output devices, the only requirement being that an operator has to key in a different set of instructions. Also they are much faster than relay-operated systems. The result is a flexible system which can be used to control systems which vary quite widely in their nature and complexity. Such systems are widely used for the implementation of logic control functions because they are easy to use and program.

PLCs are similar to computers but have certain features which are specific to their use as controllers. These are:

1   They are rugged and designed to withstand vibrations, temperature, humidity and noise.
2   The interfacing for inputs and outputs is inside the controller.
3   They are easily programmed and have an easily understood programming language. Programming is primarily concerned with logic and switching operations.

PLCs were first conceived in 1968. They are now widely used and extend from small self-contained units for use with perhaps 20 digital input/outputs to modular systems which can be used for large numbers of inputs/outputs, handle digital or analogue inputs/outputs, and also carry out PID control modes.

This chapter is a discussion of the basic structure of PLCs and how they can be used to control machines or processes. For a more detailed discussion the reader is referred to specialist texts such as *Programmable Controllers, Operation and Application* by I.G. Warnock (Prentice-Hall 1988), *Programmable Logic Controllers* by W. Bolton (Newnes 1996, 2003) or *Automation with Programmable Logic Controllers* by P. Rohner (Macmillan 1996).

19.2 Basic structure

Figure 19.2 shows the basic internal structure of a PLC. It consists essentially of a central processing unit (CPU), memory, and input/output circuitry. The CPU controls and processes all the operations within the PLC. It is supplied with a clock with a frequency of typically between 1 and 8 MHz. This frequency determines the operating speed of the PLC and provides the timing and synchronisation for all elements in the system. A bus system carries information and data to and from the CPU, memory and input/output units. There are several memory elements: a system ROM to give permanent storage for the operating system and fixed data, RAM for the user's program, and temporary buffer stores for the input/output channels.

Fig. 19.2 Architecture of a PLC

The programs in RAM can be changed by the user. However, to prevent the loss of these programs when the power supply is switched off, a battery is likely to be used in the PLC to maintain the RAM contents for a period of time. After a program has been developed in RAM it may be loaded into an EPROM memory chip and so made permanent. Specifications for small PLCs often specify the program memory size in terms of the number of program steps that can be stored. A program step is an instruction for some event to occur. A program task might consist of a number of steps and could be, for example: examine the state of switch A, examine the state of switch B, if A and B are closed then energise solenoid P which then might result in the operation of some actuator. When this happens another task might then be started. Typically the number of steps that can be handled by a small PLC is of the order of 300 to 1000, which is generally more than adequate for most control situations.

19.2.1 Input/output

The input/output unit provides the interface between the system and the outside world. Programs are entered into the input/output unit from a panel which can vary from small keyboards with liquid crystal displays to those using a visual display unit (VDU) with keyboard and screen display. Alternatively the programs can be entered into the system by means of a link to a personal computer (PC) which is loaded with an appropriate software package.

The input/output channels provide signal conditioning and isolation functions so that sensors and actuators can be generally directly connected to them without the need for other circuitry. Figure 19.3 shows the basic form of an input channel. Common input voltages are 5 V and 24 V.

Common output voltages are 24 V and 240 V. Outputs are specified as being of relay type, transistor type or triac type. With the relay type (Fig. 19.4), the signal from the PLC output is used to operate a relay and so is able to switch currents of the order of a few amperes in an external circuit. The relay isolates the PLC from the external circuit and can be used for both d.c. and a.c. switching. Relays are, however, relatively slow to operate. The transistor type of output (Fig. 19.5) uses a transistor to switch current through the external circuit. This gives a faster switching action. In Figure 19.5(a), using the conventional current flow direction as from positive to negative, an output device receives current from the output module and is referred to as *sinking*; in Figure 19.5(b) the current flows from the output module to an output load and is referred to as sourcing. Optoisolators are used with transistor switches to provide isolation between the external circuit and the PLC. The transistor output is only for d.c. switching. Triac outputs can be used to control external loads

Fig. 19.3   Input channel

Fig. 19.4   Relay type of output

which are connected to the a.c. power supply. Optoisolators are again used to provide isolation.

Fig. 19.5 Transistor outputs: (a) sinking, (b) sourcing

19.2.2 Example of a PLC

The following are some of the features of a typical small PLC, a Mitsubishi F2-20MR-ES.

Power supply	110–120 V/220–240 V a.c., single phase 50/60 Hz
Program language	Ladder logic
Programming capacity	1000 steps
Execution speed	Average 7 µs/step
Program memory	CMOS-RAM built in, EPROM can be added
Battery back-up	Lithium battery, approx. 5 years' life
Timers	0.1 s timer: 24 points, on-delay timers (0.1 to 999 s)
	0.01 s timer: 8 points, on-delay timers (0.01 to 99.9 s)
Counters (retentive)	32 points, down counter (0 to 999)
Number of inputs	12 points, all optoisolated
Input voltage	Built-in 24 V d.c., external 24 V d.c.
Number of outputs	8 points
Choice of output	Relay output: relay isolated
	Transistor output: optoisolated
	Triac output: optoisolated

19.3 Input/output processing

The basic form of programming commonly used with PLCs is *ladder programming*. This involves each program task being specified as though a rung of a ladder. Thus such a rung could specify that the state of switches A and B, the inputs, be examined and if A and B are both closed then a solenoid, the output, is energised. Ladder programming is discussed in more detail in the next section.

The sequence followed by a PLC when carrying out a program can be summarised as:

1   Scan the inputs associated with one rung of the ladder program.
2   Solve the logic operation involving those inputs.
3   Set/reset the outputs for that rung.
4   Move on to the next rung and repeat operations 1, 2, 3.
5   Move on to the next rung and repeat operations 1, 2, 3.
6   Move on to the next rung and repeat operations 1, 2, 3.

And so on until the end of the program.

Each rung of the ladder program is thus scanned in turn.

There are two methods that can be used for input/output processing:

1   *Continuous updating*
This involves the CPU scanning the input channels as they occur in the program instructions. Each input point is examined individually and its effect on the program determined. There will be a built-in delay, typically about 3 ms, when each input is examined in order to ensure that only valid input signals are read by the microprocessor. This delay enables the microprocessor to avoid counting an input signal twice, or more frequently, if there is contact bounce at a switch. A number of inputs may have to be scanned, each with a 3 ms delay, before the program has the instruction for a logic operation to be executed and an output to occur. The outputs are latched so that they retain their status until the next updating.

2   *Mass input/output copying*
Because, with continuous updating, there has to be a 3 ms delay on each input, the time taken to examine several hundred input/output points can become comparatively long. To allow a more rapid execution of a program, a specific area of RAM is used as a buffer store between the control logic and the input/output unit. Each input/output has an address in this memory. At the start of each program cycle the CPU scans all the inputs and copies their status into the input/output addresses in RAM. As the program is executed the stored input data is read, as required, from RAM and the

logic operations carried out. The resulting output signals are stored in the reserved input/output section of RAM. At the end of each program cycle all the outputs are transferred from RAM to the output channels. The outputs are latched so that they retain their status until the next updating.

19.4 Programming

PLC programming based on the use of *ladder diagrams* involves writing a program in a similar manner to drawing a switching circuit. The ladder diagram consists of two vertical lines representing the power rails. Circuits are connected as horizontal lines, i.e. the rungs of the ladder, between these two verticals. Figure 19.6 shows the basic standard symbols that are used and an example of rungs in a ladder diagram.

In drawing the circuit line for a rung, inputs must always precede outputs and there must be at least one output on each line. Each rung must start with an input or a series of inputs and end with an output.

The inputs and outputs are numbered, the notation used depending on the PLC manufacturer, e.g. the Mitsubishi F series of PLCs precedes input elements by an X and output elements by a Y and uses the following numbers:

Inputs   X400–407, 410–413

                 X500–507, 510–513   (24 possible outputs)

Outputs   Y430–437

                 Y530–537   (16 possible inputs)

Fig. 19.6   Ladder diagram

To illustrate the drawing of a ladder diagram, consider a situation where the output from the PLC is to energise a solenoid when a normally open start switch connected to the input is activated by being closed (Fig. 19.7(a)). The program required is

(a)

(b)

Fig. 19.7 Switch controlling a solenoid

shown in Figure 19.7(b)). Starting with the input, we have the normally open symbol | |. This might have an input address X400. The line terminates with the output, the solenoid, with the symbol O. This might have the output address Y430. To indicate the end of the program the end rung is marked. When the switch is closed the solenoid is activated. This might, for example, be a solenoid valve which opens to allow water to enter a vessel.

Another example might be an on-off temperature control (Fig. 19.8) in which the input goes from low to high when the temperature sensor reaches the set temperature. The output is then to go from on to off. The temperature sensor shown in the figure is a thermistor connected in a bridge arrangement with output to an operational amplifier connected as a comparator (see Section 3.2.7). The program shows the input as a normally closed pair of contacts, so giving the on signal and hence an output. When the contacts are opened to give the off signal then the output is switched off.

Such ladder programs can be entered from special keypads or selected from a monitor screen by using a mouse. They can also be specified by using a mnemonic language. However they are entered, the programs are then translated by the PLC into machine language for the benefit of the microprocessor and its associated elements.

(a)

(b)

Fig. 19.8 Temperature control system

(a)

(b)

Fig. 19.9 An AND system

19.4.1 Logic functions

The logic functions can be obtained by combinations of switches (see Section 14.3) and the following shows how we can write ladder programs for such combinations.

1   AND
    Figure 19.9(a) shows a situation where a coil is not energised unless two, normally open, switches are both closed. Switch A and switch B have both to be closed, which thus gives an

(a)

(b)

Fig. 19.10   An OR system

(a)

(b)

Fig. 19.11   A NOR system

(a)

(b)

Fig. 19.12   A NAND system

Fig. 19.13   An XOR system

AND logic situation. The ladder diagram starts with | |, labelled Input 1, to represent switch A and in series with it | |, labelled Input 2, to represent switch B. The line then terminates with O to represent the output. Figure 19.9(b) shows the line.

2   OR
Figure 19.10(a) shows a situation where a coil is not energised until either, normally open, switch A or B is closed. The situation is an OR logic gate. The ladder diagram starts with | |, labelled Input 1, to represent switch A and in parallel with it | |, labelled Input 2, to represent switch B. The line then terminates with O to represent the output. Figure 19.10(b) shows the line.

3   NOR
Figure 19.11 shows how we can represent the ladder program line for a NOR gate. Since there has to be an output when neither A nor B have an input and when there is an input to A or B the output ceases, the ladder program shows Input 1 in parallel with Input 2, with both being represented by normally closed contacts.

4   NAND
Figure 19.12 shows a NAND gate. There is no output when both A and B have an input. Thus for the ladder program line to obtain an output we require no inputs to Input 1 and to Input 2.

5   EXCLUSIVE-OR (XOR)
Figure 19.13 shows how we can draw the ladder program line for an XOR gate, there being no output when there is no input to Input 1 and Input 2 and when there is an input to both Input 1 and Input 2. Note that we have represented each input by two sets of contacts, one normally open and one normally closed.

Consider a situation where a normally open switch A must be activated and either of two other, normally open, switches B and C must be activated for a coil to be energised. We can represent this arrangement of switches as switch A in series with two parallel switches B and C (Fig. 19.14(a)). For the coil to be energised we require A to be closed and either B or C to be closed. Switch A when considered with the parallel switches gives an AND logic situation. The two parallel switches give an OR logic situation. We thus have a combination of two gates. The truth table is:

(a)

(b)

Fig. 19.14 Switches controlling a solenoid

Inputs			Output
A	B	C	
0	0	0	0
0	0	1	0
0	1	0	0
0	1	1	0
1	0	0	0
1	0	1	1
1	1	0	1
1	1	1	1

For the ladder diagram, we start the ladder diagram with | |, labelled Input 1, to represent switch A. This is in series with two | | in parallel, labelled Input 1 and Input 2, for switches B and C. The line then terminates with O to represent the output, the coil. Figure 19.14(b) shows the line.

As a simple example of a program using logic gates consider the requirement for there to be an output to the solenoid controlling the valve that will open a shop door when the shopkeeper has closed a switch to open shop and a customer approaches the door and is detected by a sensor which then gives a high signal. The truth table for this system is thus:

Shop open switch	Customer approaching sensor	Solenoid output
Off	Off	Off
Off	On	Off
On	Off	Off
On	On	On

Fig. 19.15 Shop door system

This truth table is that of an AND gate and thus the program for a PLC controlling the door is as shown in Figure 19.15.

19.4.2 Latching

There are often situations where it is necessary to hold a coil energised, even when the input which energised it ceases. The term *latch circuit* is used for the circuit used to carry out such an operation. It is a self-maintaining circuit in that, after being energised, it maintains that state until another input is received. It remembers its last state.

An example of a latch circuit is shown in Figure 19.16. When Input 1 is energised and closes, there is an output. However, when there is an output, a set of contacts associated with the output is energised and closes. These contacts OR the Input 1 contacts.

Fig. 19.16 A latch circuit

(a) Unsafe system

(a) Safe system

Fig. 19.17 Motor latching

Thus, even if Input 1 contacts open, the circuit will still maintain the output energised. The only way to release the output is by operating the normally closed contact Input 2.

As an example of the use of a latching circuit, consider the requirement for a PLC to control a motor so that when the start signal button is momentarily pressed the motor starts and when the stop switch is used the motor switches off, signal lamps indicating when the motor is off and when on. Figure 19.17 shows a possible program. With no inputs, the signal lamp for the motor indicates that it is off. The motor contacts which are normally closed have the off lamp on. When the start button is pressed, the normally open contacts are closed and the motor switched on. This is latched on by the motor contacts in parallel with the start contacts. Also, other motor contacts which are normally closed open and the off lamp goes off and motor contacts which are normally open close and the on lamp goes on. The motor is stopped by the stop switch. It might seem that a perfectly simple arrangement for a stop switch would be to use a normally-closed set of program contacts and open them to stop the motor. However, this has the disadvantage that if there is a PLC failure then the motor cannot be stopped. Using a normally-closed real switch in the output of the PLC, and then the PLC program being programmed as normally-open and being switched because it finds the contacts closed, is a much safer option.

19.4.3 Sequencing

There are often control situations where sequences of outputs are required, with the switch from one output to another being controlled by sensors. Consider the requirement for a ladder program for a pneumatic system (Fig. 19.18) with double-solenoid valves controlling two double-acting cylinders A and B if limit switches a–, a+, b–, b+ are used to detect the limits of the piston rod movements in the cylinders and the cylinder activation sequence A+, B+, A–, B– is required.

Fig. 19.18 Cylinder sequencing

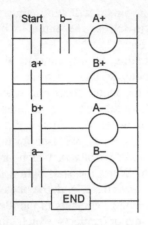

Fig 19.19   Cylinder sequencing

19.5 Mnemonics

Mitsubishi mnemonics

LD	Start a rung with an open contact.
OUT	An output.
AND	A series element and so an AND logic instruction.
OR	Parallel elements and so an OR logic instruction.
I	A NOT logic instruction.
... I	Used in conjunction with other instructions to indicate the inverse.
ORI	An OR NOT logic function.
ANI	An AND NOT logic function.
LDI	Start a rung with a closed contact.
ANB	AND used with two subcircuits.
ORB	OR used with two subcircuits.
RST	Reset shift register/counter.
SHF	Shift.
K	Insert a constant.
END	End ladder.

Fig. 19.20   An AND system

Figure 19.19 shows a possible program. A start switch input has been included in the first rung. Thus cylinder extension for A, i.e. the solenoid A+ energised, only occurs when the start switch is closed and the b– switch is closed, this switch indicating that the B cylinder is retracted. When cylinder A is extended, the switch a+, which indicates the extension of A, is activated. This then leads to an output to solenoid B+ which results in B extending. This closes the switch indicating the extension of B, i.e. the b+ switch, and leads to the output to solenoid A– and the retraction of cylinder A. This retraction closes limit switch a– and so gives the output to solenoid B– which results in B retracting. This concludes the program cycle and leads to the first rung again, which awaits the closure of the start switch before being repeated.

Each horizontal rung on the ladder in a ladder program represents a line in the program and the entire ladder gives the complete program in 'ladder language'. The programmer can enter the program into the PLC using a keyboard with the graphic symbols for the ladder elements, or using a computer screen and a mouse to select symbols, and the program panel or computer then translates these symbols into machine language that can be stored in the PLC memory.

There is an alternative way of entering a program and that is to translate the ladder program into mnemonics, each code corresponding to a ladder element, and then enter these into the programming panel or computer. These are then translated into machine language. The mnemonics used by different PLC manufacturers differ. For the Mitsubishi F series PLCs, the mnemonics used are shown opposite. In examples discussed in the rest of this chapter, where general descriptions are not used, the Mitsubishi mnemonics will be used. However, those used by other manufacturers do not differ widely from these and the principles involved in their use are the same.

The following shows how individual rungs on a ladder are entered. Using the Mitsubishi mnemonics, the AND gate shown in Figure 19.20 would be entered as:

Step	Instruction	
0	LD	X400
1	AND	X401
2	OUT	Y430

The OR gate shown in Figure 19.21 would be entered as:

Fig. 19.21   An OR system

Fig. 19.22   A NOR system

Step	Instruction	
0	LD	X400
1	OR	X401
2	OUT	Y430

The NOR gate shown in Figure 19.22 would be entered as:

Step	Instruction	
0	LDI	X400
1	ANI	X401
2	OUT	Y430

The NAND gate shown in Figure 19.23 would be entered as:

Step	Instruction	
0	LDI	X400
1	ORI	X401
2	OUT	Y430

The XOR gate shown in Figure 19.24 would be entered as:

Step	Instruction	
0	LD	X400
1	ANI	X401
2	LDI	X400
3	AND	X401
4	ORB	
5	OUT	Y430

Fig. 19.23   A NAND system

Fig. 19.24   An XOR system

After reading the first two instructions, the third instruction starts a new line. But the first line has not been ended by an output. The CPU thus recognises a parallel line is involved for the second line and reads together the listed elements until the ORB instruction is reached. The mnemonic ORB (OR branches/blocks together) indicates to the CPU that it should OR the results of steps 0 and 1 with those of steps 2 and 3.

19.6 Timers, internal relays and counters

The previous sections in this chapter have been concerned with tasks requiring the series and parallel connections of input contacts. However, there are tasks which can involve time delays and event counting. These requirements can be met by the timers and counters which are supplied as a feature of PLCs. They can be controlled by logic instructions and represented on ladder diagrams.

Fig. 19.25   Delay-on timer

Fig. 19.26   Timer

Fig. 19.27   Cascaded timers

Fig. 19.28   On-off cyclic timer

The numbering system used for the functions varies from one PLC manufacturer to another. For the Mitsubishi F series, the numbers used are:

Timers	450–457, 8 points, delay-on period 0.1–999 s
T	550–557, 8 points, delay-on period 0.1–999 s
Markers	100–107, 170–177, 200–207, 270–277,
M	128 points
	300–307, 370–377, battery backed, 64 points
Counters	460–467, 8 points, 1 to 999
C	560–567, 8 points, 1 to 999

The term *point* is used for a data point and so is a timing, marker (internal relay) or counter element. Thus, the 16 points for timers means that there are 16 timer circuits. The term *delay-on* is used to indicate that this type of timer waits for a fixed delay period before turning on (Fig. 19.25). In the data given above, this is a period which can be set between 0.1 and 999 s in steps of 0.1 s. Other time delay ranges and steps are possible.

19.6.1 Timers

A timer circuit is specified by stating the interval to be timed and the conditions or events that are to start and/or stop the timer. They are commonly regarded as relays with coils which, when energised, result in the closing or opening of input contacts after some preset time. Figure 19.26 shows part of a program involving a delay-on timer. When there is an input, the timer is energised and starts timing. After some preset time the contacts associated with the timer close and the output occurs.

Timers can be linked together, the term is *cascaded*, to give larger delay times than is possible with just one timer. Figure 19.27 shows such an arrangement. When the input contacts close, timer 1 is started. After its time delay, its contacts close and timer 2 is started. After its time delay, its contacts close and there is an output.

Figure 19.28 shows a program that can be used to cause an output to go on for 0.5 s, then off for 0.5 s, then on for 0.5 s, then off for 0.5 s, and so on. When the input contacts close, Timer 1 is started and comes on after 0.5 s, this being the time for which it was preset. After this time the Timer 1 contacts close and start Timer 2. It comes on after 0.5 s, its preset time, and opens its contacts. This results in Timer 1 being switched off. This results in its contact opening and switching off Timer 1. This then closes its contact and so starts the entire cycle again. The result is that timer contacts for Timer 1 are switched on for 0.5 s, then off for 0.5 s, on for 0.5 s, and so on. Thus the output is switched on for 0.5 s, then off for 0.5 s, on for 0.5 s, and so on.

Fig. 19.29   Delay-off timer

Fig. 19.30   An output controlled by two input arrangements

Fig. 19.31   An output controlled by two input arrangements

PLCs are generally provided with only delay-on timers, i.e. a timer which comes on after a time delay. Figure 19.29 shows how a delay-off timer, i.e. a timer which switches off an output after a time delay from being energised, can be devised. When the input contacts are momentarily closed the output is energised and the timer started. The output contacts latch the input and keep the output on. After the preset time of the timer, the timer comes on and breaks the latch circuit, so switching the output off.

19.6.2 Internal relays

The term *internal* relay, *auxiliary relay* or *marker* is used for what can be considered as an internal relay in the PLC. These behave like relays with their associated contacts, but in reality are not actual relays but simulations by the software of the PLC. Some have battery back-up so that they can be used in circuits to ensure a safe shut-down of plant in the event of a power failure. Internal relays can be very useful aids in the implementation of switching sequences.

Internal relays are often used when there are programs with multiple input conditions. Consider the situation where the excitation of an output depends on two different input arrangements. Figure 19.30 shows how we can draw a ladder diagram using internal relays. The first rung shows one input arrangement being used to control the coil of internal relay IR1. The second rung shows the other input arrangement controlling the coil of internal relay IR2. The contacts of the two relays are then put in an OR situation to control the output. The instruction sequence, using Mitsubishi notation (Fig. 19.31), is:

Step	Instruction	
0	LD	X400
1	OR	X401
2	OUT	M100
3	LD	X402
4	AND	X403
5	OUT	M101
6	LD	M100
7	OR	M101
8	OUT	Y430
9	END	

Another use of internal relays is for the starting of multiple outputs. Figure 19.32 shows such a ladder program. When the start contacts are closed, the internal relay is activated and latches the input. It also starts output 1 and makes it possible for outputs 2 and 3 to be activated.

Fig. 19.32   Starting of multiple outputs

As an illustration of this type of use of internal relays, consider the problem of devising a ladder program to control a pneumatic system with double-solenoid controlled valves and two cylinders A and B if limit switches a–, a+, b– and b+ are used to detect the limits of movement of the piston rod movements in the cylinders (Fig. 19.33(a)) and the sequence required is for the piston rod in A to extend, followed by the piston rod in B extending, then the piston in B retracting and finally the cycle is completed by the piston in A retracting. An internal relay can be used to switch between groups of outputs to give the form of control for pneumatic cylinders which is termed *cascade control* (see Section 5.5.1). Figure 19.33(b) shows a possible program. When the start switch is closed, the internal relay is activated. This energises solenoid A+ with the result that the piston in cylinder A extends. When extended it closes limit switch a+ and the piston in cylinder B extends. When this is extended it closes the limit switch b+. This activates the relay. As a result the B– solenoid is energised and the piston in B retracts. When this closes limit switch b–, solenoid A– is energised and the piston in cylinder A retracts.

Fig. 19.33   Cylinder sequencing

Another example of the use of markers/internal relays is resetting a latch. Figure 19.34 shows the ladder diagram. When the contacts of Input 1 are momentarily pressed then the output is energised. The contacts of the output are then closed and so latch the output, i.e. keeps it on even when the contacts of the input are no longer closed. The output can be unlatched by the internal relay contacts opening. This will occur if Input 2 is closed and energises the coil of the internal relay.

Fig. 19.34   Resetting a latch

Fig. 19.35   Resetting a latch

(a)

(b)

Fig. 19.36   Counter

An example of the use of a battery-backed internal relay is shown in Figure 19.35. When the contacts of Input 1 close, the coil of the battery-backed internal relay is energised. This closes the internal relay contacts and so even if contacts of the input open as a result of power failure the internal relay contacts remain closed. This means that the output controlled by the internal relay remains energised, even when there is a power failure.

19.6.3 Counters

Counters are used when there is a need to count a specified number of contact operations, e.g. where items pass along a conveyor into boxes, and when the specified number of items has passed into a box the next item is diverted into another box. Counter circuits are supplied as an internal feature of PLCs. In most cases the counter operates as a *down-counter*. This means that the counter counts down from the present value to zero, i.e. events are subtracted from the set value. When zero is reached the counter's contact changes state. An *up-counter* would count up to the preset value, i.e. events are added until the number reaches the set value. When the set value is reached the counter's contact changes state.

Figure 19.36(a) shows a basic counting program. In the ladder diagram the counter is represented by a rectangle spanning two lines. On one line is the reset which is used to reset the counter. The other line is the out line and the K10 indicates that the counter contact will change state on the 10th pulse. When the contacts of Input 1 momentarily close the counter is reset to the set value. The count will then count the number of pulses resulting from the contacts of Input 2 closing and opening. When this reaches the set value, in this case 10, the counter contact closes. Thus output is switched on after 10 pulses have been from Input 2. If the contacts of Input 1 are momentarily closed any time during the count, the counter will reset to 10. The instruction sequence, with Mitsubishi notation (Fig. 19.36(b)), is:

Step	Instruction	
0	LD	X400
1	RST	C460
2	LD	X401
3	OUT	C460
4	K	10
5	LD	C460
6	OUT	Y430

As an illustration of the use of a counter, consider the problem of the control for a machine which is required to direct 6 items along one path for packaging in a box, and then 12 items along another path for packaging in another box. Figure 19.37 shows the program that could be used. It involves two counters, one preset to count 6 and the other to count 12. Input 1 momentarily closes its contacts to start the counting cycle, resetting both counters. Input 2 contacts could be activated by a microswitch which is activated every time an item passes up to the junction in the paths. Counter 1 counts 6 items and then closes its contact. This activates the output, which might be a solenoid used to activate a flap which closes one path and opens another. Counter 1 also has contacts which close and enables counter 2 to start counting. When counter 2 has counted 12 items it resets both the counters and opens the counter 1 contacts which then results in the output becoming deactivated and items no longer directed towards the box to contain 12 items.

19.7 Shift registers

A number of internal relays can be grouped together to form a register which can provide a storage area for a series sequence of individual bits. Thus a 4-bit register would be formed by using four internal registers, an 8-bit using eight. The term *shift register* is used because the bits can be shifted along by one bit when there is a suitable input to the register. For example, with an 8-bit register we might initially have:

1	0	1	1	0	1	0	1

Then there is an input of a 0 shift pulse:

$$0 \rightarrow \boxed{0 \mid 1 \mid 0 \mid 1 \mid 1 \mid 0 \mid 1 \mid 0} \rightarrow 1$$

with the result that all the bits shift along one place and the last bit overflows.

The grouping together of a number of auxiliary registers to form a shift register is done automatically by a PLC when the shift register function is selected at the control panel. With the Mitsubishi PLC, this is done by using the programming function SFT (shift) against the auxiliary relay number that is to be the first in the register array. This then causes the block of relays, starting from that initial number, to be reserved for the shift register. Thus, if we select M140 to be the first relay then the shift register will consist of M140, M141, M142, M143, M144, M145, M146 and M147.

Shift registers have three inputs: one to load data into the first element of the register (OUT), one as the shift command (SFT) and one for resetting (RST). With OUT, a logic level 0 or 1 is loaded into the first element of the shift register. With SFT, a pulse moves the contents of the register along one bit at a time,

Fig. 19.37 Counter

Fig. 19.38   Shift register

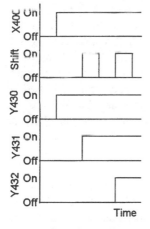

Fig. 19.39   Shift register

19.8 Master and jump controls

the final bit overflowing and being lost. With RST, a pulse of a closure of a contact resets the register contents to all 0s.

Figure 19.38 gives a ladder diagram involving a shift register when Mitsubishi notation is used; the principle is, however, the same with other manufacturers. M140 has been designated as the first relay of the register. When X400 is switched on, a logic 1 is loaded into the first element of the shift register, i.e. M140. We thus have the register with 10000000. The circuit shows that each element of the shift register has been connected as a contact in the circuit. Thus M140 contact closes and Y430 is switched on (Fig. 19.39). When contact X401 is closed, then the bits in the register are shifted along the register by one place to give 11000000, a 1 being shifted into the register because X400 is still on. Contact M141 thus closes and Y430 is switched on. As each bit is shifted along, the outputs are energised in turn. The instruction sequence for this ladder is:

Step	Instruction	
0	LD	X400
1	OUT	M140
2	LD	X401
3	SFT	M140
4	LD	X402
5	RST	M140
6	LD	M140
7	OUT	Y430
8	LD	M141
9	OUT	Y431
10	LD	M142
11	OUT	Y432
etc.		
20	LD	M147
21	OUT	Y437

Shift registers can thus be used to sequence events.

A whole block of outputs can be simultaneously turned off or on by using the same internal relay contacts in each output rung so that switching it on or off affects every one of the rungs. An alternative way of programming to achieve the same effect is to use a *master relay*. Figure 19.40 illustrates its use. We can think of it controlling the power to a length of the vertical rails of the ladder. When there is an input to close Input 1 contacts then master relay MC1 is activated and then the block of program rungs controlled by that relay follows. The end of a master relay controlled section is indicated by the reset MCR. If there is an

Fig. 19.40   Master relay

19.9 Data handling

input to Input 2 then master relay MC2 is activated and the rungs controlled by that master relay follow. If there are no inputs to either Input 1 or Input 2 then the rung that follows in the program is that for Input 9 controlled output Out 6. It is thus a branching program in that if there is Input 1 then branch to follow the MC1 controlled path, if Input 2 then branch to follow the MC2 controlled path and if neither follow the rest of the program and ignore the branches.

With a Mitsubishi PLC, an internal relay can be designated as a master control relay by programming it accordingly. Thus to program an internal relay M100 as a master control relay, the program instruction is:

    MC   M100

To indicate the end of the section controlled by a master control relay, the program instruction is:

    MCR   M100

19.8.1 Jumps

A function which is often provided with PLCs is the *conditional jump* function. Such a function enables programs to be designed so that if a certain condition exists then a section of the program is jumped. Figure 19.41 illustrates this on a flow diagram and with a section of ladder program. Following a section of program, A, the program rung is encountered with Input 1 and the conditional jump relay CJP. If Input 1 occurs then the program jumps to the rung with the end of jump relay coil EJP and so continues with that section of the program labelled as C, otherwise it continues with the program rungs labelled as program B.

With the exception of the shift register, the previous parts of this chapter have been concerned with the handling of individual bits of information, e.g. a switch being closed or not. There are, however, some control tasks where it is useful to deal with related groups of bits, e.g. a block of eight inputs, and so operate on them as a data word. Such a situation can arise when a sensor supplies an analogue signal which is converted to, say, an 8-bit word before becoming an input to a PLC.

The operations that may be carried out with a PLC on data words normally include:

1   Moving data.
2   Comparison of magnitudes of data, i.e. greater than, equal to, or less than.
3   Arithmetic operations such as addition and subtraction.

Fig. 19.41   Jump

Fig. 19.42   Move data

4   Conversions between binary coded decimal (BCD), binary and octal.

As discussed earlier, individual bits have been stored in memory locations specified by unique addresses. For example, for the Mitsubishi PLC, input memory addresses have been preceded by an A, outputs by a Y, timers by a T, auxiliary relays by an M, etc. Data instructions also require memory addresses and the locations in the PLC memory allocated for data are termed *data registers*. Each data register can store a binary word of, usually, 8 or 16 bits and is given an address such as D0, D1, D2, etc. An 8-bit word means that a quantity is specified to a precision of 1 in 256, a 16-bit a precision of 1 in 65 536.

Each instruction has to specify the form of the operation, the source of the data used in terms of its data register and the destination data register of the data.

19.9.1 Data movement

For data movement the instruction will contain the move data instruction, the source address of the data and the destination address of the data. Thus the ladder rung could be of the form shown in Figure 19.42 and the instructions could be, when data is to be moved from data register D1 to data register D2:

Step	Instruction	
0	LD	X300
1	MOV	
2		D1
3		D2

Such data transfers might be to move a constant into a data register, a time or count value to a data register, data from a data register to a timer or counter, data from a data register to an output, input data to a data register, etc.

19.9.2 Data comparison

PLCs can generally make the data comparisons of *less than* (usually denoted by < or LES), *equal to* (= or EQU), *less than or equal to* (≤ or <= or LEQ), *greater than* (> or GRT), *greater than or equal to* (≥, >= or GEQ) and *not equal to* (≠ or <> or NEQ). To compare data, the program instruction will contain the comparison instruction, the source address of the data and the destination address. Thus to compare the data in data register D1 to see if it is greater than data in data register D2, the ladder program rung would be of the form shown in Figure 19.43 and the instructions would be:

Fig. 19.43   Compare data

Fig. 19.44   Temperature alarm

Fig. 19.45   Add data

Fig. 19.46   BCD to binary

Step	Instruction	
0	LD	X300
1	>	
2		D1
3		D2

Such a comparison might be used when the signals from two sensors are to be compared by the PLC before action is taken. For example, an alarm might be required to be sounded if a sensor indicates a temperature above 80°C and remain sounding until the temperature falls below 70°C. Figure 19.44 shows the ladder program that could be used. The input temperature data is inputted to the source address and the destination address contains the set value. When the temperature rises to 80°C, or higher, the data value in the source address becomes ≥ the destination address value and there is an output to the alarm which latches the input. When the temperature falls to 70°C, or lower, the data value in the source address becomes ≤ to the destination address value and there is an output to the relay which then opens its contacts and so switches the alarm off.

19.9.3 Arithmetic operations

Some PLCs can carry out just the arithmetic operations of addition and subtraction, others have even more arithmetic functions. The instruction to add or subtract generally states the instruction, the register containing the address of the value to be added or subtracted, the address of the value to which the addition or from which the subtraction is to be made and the register where the result is to be stored. Figure 19.45 shows the form used for the ladder symbol for addition with OMRON.

Addition or subtraction might be used to alter the value of some sensor input value, perhaps a correction or offset term, or alter the preset values of timers or counters.

19.9.4 Code conversions

All the internal operations in the CPU of a PLC are carried out using binary numbers. Thus, when the input is a signal which is decimal, conversion to binary coded decimal (BCD) is used. Likewise, where a decimal output is required, conversion to decimal is required. Such conversions are provided with most PLCs. For example, with Mitsubishi, the ladder rung to convert BCD to binary is of the form shown in Figure 19.46. The data at the source address is in BCD and converted to binary and placed at the destination address.

19.10 Analogue input/output

Many sensors generate analogue signals and many actuators require analogue signals. Thus, some PLCs may have an analogue-to-digital converter module fitted to input channels and a digital-to-analogue converter module fitted to output channels. An example of where such an item might be used is for the control of the speed of a motor so that its speed moves up to its steady value at a steady rate (Fig. 19.47). The input is an on/off switch to start the operation. This opens the contacts for the data register and so it stores zero. Thus the output from the controller is zero and the analogue signal from the DAC is zero and hence motor speed is zero. The closing of the start contacts gives outputs to the DAC and the data register. Each time the program cycles though these rungs on the program the data register is incremented by 1 and so the analogue signal is increased and hence the motor speed. Full speed is realised when the output from the data register is the word 11111111. The timer function of a PLC can be used to incorporate a delay between each of the output bit signals.

Fig. 19.47 Ramping the speed of a motor

A PLC equipped with analogue input channels can be used to carry out a continuous control function, i.e. PID control (see Chapter 13). Thus, for example, to carry out proportional control on an analogue input the following set of operations can be used:

1 Convert the sensor output to a digital signal.
2 Compare the converted actual sensor output with the required sensor value, i.e. the set point, and obtain the difference. This difference is the error.
3 Multiply the error by the proportional constant K_P.
4 Move this result to the digital-to-analogue converter output and use the result as the correction signal to the actuator.

An example of where such a control action might be used is with a temperature controller. Figure 19.48 shows a possibility. The input could be from a thermocouple, which after amplification is fed through an analogue-to-digital converter into the PLC. The PLC is programmed to give an output proportional to the error between the input from the sensor and the required temperature.

Fig. 19.48 Proportional control of temperature

The output word is then fed through a digital-to-analogue converter to the actuator, a heater, in order to reduce the error.

With the ladder program shown, rung 0 reads the ADC and stores the temperature value in data register DR1. With rung 1, the data register DR2 is used to store the set point temperature. Rung 2 uses the subtract function to subtract the values held in data registers DR1 and DR2 and store the result in data register DR3, i.e. this data register holds the error value. With rung 3 a multiply function is used, in this case to multiply the value in data register DR3 by the proportional gain of 4. Rung 4 uses an internal relay which can be programmed to switch off DR3 if it takes a negative value. With rung 5 the data register DR3 is reset to zero when the input is switched off.

Some PLCs have add-on modules which more easily enable PLC control to be used, without the need to write lists of instructions in the way outlined above.

19.11 Selection of a PLC

In considering the size and type of PLC required for a particular task or tasks, criteria that need to be considered are:

1 What input/output capacity is required, i.e. the number of inputs/outputs, capability of expansion for future needs?
2 What types of inputs/outputs are required, i.e. isolation, on-board power supply for inputs/outputs, signal conditioning?
3 What size of memory is required? This is linked to the number of inputs/outputs and the complexity of program used.
4 What speed and power is required of the CPU? This is linked to the number of types of instruction that can be handled by a PLC. As the number of types increases so a faster CPU is required. Likewise, the greater the number of inputs/outputs to be handled the faster the CPU required.

Problems

1 What are the logic functions used for switches (a) in series, (b) in parallel?
2 Draw the ladder rungs to represent:

Fig. 19.49   Problem 3

Fig. 19.50   Problem 4

Fig. 19.51   Problem 5

Fig. 19.52   Problem 11

(a) Two switches are normally open and both have to be closed for a motor to operate.

(b) Either of two, normally open, switches have to be closed for a coil to be energised and operate an actuator.

(c) A motor is switched on by pressing a spring-return push-button start switch, and the motor remains on until another spring-return push-button stop switch is pressed.

3 Write the program instructions corresponding to the latch program shown in Figure 19.49.

4 Write the program instructions for the program in Figure 19.50 and state how the output varies with time.

5 Write the program instructions corresponding to the program in Figure 19.51 and state the results of inputs to the PLC.

6 Devise a timing circuit that will switch an output on for 1 s then off for 20 s, then on for 1 s, then off for 20 s, and so on.

7 Devise a timing circuit that will switch an output on for 10 s then switch it off.

8 Devise a circuit that can be used to start a motor and then after a delay of 100 s start a pump. When the motor is switched off there should be a delay of 10 s before the pump is switched off.

9 Devise a circuit that could be used with a domestic washing machine to switch on a pump to pump water for 100 s into the machine, then switch off and switch on a heater for 50 s to heat the water. The heater is then switched off and another pump is to empty the water from the machine for 100 s.

10 Devise a circuit that could be used with a conveyor belt which is used to move an item to a work station. The presence of the item at the work station is detected by means of breaking a contact activated by a beam of light to a photosensor. There it stops for 100 s for an operation to be carried out before moving on and off the conveyor. The motor for the belt is started by a normally open start switch and stopped by a normally closed switch.

11 How would the timing pattern for the shift register in Figure 19.38 change if the data input X400 was of the form shown in Figure 19.52?

12 Explain how a PLC can be used to handle an analogue input.

13 Devise a system, using a PLC, which can be used to control the movement of a piston in a cylinder so that when a switch is momentarily pressed, the piston moves in one direction and when a second switch is momentarily pressed the piston moves in the other direction. Hint: you might consider using a 4/2 solenoid controlled valve.

14 Devise a system, using a PLC, which can be used to control the movement of a piston in a cylinder using a 4/2 solenoid operated pilot valve. The piston is to move in one direction when a proximity sensor at one end of the stroke closes contacts and in the other direction when a proximity sensor at the other end of the stroke indicates its arrival there.

20 Communication systems

20.1 Digital communications

An *external bus* is a set of signal lines that interconnects microprocessors, microcontrollers, computers and PLCs and also connects them with peripheral equipment. Thus a computer needs to have a bus connecting it with a printer if its output is to be directed to the printer and printed. Multiprocessor systems are quite common. For example, in a car there are likely to be several microcontrollers with each controlling a different part of the system, e.g. engine management, braking and instrument panel, and communication between them is necessary. In automated plant there is not only a need for data to pass between programmable logic controllers, displays, sensors, and actuators and allow for data and programs to be inputted by operator, but there can also be data communications with other computers. There may, for example, be a need to link a PLC into a control system involving a number of PLCs and computers. Computer integrated manufacturing (CIM) is an example of a large network which can involve large numbers of machines linked together. This chapter is a consideration of how such data communications between computers can take place, whether it is just simply machine-to-machine or a large network involving large numbers of machines linked together, and the forms of standard communication interfaces.

For more details concerning digital data communications, the reader is referred to texts such as *Data Communication for Engineers* by C.G. Freer (Macmillan 1992) or *Advanced Electronic Communications Systems* by W. Tomasi (Prentice-Hall 2000).

20.2 Centralised, hierarchical and distributed control

Centralised computer control involves the use of one central computer to control an entire plant. This has the problem that failure of the computer results in the loss of control of the entire plant. This can be avoided by the use of dual computer systems. If one computer fails the other one takes over. Such centralised systems were common in the 1960s and 1970s. The development of the microprocessor and the ever reducing costs of computers has led to multi-computer systems becoming more common and the development of hierarchical and distributed systems.

With the *hierarchical system*, there is a hierarchical system of computers according to the tasks they carry out. The computers handling the more routine tasks are supervised by computers which have a greater decision-making role. For example, the computers which are used for direct digital control of systems are subservient to a computer which performs supervisory control of the entire system. The work is divided between the computers according to the function involved. There is specialisation of computers with some computers only receiving some information and others different information.

With the *distributed system*, each computer system carries out essentially similar tasks to all the other computer systems. In the event of a failure of one, or overloading of a particular computer, work can be transferred to other computers. The work is spread across all the computers and not allocated to specific computers according to the function involved. There is no specialisation of computers. Each computer thus needs access to all the information in the system.

In most modern systems there is generally a mixture of distributed and hierarchical systems. For example, the work of measurement and actuation may be distributed among a number of microcontrollers/computers which are linked together and provide the data base for the plant. These may be overseen by a computer used for direct digital control or sequencing and this in turn may be supervised by one used for supervisory control of the plant as a whole. Typical levels in such a scheme are:

Level 1 Measurement and actuators
Level 2 Direct digital and sequence control
Level 3 Supervisory control
Level 4 Management control and design

Distributed/hierarchical systems have the advantage of allowing the task of measurement scanning and signal conditioning in control systems to be carried out by sharing it between a number of microprocessors. This can involve a large number of signals with a high frequency of scanning. If extra measurement loops are required, it is a simple matter to increase the capacity of the system by adding microprocessors. The units can be quite widely dispersed, being located near the source of the measurements. Failure of one unit does not result in failure of the entire system.

20.2.1 Parallel and serial data transmission

Data communication can be via parallel or serial transmission links.

1 *Parallel data transmission*

Within computers, data transmission is usually by *parallel data paths*. Parallel data buses transmit 8, 16 or 32 bits simultaneously, having a separate bus wire for each data bit and the control signals. Thus, if there are 8 data bits to be transmitted, e.g. 11000111, then 8 data wires are needed. The entire 8 data bits are transmitted in the same time as it takes to transmit 1 data bit because each bit is on a parallel wire. Handshaking (see Section 18.3.2) lines are also needed, handshaking being used for each character transmitted with lines needed to indicate that data is available for transmission and that the receiving terminal is ready to receive. Parallel data transmission permits high data transfer rates but is expensive because of the cabling and interface circuitry required. It is thus normally only used over short distances or where high transfer rates are essential.

2 *Serial data transmission*

This involves the transmission of data which, together with control signals, is sent bit by bit in sequence along a single line. Only two conductors are needed, to transmit data and to receive data. Since the bits of a word are transmitted sequentially and not simultaneously, the data transfer rate is considerably less than with parallel data transmission. However, it is cheaper since far fewer conductors are required. For example, with a car when a number of microcontrollers are used, the connections between them are by serial data transmission. Without the use of serial transmission the number of wires involved would be considerable. In general, serial data transmission is used for all but the shortest peripheral connections.

Consider the problem of sending a sequence of characters along a serial link. The receiver needs to know where one character starts and stops. Serial data transmission can be either asynchronous or synchronous transmission. *Asynchronous transmission* implies that both the transmitter and receiver computers are not synchronised, each having its own independent clock signals. The time between transmitted characters is arbitrary. Each character transmitted along the link is thus preceded by a start bit to indicate to the receiver the start of a character, and followed by a stop bit to indicate its completion. This method has the disadvantage of requiring extra bits to be transmitted along with each character and thus reduces the efficiency of the line for data transmission. With *synchronous transmission* there is no need for start and stop bits since the transmitter and receiver have a common clock signal and thus characters automatically start and stop always at the same time in each cycle.

The *rate of data transmission* is measured in bits per second. If a group of *n* bits form a single symbol being transmitted and the symbol has a duration of T seconds then the data rate of transmission is n/T. The *baud* is the unit used. The baud rate is only the same as the number of bits per second transmitted if each character is represented by just one symbol. Thus a system which does not use start and stop pulses has a baud rate equal to the bit rate, but this will not be the case when there are such bits.

20.2.2 Serial data communication modes

Serial data transmission occurs in one of three modes:

1 *Simplex mode*

Transmission is only possible in one direction, from device A to device B, where device B is not capable of transmitting back to device A (Fig. 20.1(a)). You can think of the connection between the devices being like a one-way road. This method is usually only used for transmission to devices such as printers which never transmit information.

2 *Half-duplex mode*

Data is transmitted in one direction at a time but the direction can be changed (Fig. 20.1(b)). Terminals at each end of the link can be switched from transmit to receive. Thus device A can transmit to device B and device B to device A but not at the same time. You can think of this being like a two lane road under repair with traffic from one lane being stopped by a traffic control to allow the traffic for the other lane through. Citizens Band (CB) radio is an example of half-duplex mode, a person can receive or talk but not do both simultaneously.

3 *Full-duplex mode*

Data may be transmitted simultaneously in both directions between devices A and B (Fig. 20.1(c)). This is like a two lane highway in which traffic can occur in both directions simultaneously. The telephone system is an example of full-duplex mode in that a person can talk and receive at the same time.

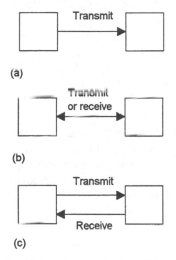

Fig. 20.1 Communication modes

20.3 Networks

The term *network* is used for a system which allows two or more computers/microprocessors to be linked for the interchange of data. The logical form of the links is known as the network *topology*. The term *node* is used for a point in a network where one or more communication lines terminate or a unit is connected to the communication lines. Commonly used forms are:

(a)

(b)

(c)

(d)

(e)

Fig. 20.2 Network topologies:
(a) data bus, (b) star, (c) hierarchy,
(d) ring, (e) mesh

1   *Data bus*
    This has a linear bus (Fig. 20.2(a)) into which all the stations
    are plugged. This system is often used for multipoint
    terminal clusters. It is generally the preferred method for
    distances between nodes of more than 100 m.

2   *Star*
    This has dedicated channels between each station and a
    central switching hub (Fig. 20.2(b)) through which all
    communications must pass. This is the type of network used
    in the telephone systems (private branch exchanges (PBXs))
    in many companies, all the lines passing through a central
    exchange. This system is also often used to connect remote
    and local terminals to a central mainframe computer. There
    is a major problem with this system in that if the central hub
    fails then the entire system fails.

3   *Hierarchy* or *tree*
    This consists of a series of branches converging indirectly to
    a point at the head of the tree (Fig. 20.2(c)). With this system
    there is only one transmission path between any two stations.
    This arrangement may be formed from a number of linked
    data bus systems. Like the bus method, it is often used for
    distances between nodes of more than 100 m.

4   *Ring*
    This is a very popular method for local area networks,
    involving each station being connected to a ring (Fig.
    20.2(d)). The distances between nodes is generally less than
    100 m. Data put into the ring system continues to circulate
    round the ring until some system removes it. The data is
    available to all the stations.

5   *Mesh*
    This method (Fig. 20.2(e)) has no formal pattern to the
    connections between stations and there will be multiple data
    paths between them.

The term *local area network* (LAN) is used for a network over a
local geographic area such as a building or a group of buildings
on one site. The topology is commonly bus, star or ring. A *wide
area network* is one that interconnects computers, terminals and
local area networks over a national or international level. This
chapter is primarily concerned with local area networks.

20.3.1 Network access control

Access control methods are necessary with a network to ensure
that only one user of the network is able to transmit at any one
time. The following are methods used.

With ring-based local area networks, two commonly used methods are:

1 *Token passing*

With this method a token, a special bit pattern, is circulated. When a station wishes to transmit it waits until it receives the token, then transmits the data with the token attached to its end. Another station wishing to transmit removes the token from the package of data and transmits its own data with the token attached to its end.

2 *Slot passing*

This method involves empty slots being circulated. When a station wises to transmit data it deposits it in the first empty slot that comes along.

With bus or tree networks a method that is often used is:

3 *Carrier sense multiple access with collision detection access (CSMA/CD)*

This method is generally identified with the *Ethernet LAN bus*. With the CSMA/CD access method, stations have to listen for other transmissions before transmitting with any station being able to gain control of the network and transmit, hence the term multiple access. If no activity is detected then transmission can occur. If there is activity then the system has to wait until it can detect no further activity. Despite this listening before transmission, it is still possible for two or more systems to start to transmit at the same time. If such a situation is detected, both stations cease submitting and wait a random time before attempting to retransmit.

20.3.2 Broadband and baseband

The term *broadband transmission* is used for a network in which information is modulated onto a radio frequency carrier which passes through the transmission medium such as a coaxial cable. Typically the topology of broadband local area networks is a bus with branches. Broadband transmission allows a number of modulated radio frequency carriers to be simultaneously transmitted and so offers a multichannel capability. The term *baseband transmission* is used when digital information is passed directly through the transmission medium. Baseband transmission networks can only support one information signal at a time. A LAN may be either baseband or broadband.

20.4 Protocols

Transmitted data will contain two types of information. One is the data which one computer wishes to send to another, the other is information termed *protocol data* and is used by the interface between a computer and the network to control the transfer of the data into the network or from the network into the computer. A protocol is a formal set of rules governing data format, timing, sequencing, access control and error control. The three elements of a protocol are:

1 *Syntax*, which defines data format, coding and signal levels.

2 *Semantics*, which deals with synchronisation, control and error handling.

3 *Timing*, which deals with the sequencing of data and the choice of data rate.

When a sender communicates with a receiver then both must employ the same protocol, e.g. two microcontrollers with data to be serially transmitted between them. With simplex communication the data block can be just sent from sender to receiver. However, with half-duplex, each block of transmitted data, if valid, must be acknowledged (ACK) by the receiver before the next block of data can be sent (Fig. 20.3(a)); if invalid a NAK, negative acknowledgement, signal is sent. Thus a continuous stream of data cannot be transmitted. The CRC bits, *cyclic redundancy checks*, are a means of error detection and are transmitted immediately after a block of data. The data is transmitted as a binary number and at the transmitter the data is divided by a number and the remainder is used as the cyclic check code. At the receiver the incoming data, including the CRC, is divided by the same number and will give zero remainder if the signal is error free. With full-duplex mode (Fig. 20.3(b)), data can be continuously sent and received.

Fig. 20.3   Protocols:
(a) half-duplex, (b) full-duplex

Within a package being sent, there is a need to include protocol information. For example, with asynchronous transmission there may be characters to indicate the start and end of data. With

Fig. 20.4   (a) Bisync, (b) HDLC

synchronous transmission and the *Bisync protocol*, a data block is preceded by a synchronising sequence of bits, usually the ASCII character SYN (Fig. 20.4(a)). The SYN characters are used by the receiver to achieve character synchronisation, preparing the receiver to receive data in 8-bit groupings. The Motorola MC6852 is a synchronous serial data adapter (SSDA) that is designed for use with 6800 microprocessors to provide a synchronous serial communications interface using the Bisync protocol. It is similar to the asynchronous communications interface adapter described in Section 18.5. Another protocol is the *High-level Data Link Control* (HDLC). This is a full-duplex protocol with the beginning and end of a message being denoted by the bit pattern 01111110. Address and control fields follow the start flag. The address identifies the address of the destination station, the control field defines whether the frame is supervisory, information or unnumbered. Following the message is a 16-bit frame check sequence which is used to give a cyclic redundancy check (CRC). The Motorola 6854 is an example of a serial interface adapter using this HDLC protocol.

20.5 Open Systems Interconnection communication model

Communication protocols have to exist on a number of levels. The International Organisation for Standardisation (ISO) has defined a seven-layer standard protocol system known as the *Open Systems Interconnection* (OSI) model. The model is a framework for developing a co-ordinated system of standards. The layers are:

1 *Physical layer*
    This layer describes the means for bit transmission to and from physical components of the network. It deals with hardware issues, e.g. the types of cable and connectors to be used, synchronising data transfer and signal levels. Commonly used LAN systems defined at the physical layer are Ethernet and token ring.

2 *Data link layer*
    This layer defines the protocols for sending and receiving messages, error detection and correction and the proper sequencing of transmitted data. It is concerned with packaging data into packets and placing them on the cable and then taking them off the cable at the receiving end. Ethernet and token ring are also defined at this level.

3 *Network layer*
    This deals with communication paths and the addressing, routing and control of messages on the network and thus making certain that the messages get to the right destinations. Commonly used network layer protocols are

Internet Protocol (IP) and Novell's Internetwork Packet Exchange (IPX).

4 *Transport layer*
This provides for reliable end-to-end message transport. It is concerned with establishing and maintaining the connection between transmitter and receiver. Commonly used transport layer protocols are Internet Transmission Control protocol (TCP) and Novell's Sequenced Packet Exchange (SPX).

5 *Session layer*
This layer is concerned with the establishment of dialogues between application processes which are connected together by the network. It is responsible for determining when to turn a communication between two stations on or off.

6 *Presentation layer*
This layer is concerned with allowing the encoded data transmitted to be presented in a suitable form for user manipulation.

7 *Application layer*
This layer provides the actual user information processing function and application specific services. It provides such functions as file transfer or electronic mail which a station can use to communicate with other systems on the network.

20.5.1 Network standards

There are a number of network standards, based on the OSI layer model, that are commonly used. The following are examples.

In the United States, General Motors realised that the automation of their manufacturing activities posed a problem of equipment being supplied with a variety of non-standard protocols. They thus developed a standard communication system for factory automation applications. The standard is referred to as the *manufacturing automation protocol* (MAP) (Figure 20.5). The choice of protocols at the different layers reflects the requirement for the system to fit the manufacturing environment. Layers 1 and 2 are implemented in hardware electronics and levels 3 to 7 using software. For the physical layer broadband transmission is used. The broadband method allows the system to be used for services in addition to those required for MAP communications. For the data link layer the token system with a bus is used with logical link control (LLC) to implement such functions as error checking, etc. For the other layers ISO standards are used. At layer 7, MAP includes manufacturing message services (MMS), an application relevant to factory floor communications which defines interactions between program-

Fig. 20.5 MAP

OSI		SNI	
7	Application	Transaction services	7
6	Presentation	Presentation services	6
5	Session	Data flow control	5
4	Transport	Transmission control	4
3	Network	Path control	3
2	Data link	Data link control	2
1	Physical	Physical control	1

Fig. 20.6 SNA

OSI		Netware	
7	Application	Netware core protocol	
6	Presentation		
5	Session	Named pipes	NetBIOS
4	Transport	Sequenced packet exchange	
3	Network	Internetwork packet exchange	
2	Data link	LAN drivers	
		ODI	NDIS
1	Physical	Physical	

Fig. 20.7 Netware

mable logic controllers and numerically controlled machines or robots.

Technical and office protocol (TOP) is a standard that was developed by Boeing Computer Services. It has much in common with MAP but can be implemented at a lower cost because it is a baseband system. It differs from MAP in layers 1 and 2, using either the token with a ring or the carrier sense multiple access and collision detection (CSMA/CD) access method with a bus network. Also, at layer 7, it specifies application protocols that concern office requirements, rather than factory floor requirements. With the CSMA/CD access method, stations have to listen for other transmissions before transmitting. TOP and MAP networks are compatible and a gateway device can be used to connect TOP and MAP networks. This device carries out the appropriate address conversions and protocol changes.

Systems network architecture (SNA) is a system developed by IBM as a design standard for IBM products. SNA is divided into seven layers; it, however, differs to some extent from the OSI layers (Fig. 20.6). The data link control layer provides support for the token ring protocol for LANs. Five of the SNA layers are consolidated in two packages: the path control network for layers 2 and 3 and the network addressable units for layers 4, 5 and 6.

Novell Netware describes a suite of local area network operating systems, this includes Netware Lite for peer-to-peer networks with two to about two dozen users, Netware 3.x for single-server LANs supporting hundreds of users and Netware 4.x which is an enterprise network operating scheme. Figure 20.7 shows how the Netware protocol suite relates to the OSI layers. The Netware core protocol defines the procedures used for transmitting information to a server and its clients on a Netware LAN. Named pipes provide an interface between processes running on separate computers. The Network Basic Input Output System (NetBIOS) is an application program interface. Sequenced packet exchange (SPX) defines protocols for sending messages from one end of the network to the other. Internetwork packet exchange (IPX) is a peer-to-peer networking protocol. Open data link interface (ODI) and Network driver interface specifications (NDIS) define protocols for sending and receiving information between units that are directly connected to each other. In the physical layer, Netware supports such technologies as Ethernet and token ring.

With PLC systems, it is quite common for the system used to be that marketed by the PLC manufacturer. For example, Allen Bradley have the *Allen Bradley data highway* which uses token passing to control message transmission; Mitsubishi have Melsec-Net and Texas Instruments have TIWAY. A commonly used system with PLC networks is the *Ethernet*. This is a single bus system with CSMA/CD used to control access and is widely used with systems involving PLCs communicating with computers. The problem with using CSMA/CD is that though this method works well when traffic is light, as network traffic increases the number of collisions and corresponding back-off of transmitters increases. Network throughput can thus slow down quite dramatically.

20.6 Communication interfaces

Digital data can be communicated between devices by serial or parallel communication. With serial communications the main standard is *RS-232*. With parallel communications commonly used standards are the *Centronics parallel interface* and the *general purpose interface bus* (GPIB) (IEEE-488).

20.6.1 Serial communication interface

The most popular serial interface is RS-232; this was first defined by the American Electronic Industries Association (EIA) in 1962. The standard relates to data terminal equipment (DTE) and data circuit-terminating equipment (DCE). Data terminal equipment can send or receive data via the interface, e.g. a microcontroller. Data circuit-terminating equipment is devices which facilitate communication; a typical example is a modem. This forms an essential link between a microcomputer and a conventional analogue telephone line.

RS-232 signals can be grouped into three categories:

1 *Data*
RS-232 provides two independent serial data channels, termed primary and secondary. Both these channels are used for full-duplex operation.

2 *Handshake control*
Handshaking signals are used to control the flow of serial data over the communication path.

3 *Timing*
For synchronous operation it is necessary to pass clock signals between transmitters and receivers.

Table 20.1 gives the RS-232C connector pin numbers and signals for which each is used; not all the pins and signals are necessarily used in a particular set-up. The signal ground wire allows for a return path.

The connector to a RS-232C serial port is via a 25-pin D-type connector; usually a male plug is used on cables and a female socket on the DCE or DTE.

For the simplest bi-directional link only the two lines 2 and 3 for transmitted data and received data, with signal ground (7) for the return path of these signals, are needed (Fig. 20.8(a)). Thus the minimum connection is via a three-wire cable. For a simple set-up involving a personal computer (PC) being linked with a visual display unit (VDU) pins 1, 2, 3, 4, 5, 6, 7 and 20 are

(a)

(b)

Fig. 20.8 RS-232 connections: (a) minimum configuration, (b) PC connection

Table 20.1 RS-232 pin assignments

Pin	Abbreviation	Direction: To	Signal/function
1	FG		Frame ground
2	TXD	DCE	Transmitted data
3	RXD	DTE	Received data
4	RTS	DCE	Request to send
5	CTS	DTE	Clear to send
6	DSR	DTE	DCE ready
7	SG		Signal ground/common return
8	DCD	DTE	Received line detector
12	SDCD	DTE	Secondary received line signal detector
13	SCTS	DTE	Secondary clear to send
14	STD	DCE	Secondary transmitted data
15	TC	DTE	Transmit signal timing
16	SRD	DTE	Secondary received data
17	RC	DTE	Received signal timing
18		DCE	Local loop back
19	SRTS	DCE	Secondary request to send
20	DTR	DCE	Data terminal ready
21	SQ	DEC/DTE	Remote loop back/signal quality detector
22	RI	DTE	Ring indicator
23		DEC/DTE	Data signal rate selector
24	TC	DCE	Transmit signal timing
25		DTE	Test mode

involved (Fig. 20.8(b)). The signals sent through pins 4, 5, 6 and 20 are used to check that the receiving end is ready to receive a signal; the transmitting end is ready to send and the data is ready to be sent.

RS-232 is limited concerning the distance over which it can be used as noise limits the transmission of high numbers of bits per second when the length of cable is more than about 15 m. The maximum data rate is about 20 kbits/s. Other standards such as RS-422 and RS-485 are similar to RS-232 and can be used for higher transmission rates and longer distances.

RS-422 uses a pair of lines for each signal and can operate up to about 1220 metres or at higher transmission speeds up to 100 bits/s and in noisier environments; maximum speed and maximum distance cannot, however, be achieved simultaneously. RS-485 can be used up to about 1220 m with speeds of 100 kbits/s.

The serial communications interface of the Motorola micro-controller MC68HC11 is capable of full-duplex communications at a variety of baud rates. However, the input and output of this system use transistor–transistor logic (TTL) for which logic 0 is 0 V and logic 1 is +5 V. The RS-232C standards are +12 V for logic 0 and –12 V for logic 1. Thus conversion in signal levels is necessary. This can be achieved by using integrated circuit devices such as MC1488 for TTL to RS-232C conversion and MC1489 for RS-232C to TTL conversion (Fig. 20.9).

Fig. 20.9 Level conversion

20.6.2 Centronics parallel interface

The Centronics parallel interface is commonly used for the parallel interface to a printer. Table 20.2 shows the pin assignments, an Amphenol 36-way connector being used.

The computer, or microcontroller, sends a strobe pulse every time it sends data to the printer (Fig. 20.10). Handshaking involves two printer signals, ACK and BUSY. When the printer receives the strobe pulse it sets its BUSY line high and after the data has been received it sends back an acknowledge pulse to indicate that it is ready for more data and sets BUSY low. The signal levels used for the Centronics interface are transistor–transistor logic (TTL).

Data

STROBE

BUSY

ACK

Fig. 20.10 Handshaking signals

If the Centronics interface is used to send characters from a microcontroller to a printer then a subroutine program would be used. The data pins (2 to 9) of the Centronics interface would be connected to an output port of the microcontroller. An input/output port would be used for status signals, i.e. ERROR, PO, SLCT, BUSY and ACK. STROBE is connected to STRB. The subroutine program would save the register and accumulator settings of the main program and then switch to the subroutine. This would then read the ERROR input to determine if the printer is ready for printing. If this signal is not OK then the subroutine program does not proceed. If it is OK then with a M68HC11 microcontroller, we might set port C as an output by writing $FF into the DDRC register, and output the character to the printer. A time delay is then inserted to allow for the set-up time of the printer. STRB can be used for handshaking signals and thus, after setting the PIOC (parallel input/output control) register at address $1002 to select the conditions for operation of STRB so that it is active when low, is used to send a falling edge signal to indicate data is being sent and, after a suitable time delay, is made a rising edge signal to show the end of transmission. The microcontroller then waits for the ACK signal from the printer before completing the subroutine and returning to the main program.

Table 20.2 Centronics pin assignments

Signal pin	Return pin	Signal	Function
1	19	STROBE	Strobe pulse to read data in
2	20	DATA 1	Data bit 1 (LSB)
3	21	DATA 2	Data bit 2
4	22	DATA 3	Data bit 3
5	23	DATA 4	Data bit 4
6	24	DATA 5	Data bit 5
7	25	DATA 6	Data bit 6
8	26	DATA 7	Data bit 7
9	27	DATA 8	Data bit 8 (MSB)
10	28	ACK	Acknowledge pulse to indicate data has been received and printer ready for new data
11	29	BUSY	Printer busy, a high signal indicating printer cannot receive data
12		PO	Goes high when printer out of paper
13		SLCT	Select status, high when printer can communicate, low when not
14		AUTO FEED	Auto linefeed, if low a linefeed is added to a carriage return
16		SG	Signal ground
17		FG	Frame ground
18		+5 V	
31	30	PRIME	Used to initialise printer, when low the printer resets
32		ERROR	Error status line, low when printer detects a fault
33		SG	Signal ground

Fig. 20.11 GPIB bus structure

20.6.3 General purpose instrument bus

The standard interface commonly used for general parallel communications is the *General purpose instrument bus* (GPIB), the IEEE-488 standard, originally developed by Hewlett Packard to link their computers and instruments and thus often referred to as the *Hewlett Packard Instrumentation Bus*. Each of the devices connected to the bus is termed a listener, talker or controller. Listeners are devices that accept data from the bus, talkers place data, on request, on the bus and controllers manage the flow of data on the bus by sending commands to talkers and listeners and carries out polls to see which devices are active. There is a total of 24 lines with the interface (Fig. 20.11):

1 Eight bi-directional lines to carry data and commands between the various devices connected to the bus.

2 Five lines for control and status signals.

3 Three lines for handshaking between devices.

4 Eight lines are ground return lines.

Table 20.3 lists the functions of the lines and their pin numbers in a 25-way D-type connector. Up to 15 devices can be attached to the bus at any one time, each device having its own address.

The 8-bit parallel data bus can transmit data as one 8-bit byte at a time. Each time a byte is transferred the bus goes through a handshake cycle. Each device on the bus has its own address.

Table 20.3 IEEE-488 bus system

Pin	Signal group	Abbreviation	Function
1	Data	D101	Data line 1
2	Data	D102	Data line 2
3	Data	D103	Data line 3
4	Data	D104	Data line 4
5	Management	EOI	End Or Identify. This is used either to signify the end of a message sequence from a talker device or is used by the controller to ask a device to identify itself
6	Handshake	DAV	Data Valid. When the level is low on this line then the information on the data bus is valid and acceptable
7	Handshake	NRFD	Not Ready For Data. This line is used by listener devices taking it high to indicate that they are ready to accept data
8	Handshake	NDAC	Not Data Accepted. This line is used by listeners taking it high to indicate that data is being accepted
9	Management	IFC	Interface Clear. This is used by the controller to reset all the devices of the system to the start state
10	Management	SRQ	Service Request. This is used by devices to signal to the controller that they need attention
11	Management	ATN	Attention. This is used by the controller to signal that it is placing a command on the data lines
12		SHIELD	Shield
13	Data	D105	Data line 5
14	Data	D106	Data line 6
15	Data	D107	Data line 7
16	Data	D108	Data line 8
17	Management	REN	Remote Enable. This enables a device to indicate that it is to be selected for remote control rather than by its own control panel
18		GND	Ground/common (twisted pair with DAV)
19		GND	Ground/common (twisted pair with NRFD)
20		GND	Ground/common (twisted pair with NDAC)
21		GND	Ground/common (twisted pair with IFC)
22		GND	Ground/common (twisted pair with SRG)
23		GND	Ground/common (twisted pair with ATN)
24		GND	Signal ground

Commands from the controller are signalled by taking the Attention Line (ATN) low. Commands are then directed to individual devices by placing addresses on the data lines; device addresses are sent via the data lines as a parallel 7-bit word with the lowest 5 bits providing the device address and the other 2 bits control information. If both these bits are 0 then the commands are sent to all addresses; if bit 6 is 1 and bit 7 a 0 the addressed device is switched to be a listener; if bit 6 is 0 and bit 7 is 1 then the device is switched to be a talker.

Handshaking uses the lines DAV, NRFD and NDAC, the three lines ensuring that the talker will only talk when it is being

Fig. 20.12 Handshaking

Fig. 20.13 GPIB hardware

listened to by listeners (Fig. 20.12). When a listener is ready to accept data, NRFD is made high. When data has been placed on the line, DAV is made low to notify devices that data is available. When a device accepts a data word it sets NDAC high to indicate that it has accepted the data and NRFD low to indicate that it is now not ready to accept data. When all the listeners have set NDAC high, then the talker cancels the data valid signal, DAV going high. This then results in NDAC being set low. The entire process can then be repeated for another word being put on the data bus.

The GPIB is a bus which is used to interface a wide range of instruments, e.g. digital multimeters and digital oscilloscopes, via plug-in boards (Fig. 20.13) to computers with standard cables used to link the board o the instruments via interfaces.

20.6.4 Personal computer buses

A computer bus is the bus used to connect the CPU to the input/output ports or other devices, the form of the bus depending on the microprocessor used. For example:

1 The *XT computer bus* was introduced in 1983 for 8-bit data transfers with IBM PC/XT and compatible computers.

2 The *AT bus*, also referred to as the *industry standard architecture* (ISA) *bus* was later introduced for use with 16-bit transfers with IBM PC and other compatible computers using 80286 and 80386 microprocessors. The AT bus is compatible with the XT bus so that plug-in XT boards can be used in AT bus slots.

3 The *extended industry standard architecture* (EISA) *bus* was developed to cope with 32-bit data transfers with IBM PC and other compatible computers using 80386 and 80486 microprocessors.

4 The *Micro Channel architecture* (MCA) *bus* is a 16-bit or 32-bit data transfer bus designed for use with IBM personal System/1 (PS/2) computers. Boards for use with this bus are not compatible with PC/XT/AT boards.

5 The *NuBus* is the 32-bit bus used in Apple's Macintosh II computers.

6 The *S-bus* is the 32-bit bus used in Sun Microsystem's SPARC stations,

7 The *TURBOchannel* is the 32-bit bus used in DECstation 5000 workstations.

8 The *VME bus* is the bus designed by Motorola for use with its 32-bit 68000-microprocessor-based system. Such a bus is now, however, widely used with other computer systems as the bus for use with instrumentation systems.

The above are called *backplane buses*, the term backplane being for the board (Figure 20.14) on which connectors are mounted and into which printed circuit boards containing a particular function, e.g. memory, can be plugged. The backplane provides the data, address and control bus signals to each board, so enabling systems to be easily expanded by the use of off-the-shelf boards. It is into such computer buses that data acquisition boards and boards used for interfacing instruments and other peripherals have to interface. Data acquisition and instrument boards are usually available in various configurations, depending on the computer with which they are to be used.

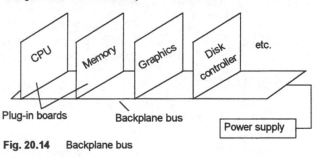

Fig. 20.14 Backplane bus

20.6.5 VXIbus

The *VME bus* is the bus designed by Motorola for use with its 32-bit 68000-microprocessor-based system. The *VXIbus* (VME Extensions for Instrumentation) is an extension of the specification of the VMEbus which has been designed for instrumentation applications such as automatic test equipment where higher speed communications are required than possible with the GPIB bus. It also provides better synchronisation and triggering and has been developed by a consortium of instrument manufacturers so that interoperability is possible between the products of different companies. The system involves VXI boards plugging into a mainframe. Figure 20.15 shows a number of possible system configurations that can be used. In Figure 20.15(a) a VXI mainframe is linked to an external controller, a computer, via a GPIB link. The controller talks across this link using a GPIB protocol to an interface board in the chassis which translates the GPIB protocol into the VXI protocol. This makes VXI instruments appear to the controller to be GPIB instruments and enables them to be programmed using GPIB methods. Figure 20.15(b) shows the complete computer embedded in the VXI chassis. This option offers the smallest possible physical size for the system and allows the computer to directly use the VXI

Fig. 20.15 VXI options

backplane bus. Figure 20.15(c) uses a special high-speed system-bus-on-cable, the MXIbus, to link a computer and the VXI chassis, the MXI being 20 times faster than the GPIB.

20.6.6 I²C bus

The *Inter-IC-Communication bus*, referred to as the I²C bus, is a data bus designed by Philips for use for communications between integrated circuits or modules. The bus allows data and instructions to be exchanged between devices by means of just two wires. This results in a considerable simplification of circuits.

The two lines are a bi-directional data line (SDA) and a clock line (SCL). Both lines are connected to the positive power supply via resistors (Fig. 20.16). The device generating the message is the transmitter and the device receiving the message the receiver. The device that controls the bus operation is the master and the devices which are controlled by the master are the slaves.

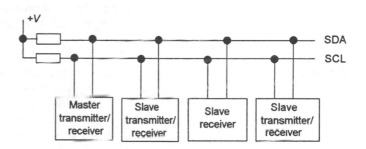

Fig. 20.16   I²C bus

The following is the protocol used: a data transfer may only be initiated when the bus is not busy and during the data transfer, when the clock line is high, the data line must remain. Changes in the data line when the clock line is high are interpreted as control signals.

1   When both the data and clock lines are high the bus is not busy.
2   A change in the state of the data line from high to low while the clock is high is used to define the start of data transfer.
3   A change in the state of the data line from low to high while the clock is high defines the stop of data transfer.
4   Data is transferred between start and stop conditions.
5   After a start of data transfer, the data line is stable for the duration of the high periods of the clock signal, being able to change during the low periods of the clock signal.
6   There is one clock pulse per data bit transferred with no limit on the number of data bytes that can be transferred between the start and stop conditions; after each byte of data the receiver acknowledges with a ninth bit.
7   The acknowledge bit is a high level put on the bus by the transmitter, a low level by a receiver.

Fig. 20.17 Bus conditions

Figure 20.17 illustrates the above by showing the form of the clock signal and the outputs by a transmitter and a receiver.

Problems

1 Explain the difference between centralised and distributed communication systems.
2 Explain the forms of bus/tree and ring networks.
3 A LAN is required with a distance between nodes of more than 100 m. Should the choice be bus or ring topology?
4 A multichannel LAN is required. Should the choice be broadband or baseband transmission?
5 What are MAP and TOP?
6 Explain what is meant by communication protocol.
7 Briefly explain the two types of multiple-access control used with LANs.
8 A microcontroller M68HC11 is a 'listener' to be connected to a 'talker' via a GPIB bus. Indicate the connections to be made if full handshaking is to be used.
9 What problem has to be overcome before the serial data communications interface of the microcontroller M68HC11 can output data through an RS-232C interface?
10 What is a backplane bus?

21 Fault finding

21.1 Fault-detection techniques

This chapter is a brief consideration of the problems of fault detection with measurement, control and data communication systems. For details of the fault-finding checks required for specific systems or components, the manufacturer's manuals should be used.

There are a number of techniques that can be used to detect faults:

1 *Replication checks*
   This involves duplicating or replicating an activity and comparing the results. In the absence of faults it is assumed that the results should be the same. It could mean, with transient errors, just repeating an operation twice and comparing the results or it could involve having duplicate systems and comparing the results given by the two. This can be an expensive option.

2 *Expected value checks*
   Software errors are commonly detected by checking whether an expected value is obtained when a specific numerical input is used. If the expected value is not obtained then there is a fault.

3 *Timing checks*
   This involves the use of timing checks that some function has been carried out within a specified time. These checks are commonly referred to as *watchdog timers*. For example, with a PLC, when an operation starts a timer is also started and if the operation is not completed within the specified time a fault is assumed to have occurred. The watchdog timer trips, sets off an alarm and closes down part or the entire plant.

4　*Reversal checks*

Where there is a direct relationship between input and output values, the value of the output can be taken and the input which should have caused it computed. This can then be compared with the actual input.

5　*Parity and error coding checks*

This form of checking is commonly used for detecting memory and data transmission errors. Communication channels are frequently subject to interference which can affect data being transmitted. To detect whether data has been corrupted a parity bit is a bit added to the transmitted data word. The parity bit is chosen to make the resulting number of 1s in the group either odd (odd parity) or even (even parity). If odd parity then the word can be checked after transmission to see if it is still odd. Other forms of checking involve codes added to transmitted data in order to detect corrupt bits.

6　*Diagnostic checks*

Diagnostic checks are used to test the behaviour of components in a system. Inputs are applied to a component and the outputs compared with those which should occur.

21.2 Watchdog timer

A watchdog timer is basically a timer that the system must reset before it times out. If the timer is not reset in time then an error is assumed to have occurred.

As an illustration of such a timer, Figure 21.1 shows a simple ladder program that can be used to provide a PLC with a watchdog timer for an operation involving the movement of a piston in a cylinder. When the start switch is closed the solenoid A+ is activated and starts the movement of the piston in the cylinder. It also starts a timer. When the piston is fully extended it opens the limit switch a+. This stops the timer. However, if a+ is not opened before the timer has timed out then the timer contacts close and an alarm is sounded. Thus the timer might be set for, say, 4 s on the assumption that the piston will be fully extended within that time. If, however, it sticks and fails to meet this deadline then the alarm sounds.

When a microprocessor executes instructions from its memory, a nearby electrical disturbance might momentarily upset the processor data bus and the wrong byte is accessed. Alternatively a software bug can result in the processor getting into problems when it returns from a subroutine. The consequence of such faults is that the system may crash with possibly dangerous consequences for actuators controlled by the microprocessor. To avoid this happening with critical systems a watchdog timer is used to reset the microprocessor.

Fig. 21.1　Watchdog timer program

As an illustration of the provision in microprocessor-based systems of internal watchdog timers, consider the microcontroller MC68HC11 which includes an internal watchdog timer, called *Computer Operating Properly* (COP), to detect software processing errors. When the COP timer has been started it is necessary for the main program to reset the COP periodically before it times out. If the watchdog timer times out before being set to start timing all over again then a COP failure reset occurs. The COP timer can be reset to zero time by writing a $55 (0x55 in C language) to the COP reset register (COPRST) at address $103A (0x103A) followed later in the program by writing a $AA (0xAA) to clear the COP timer. If the program hangs up in between the two instructions then COP times out and results in the COP failure reset routine being executed. The program, in assembly language, thus has the following lines:

```
LDAA      #$55       ; reset timer
STAA      $103A      ; writing $55 to COPRST
                     ; other program lines
LDAA      #$AA       ; clearing timer
STAA      $103A      ; writing $AA to COPRST
```

The COP operating period is set by setting CR1 and CR2 in the OPTION register, address $1039 (0x1039), to either 0 or 1. For example, with CR1 set to 0 and CR2 set to 0 a time-out of 16.384 ms might be given, whereas with CR1 set to 1 and CR2 set to 0 a time-out of 262.14 ms is given.

21.3 Parity and error coding checks

In order to try and detect when a data signal has been corrupted and has an error as a result of noise, error detection techniques are used. One such technique is *parity checks*.

In Section 14.2.3 a brief account was given of the parity method for error detection. With such a method an extra bit is added to a message to make the total number of 1s an even number when even parity is used or an odd number when odd parity is used. For example, the character 1010000 would have a parity bit placed after the most significant bit of a 0 with an even-parity system, i.e. 01010000, or a 1 with odd parity, i.e. 11010000.

Such a method can detect the presence of a single error in the message but not the presence of two errors which result in no change in parity, e.g. with even parity and the above number a single error in, say, the third bit would be detected in 1101100 because the parity check bit would be wrong but not if also there was an error in the first bit since 1101110 would have the correct parity bit.

If no error is detected the signal is acknowledged as being error free by the return of the ACK character to the sending terminal; if an error is detected the signal NAK is used. This is called an *automatic repeat request* (ARQ). The NAK signal then results in the retransmission of the message.

The efficiency of error detection can be increased by the use of *block parity*. The message is divided into a number of blocks and each block has a block check character added at the end of the block. For example, with the following block, a check bit for even parity is placed at the end of each row and a further check bit at the foot of each column.

	Information bits				Check bit
First symbol	0	0	1	1	0
Second symbol	0	1	0	0	1
Third symbol	1	0	1	1	1
Fourth symbol	0	0	0	0	0
Block check bits	1	1	0	0	0

At the receiver the parity of each row and each column is checked and any single error is detected by the intersection of the row and column containing the error check bit.

Another form of error detection is *cyclic redundancy checking* (CRC). At the transmitting terminal the binary number representing the data being transmitted is divided by a predetermined number using modulo-2 arithmetic. The remainder from the division is the CRC character which is transmitted with the data. At the receiver the data plus the CRC character are divided by the same number. If no transmission errors have occurred there is no remainder.

A common CRC code is CRC-16, 16 bits being used for the check sequence. The 16 bits are considered to be the coefficients of a polynomial with the number of bits equal to the highest power of the polynomial. The data block is first multiplied by the highest power of the polynomial, i.e. x^{16} and then divided by the CRC polynomial

$$x^{16} + x^{12} + x^5 + 1$$

using modulo-2 arithmetic, i.e. $x = 2$ in the polynomial. The CRC polynomial is thus 10001000000100001. The remainder of the division by this polynomial is the CRC.

As an illustration, suppose we have the data 10110111 or polynomial

$$x^7 + x^5 + x^4 + x^2 + x^1 + 1$$

and a CRC polynomial of

$$x^5 + x^4 + x^1 + 1$$

or 110011. The data polynomial is first multiplied by x^5 to give

$$x^{12} + x^{10} + x^9 + x^7 + x^6 + x^5$$

and so 1011011100000. Dividing this by the CRC polynomial gives:

```
                   11010111
         110011 |1011011100000
                 110011
                  110011
                  110011
                   100100
                   110011
                    101110
                    110011
                     111010
                     110011
                      01001
```

and so a remainder of 01001 which then becomes the CRC code transmitted with the data.

21.4 Common hardware faults

The following are some of the commonly encountered faults that can occur with specific types of components and systems.

21.4.1 Sensors

If there are faults in a measurement system then the sensor might be at fault. A simple test is to substitute the sensor with a new one and see what effect this has on the results given by the system. If the results change then it is likely that the original sensor was faulty; if the results do not change then the fault is elsewhere in the system. It is also possible to check that the voltage/current sources are supplying the correct voltages/currents, whether there is electrical continuity in connecting wires, that the sensor is correctly mounted and used under the conditions specified by the manufacturer's data sheet, etc.

21.4.2 Switches and relays

Dirt and particles of waste material between switch contacts is a common source of incorrect functioning of mechanical switches. A voltmeter used across a switch should indicate the applied voltage when the contacts are open and very nearly zero when they are closed. Mechanical switches used to detect the position of some item, e.g. the presence of a work piece on a conveyor, can fail to give the correct responses if the alignment is incorrect or if the actuating lever is bent.

Inspection of a relay can disclose evidence of arcing or contact welding. The relay should then be replaced. If a relay fails to operate then a check can be made for the voltage across the coil. If the correct voltage is present then coil continuity can be checked with an ohmmeter. If there is no voltage across the coil then the fault is likely to be the switching transistor used with the relay.

21.4.3 Motors

Maintenance of both d.c. and a.c. motors involves correct lubrication. With d.c. motors the brushes wear and can require changing. Setting of new brushes needs to be in accordance with the manufacturer's specification. A single-phase capacitor start a.c. motor that is sluggish in starting probably needs a new starting capacitor. The three-phase induction motor has no brushes, commutator, slip rings or starting capacitor and short of a severe overload, the only regular maintenance that is required is periodic lubrication.

21.4.4 Hydraulic and pneumatic systems

A common cause of faults with hydraulic and pneumatic systems is dirt. Small particles of dirt can damage seals, block orifices, cause valve spools to jam, etc. Thus filters should be regularly checked and cleaned, components should only be dismantled in clean conditions, and oil should be regularly checked and changed. With an electrical circuit a common method of testing a circuit is to measure the voltages at a number of test points. Likewise, with a hydraulic and pneumatic system there needs to be points at which pressures can be measured. Damage to seals can result in hydraulic and pneumatic cylinders leaking, beyond that which is normal, and result in a drop in system pressure when the cylinder is actuated. This can be remedied by replacing the seals in the cylinders. The vanes in vane-type motors are subject to wear and can then fail to make a good seal with the motor housing with the result of a loss of motor power. The vanes can be replaced. Leaks in hoses, pipes and fittings are common faults.

21.5 Microprocessor systems

Typical faults in microprocessor systems are:

1   *Chip failure*
    Chips are fairly reliable but occasionally there can be failure.

2   *Passive component failure*
    Microprocessor systems will usually include passive components such as resistors and capacitors. Failure of any of these can cause system malfunction.

3   *Open circuits*
Open circuits can result in a break in a signal path or in a power line. Typical reasons for such faults are unsoldered or faulty soldered joints, fracture of a printed circuit track, a faulty connection on a connector and breaks in cables.

4   *Short circuits*
Short circuits between points on a board which should not be connected often arise as a result of surplus solder bridging the gaps between neighbouring printed circuit tracks.

5   *Externally introduced interference*
Externally induced pulses will affect the operation of the system since they will be interpreted as valid digital signals. Such interference can originate from the mains supply having 'spikes' as a result of other equipment sharing the same mains circuit and being switched on or off. Filters in the mains supply to the system can be used to remove such 'spikes'.

6   *Software faults*
Despite extensive testing it is still quite feasible for software to contain bugs and under particular input or output conditions cause a malfunction.

21.5.1 Fault-finding techniques

Fault-finding techniques that are used with microprocessor-based systems include:

1   *Visual inspection*
Just carefully looking at a faulty system may reveal the source of a fault, e.g. an integrated circuit which is loose in its holder or surplus solder bridging tracks on a board.

2   *Multimeter*
This is of limited use with microprocessor systems but can be used to check for short- or open-circuit connections and the power supplies.

3   *Oscilloscope*
The oscilloscope is essentially limited to where repetitive signals occur and the most obvious such signal is the clock signal. Most of the other signals with a microprocessor system are not repetitive and depend on the program being executed.

4   *Logic probe*
The logic probe is a hand-held device (Fig. 21.2), shaped like a pen, which can be used to determine the logic level at any

Fig. 21.2   Logic probe

point in the circuit to which it is connected. The selector switch is set for TTL or CSMOS operations and when the probe tip is touched to the point in question, the indicator lamp indicates whether it is below the logic level 0 threshold, above the logic 1 threshold or a pulsating signal. A pulse stretching circuit is often included with the probe in order to stretch the duration of a pulse to allow sufficient time for it to operate the indicator lamp and the effect to be noticed. A memory circuit can be used for detecting a single pulse, the memory clear button being pressed to turn off the indicator lamp and then any change in logic level is registered by the lamp.

5 *Logic pulser*
   The logic pulser is a hand-held pulse generator, shaped like a pen, that is used to inject controlled pulses into circuits. The pulser probe tip is pressed against a node in the circuit and the button on the probe pressed to generate a pulse. It is often used with the logic probe to check the functions of logic gates.

Current trace falls
to zero at IC short circuit

Fig. 21.3   Use of a current tracer

6 *Current tracer*
   The current tracer is similar to the logic probe but it senses pulsing current in a circuit rather than voltage levels. The tip of the current tracer is magnetically sensitive and is used to detect the changing magnetic field near a conductor carrying a pulsing current. The current tracer tip is moved along printed circuit tracks to trace out the low-impedance paths along which current is flowing (Fig. 21.3).

7 *Logic clip*
   A logic clip is a device which clips to an integrated circuit and makes contact with each of the integrated circuit pins. The logic state of each pin is then shown by LED indicators, there being one for each pin.

8 *Logic comparator*
   The logic comparator tests integrated circuits by comparing them to a good, reference, integrated circuit (Fig. 21.4). Without removing the integrated circuit being tested from its circuit, each input pin is connected in parallel with the corresponding input pin on the reference integrated circuit; likewise each output pin is connected with the corresponding output pin on the reference integrated circuit. The two outputs are compared with an EXCLUSIVE-OR gate, which then gives an output when the two outputs differ. The pulse stretcher is used to extend the duration of the signal fed to the indicator so that very short duration pulses will result in the indicator being on for a noticeable period.

Fig. 21.4   Logic comparator

9   *Signature analyser*

With analogue systems, fault finding usually involves tracing through the circuitry and examining the waveforms at various nodes, comparison of the waveforms with what would be expected enabling faults to be identified and located. With digital systems the procedure is more complex since trains of pulses at nodes all look very similar. To identify whether there is a fault the sequence of pulses is converted into a more readily identifiable form, e.g. 258F, this being termed the *signature*. The signature obtained at a node can then be compared with that which should occur. When using the signature analyser with a circuit, it is often necessary for the circuit to have been designed so that data bus feedback paths can be broken easily for the test to stop faulty digital sequences being fed back during the testing. A short program, which is stored in ROM, is activated to stimulate nodes and enable signatures to be obtained. The microprocessor itself can be tested if the data bus is broken to isolate it from memory and it is then made to 'free-run' and give a 'no operation' (NO) instruction to each of its addresses in turn. The signatures for the microprocessor bus in this state can then be compared with those expected.

10   *Logic analyser*

The logic analyser is used to simultaneously sample and store in a 'first-in-first-out' (FIFO) memory the logic levels of bus and control signals in a unit under test. The point in the program at which the data capture starts or finishes is selected by the use of a 'trigger word'. The analyser compares its trigger word with the incoming data and only starts to store data when the word occurs in the program. Data capture then continues for a predetermined number of clock pulses and is then stopped. The stored data may then be displayed as a list of binary, octal, decimal or hexadecimal codes, or as a time display in which the waveforms are displayed as a function of time, or as a mnemonic display.

21.5.2 Systematic fault-location methods

Systematic fault-location methods are:

1   *Input to output*
    A suitable input signal is injected into the first block of the system and then measurements are made in sequence, starting from the first block, at the output of each block in turn until the faulty block is found.

2   *Output to input*
    A suitable input signal is injected into the first block of the system and then measurements are made in sequence, starting from the last block, at the output of each block in turn until the faulty block is found.

3   *Half-split*
    A suitable input signal is injected into the first block of the system. The blocks constituting the system are split in half and each half tested to determine in which half the fault lies. The faulty half is then split into half and the procedure repeated.

21.5.3 Self-testing

Software can be used by a microprocessor-based system to institute a self-test program for correct functioning. Such programs are often initiated during the start-up sequence of a system when it is first switched on. For example, printers include microprocessors in their control circuits and generally the control program stored in ROM also includes test routines. Thus when first switched on it goes through these test routines and is not ready to receive data until all tests indicate the system is fault-free.

A basic ROM test involves totalling all the data bytes stored in each location in ROM and comparing the sum against that already stored (the so-called *checksum test*). If there is a difference then the ROM is faulty; if they agree it is considered to be fault-free. A basic RAM test involves storing data patterns in which adjacent bits are at opposite logic levels, i.e. hex 55 and AA, into every memory location and then reading back each value stored to check that it corresponds to the data sent (the so-called '*checker board test*').

21.6 Emulation and simulation

An *emulator* is a test board which can be used to test a microcontroller and its program. The board contains:

1 The microcontroller.
2 Memory chips for the microcontroller to use as data and program memory.
3 An input/output port to enable connections to be made with the system under test.
4 A communications port to enable program code to be downloaded from a computer and the program operation to be monitored.

The program code can be written in a host computer and then downloaded through either a serial or a parallel link into the memory on the board. The microcontroller then operates as though this program was contained within its own internal memory. Figure 21.5 shows the general arrangement.

Fig. 21.5 Using an emulator

The input/output lines of the microcontroller are connected via an input/output port on the board to a plug-in device on the system under test so that it operates as though it had the microcontroller plugged into it. The board is already programmed with a monitor system which enables the operation of the program to be monitored and the contents of memory, registers, input/output ports to be checked and modified.

Figure 21.6 shows the basic elements of the evaluation board MC68HC11EVB that is provided by Motorola. This uses a monitor program called *Bit User Fast Friendly Aid to Logical Operations* (BUFFALO). The 8K EPROM contains the Buffalo monitor. An MC6850 Asynchronous Communications Interface Adapter (ACIA) (see Section 18.5) is used for interfacing parallel and serial lines. A partial RS-232 interface is supplied with each of the two serial ports for connection to the host computer and the monitoring terminal.

I/O connector to the plug-in connection to the system under test so that its pins replicate those of MC68HC11 and it is as though the microcontroller was directly connected into the system

Fig. 21.6   MC68HC11EVB

For further reading on this evaluation board, the reader is referred to *M68HC11EVB Evaluation Board User's Manual* (Motorola 1986) or *Software and Hardware Engineering: Motorola M68HC11* by F.M. Cady (OUP 1997).

21.6.1 Simulation

Instead of testing a program by running it with an actual microcontroller, you can test it by running it with a computer program that simulates the microcontroller. Such simulation can assist in the debugging of the program code. The display screen is divided into a number of windows in which information such as the source code is displayed as it is executed, the CPU registers and flags with their current states, the input/output ports, registers and timers, and the memory situation. Figure 21.7 shows the typical form of display on the computer screen.

Fig. 21.7   Simulation screen

21.7 PLC systems

Programmable logic controllers have a high reliability. The PLC is electrically isolated by optoisolator or relays from potentially damaging voltages and currents at input/output ports; battery-backed RAM protects the application software from power failures or corruption; and the construction is so designed that the PLC can operate reliably in industrial conditions for long periods of time. PLCs generally have several built-in fault procedures. Critical faults cause the CPU to stop, while other less critical faults cause the CPU to continue running but display a fault code on a display. The PLC manual will indicate the remedial action required when a fault code is displayed.

21.7.1 Program testing

The software checking program checks through a ladder program for incorrect device addresses, and provides a list on a screen, or as a printout, of all the input/output points used, counter and timer settings, etc. with any errors detected. Thus the procedure might involve:

1   Opening and displaying the ladder program concerned.
2   Selecting from the menu on the screen Ladder Test.
3   The screen might then display the message: Start from beginning of program (Y/N)?
4   Type Y and press Enter.
5   Any error message is then displayed or the message of 'No errors found'.

For example, there might be a message for a particular output address that it is used as an output more than once in the program, a timer or counter is being used without a preset value, a counter is being used without a reset, there is no END instruction, etc. As a result of such a test, there may be a need to make changes to the program. Changes needed to rectify the program might be made by selecting from the menu displayed on screen Exchange and following through the set of displayed screen messages.

21.7.2 Testing inputs and outputs

Most PLCs have the facility for testing inputs and outputs by what is termed *forcing*. Software is used to 'force' inputs and outputs on or off. To force inputs or outputs, a PLC has to be switched into the forcing or monitor mode by perhaps pressing a key marked FORCE or selecting the MONITOR mode on a screen display. As a result of forcing an input we can check that the consequential action of that input being on occurs. The installed program can thus be run and inputs and outputs simulated so that they, and all preset values, can be checked. However, care must be exercised with forcing in that an output might be forced that can result in a piece of hardware moving in an unexpected and dangerous manner.

As an illustration of the type of display obtained when forcing, Figure 21.8 shows how inputs might appear in the ladder program display when open and closed, and outputs when not energised and energised and Figure 21.9(a) a selected part of a displayed ladder program and Figure 21.9(b) what happens when forcing occurs. Initially, Figure 21.9(a) shows rung 11, with inputs to X400, X401 and M100, but not X402 and with no output from Y430. For rung 12, the timer T450 contacts are closed, the display at the bottom of the screen indicating that there is no time left to run on T450. Because Y430 is not energised the Y430 contacts are open and so there is no output from Y431. If we now force an input to X402 then the screen display changes to that shown in Figure 21.9(b) and Y430 is energised and consequently Y431 energised.

Open   Closed

Not energised   Energised

Fig. 21.8   Monitor mode symbols

Fig. 21.9 Forcing an input

21.7.3 PLC as a monitor of systems

The PLC can also be used as a monitor of the system being controlled. It can be used to sound an alarm or light up a red light if inputs move outside prescribed limits, using the greater than, equal to, or less than functions, or its operations take longer than a prescribed time. See Figure 21.1 for an illustration of how a PLC ladder program might be used as a watchdog timer for some operation.

Often with PLC systems status lamps are used to indicate the last output that has been set during a process and so, if the system stops, where the fault has occurred. The lamps are built into the program so that as each output occurs a lamp comes on and turns off the previous output status lamps. Figure 21.10 illustrates this.

Problems

1 Explain what is meant by (a) replication checks, (b) expected value checks, (c) reversal checks, (d) parity checks.
2 Explain how a watchdog timer can be used with a PLC controlled plant in order to indicate the presence of faults.
3 Explain the function of COP in the microcontroller MC68HC11.
4 The F2 series Mitsubishi PLC is specified as having:

Diagnosis: Programmable check (sum, syntax, circuit check), watchdog timer, battery voltage, power supply voltage

Explain the significance of the terms.
5 Explain how self-testing can be used by a microprocessor-based system to check its ROM and RAM.

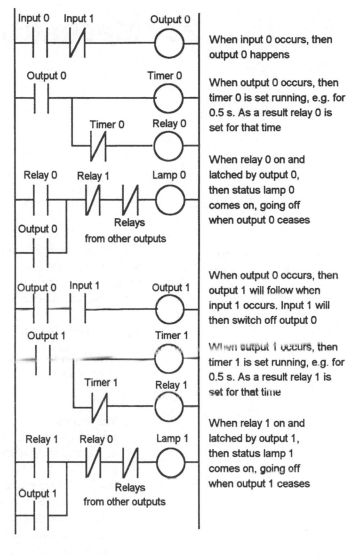

When input 0 occurs, then output 0 happens

When output 0 occurs, then timer 0 is set running, e.g. for 0.5 s. As a result relay 0 is set for that time

When relay 0 on and latched by output 0, then status lamp 0 comes on, going off when output 0 ceases

When output 0 occurs, then output 1 will follow when input 1 occurs. Input 1 will then switch off output 0

When output 1 occurs, then timer 1 is set running, e.g. for 0.5 s. As a result relay 1 is set for that time

When relay 1 on and latched by output 1, then status lamp 1 comes on, going off when output 1 ceases

Fig. 21.10   Last output set diagnostic program

22 Mechatronics systems

22.1 Traditional and mechatronics designs

This chapter brings together many of the topics discussed in this book in the consideration of both traditional and mechatronics solutions to design problems and gives mechatronics case studies. It also includes a brief review of the design process.

22.1.1 The design process

The design process for any system can be considered as involving a number of stages:

1 *The need*
   The design process begins with a need from, perhaps, a customer or client. This may be identified by market research being used to establish the needs of potential customers.

2 *Analysis of the problem*
   The first stage in developing a design is to find out the true nature of the problem, i.e. analysing it. This is an important stage in that not defining the problem accurately can lead to wasted time on designs that will not fulfil the need.

3 *Preparation of a specification*
   Following the analysis a specification of the requirements can be prepared. This will state the problem, any constraints placed on the solution, and the criteria which may be used to judge the quality of the design. In stating the problem, all the functions required of the design, together with any desirable features, should be specified. Thus there might be a statement of mass, dimensions, types and range of motion required, accuracy, input and output requirements of elements, interfaces, power requirements, operating environment, relevant standards and codes of practice, etc.

4   *Generation of possible solutions*

This is often termed the *conceptual stage*. Outline solutions are prepared which are worked out in sufficient detail to indicate the means of obtaining each of the required functions, e.g. approximate sizes, shapes, materials and costs. It also means finding out what has been done before for similar problems; there is no sense in reinventing the wheel.

5   *Selections of a suitable solution*

The various solutions are evaluated and the most suitable one selected.

6   *Production of a detailed design*

The detail of the selected design has now to be worked out. This might require the production of prototypes or mock-ups in order to determine the optimum details of a design.

7   *Production of working drawings*

The selected design is then translated into working drawings, circuit diagrams, etc. so that the item can be made.

It should not be considered that each stage of the design process just flows on stage by stage. There will often be the need to return to an earlier stage and give it further consideration. Thus when at the stage of generating possible solutions there might be a need to go back and reconsider the analysis of the problem.

22.1.1 Traditional and mechatronics designs

Engineering design is a complex process involving interactions between many skills and disciplines. The basis of the mechatronics approach is considered to lie in the inclusion of the disciplines of electronics, computer technology and control engineering. For example, the design of bathroom scales might be considered only in terms of the compression of springs and a mechanism used to convert the motion into rotation of a shaft and hence movement of a pointer across a scale; a problem that has to be taken into account in the design is that the weight indicated should not depend on the person's position on the scales. With mechatronics, other possibilities can be considered. For example, the springs might be replaced by load cells with strain gauges and the output from them used with a microprocessor to provide a digital readout of the weight on an LED display. The resulting scales might be mechanically simpler, involving fewer components and moving parts. The complexity has, however, been transferred to the software.

The traditional design of the temperature control for a domestic central heating system has been the bimetallic thermostat in a closed-loop control system. The bending of the bimetallic strip changes as the temperature changes and is used to operate an on/off switch for the heating system. A mechatronics solution to the problem might be to use a microprocessor-controlled system employing perhaps a thermodiode as the sensor. Such a system has many advantages over the bimetallic thermostat system. The bimetallic thermostat is comparatively crude and the temperature is not accurately controlled; also devising a method for having different temperatures at different times of the day is complex and not easily achieved. The microprocessor-controlled system can, however, cope easily with giving precision and programmed control. The system is much more flexible. This improvement in flexibility is a common characteristic of mechatronics systems when compared with traditional systems.

22.1.2 Embedded systems

Microprocessors and microcontrollers are often 'embedded' in systems so that control can be exercised. For example, a modern domestic washing machine has an embedded microcontroller which has been programmed with the different washing programs; all that the machine operator has to do is select the required washing program by means of a switch and the required program is implemented. The operator does not have to program the microcontroller. The term *embedded system* is used for a microprocessor-based system that is designed to control a function or range of functions and is not designed to be programmed by the system user. The programming has been done by the manufacturer and has been 'burnt' into the memory system and cannot be changed by the system user.

22.1.3 Embedding programs

In an embedded system the manufacturer makes a ROM containing the program. This is only economical if there is a need for a large number of these chips. Alternatively, for prototyping or low volume applications, a program could be loaded into the EPROM/EEPROM of the application hardware. The following illustrates how the EPROM/EEPROM of microcontrollers can be programmed.

For example, to program the EPROM of the Intel 8051 microcontroller, the arrangement shown in Figure 22.1 is required. There must be a 4–6 MHz oscillator input. The procedure is:

1 The address of an EPROM location, to be programmed in the range 0000H to 0FFFH, is applied to port 1 and pins P2.0 and P2.1 of port 2; at the same time, the code byte to be programmed into that address is applied to port 0.

2 Pins P2.7, RST and ALE should be held high, pins P2.6 and PSEN low. For pins P2.4 and P2.5 it does not matter whether they are high or low.

ig. 22.1 Programming the 8051   **Fig. 22.2** Verification with 8051

3 Pin EA/V$_{pp}$ is held at a logic high until just before ALE is to be pulsed, then it is raised to +21 V, ALE is pulsed low for 50 ms to program the code byte into the addressed location, and then EA is returned to a logic high.

Verification of the program, i.e. reading out of the program, is achieved by the arrangement shown in Figure 22.2.

1 The address of the program location to be read is applied to port 1 and pins P2.0 to P2.3 of port 2.

2 Pins EA/V$_{pp}$, RST and ALE should be held high, pins P2.7, P2.6 and PSEN low. For pins P2.4 and P2.5 it does not matter whether they are high or low.

3 The contents of the addressed location come out on port 0.

A security bit can be programmed to deny electrical access by any external means to the on-chip program memory. Once this bit has been programmed, it can only be cleared by the full erasure of the program memory. The same arrangement is used as for programming (Fig. 22.1) but P2.6 is held high. Erasure is by exposure to ultraviolet light. Since sunlight and fluorescent lighting contain some ultraviolet light, prolonged exposure (about 1 week in sunlight or 3 years in room-level fluorescent lighting) should be avoided and the chip window should be shielded by an opaque label.

The Motorola 68HC11 microcontroller is available with an internal electrically erasable programmable read-only memory (EEPROM). The EEPROM is located at addresses $B600 to

$103F

EEON

Fig. 22.3 CONFIG

$B7FF. Like an EPROM, a byte is erased when all the bits are 1 and programming involves making particular bits 0. The EEPROM is enabled by setting the EEON bit in the CONFIG register (Fig. 22.3) to 1 and disabled by setting it to 0. Programming is controlled by the EEPROM programming register (PPROG) (Fig. 22.4).

Fig. 22.4 PPROG

The procedure for programming is:

1 Write to the PPROG register to set the EELAT bit to 1 for programming.

2 Write data to the EEPROM address selected. This latches in the address and data to be programmed.

3 Write to the PPROG register to set the EEPGM bit to 1 to turn on the programming voltage.

4 Delay for 10 ms.

5 Write to the PPROG register to turn off, i.e. to 0, all the bits.

Here, in assembly language, is a programming sub-routine for use with the MC68HC11:

```
EELAT     EQU    %00000010    ; EELAT bit
EEPGM     EQU    %00000001    ; EEPGM bit
PPROG     EQU    $1028        ; address of PPROG
                             ; register

EEPROG
          PSHB
          LDAB   #EELAT
          STAB   PPROG        ; set EELAT = 1 and
                             ; EEPGM = 0
          STAA   0,X          ; store data X to
                             ; EEPROM address
```

```
            LDAB    #%00000011
            STAB    PPROG          ; set EELAT = 1 and
                                   ; EEPGM = 1
            JSR     DELAY_10       ; jump to delay 10 ms
                                   ; subroutine
            CLR     PPROG          ; clear all the PPROG
                                   ; bits and return to the
                                   ; read mode
            PULB
            RTS
```

```
; subroutine for approximately 10 ms delay
DELAY_10
            PSHX
            LDX     #2500          ; count for 20 000
                                   ; cycles
DELAY       DEX
            BNE     DELAY
            PULX
            RTS
```

The procedure for erasure is:

1 Write to the PPROG register to select for erasure of a byte, row or the entire EEPROM.

2 Write to an EEPROM address within the range to be erased.

3 Write a 1 to the PPROG register to turn on the EEPGM bit and hence the erase voltage.

4 Delay for 10 ms.

5 Write 0s to the PPROG register to turn off all the bits.

With the built-in EEPROM with a PIC microcontroller, a program to write data into it is (Fig. 22.5):

File 88h

EECON1

EECON2 at 89h requires the writing of 01010101 and then 10101010 with no interruption to unlock the write cycle.

Fig. 22.5 EECON registers

```
bcf     STATUS, RP0     ; Change to Bank 0 for
                        ; the data
mov.f   Data, w         ; Load data to be
                        ; written
movwf   EEDATA
movf    Addr, w         ; Load address of write
                        ; data
movwf   EEADR
bsf     STATUS, RP0     ; Change to Bank 1
bcf     INTCON, GIE     ; Disable interrupts
bsf     EECON1, WREN    ; Enable for writing
movlw   55h             ; Special sequence to
                        ; enable writing
movwf   EECON2
movlw   0AAh
movwf   EECON2
bsf     EECON1, WR      ; Initiate write cycle
```

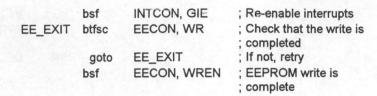

	bsf	INTCON, GIE	; Re-enable interrupts
EE_EXIT	btfsc	EECON, WR	; Check that the write is ; completed
	goto	EE_EXIT	; If not, retry
	bsf	EECON, WREN	; EEPROM write is ; complete

22.2 Possible mechatronics design solutions

Consider possible solutions to the following requirements for systems to carry out simple tasks.

22.2.1 Timed switch

Consider a simple requirement for a device which switches on some actuator, e.g. a motor, for some prescribed time.

A mechanical solution could involve a rotating cam (Fig. 22.6) (see Section 6.4). The cam would be rotated at a constant rate and the cam follower used to actuate a switch, the length of time for which the switch is closed depending on the shape of the cam.

A PLC solution could involve the arrangement shown in Figure 22.7 with the given ladder program. This would have the advantage over the rotating cam of having off and on times which can be adjusted by purely changing the timer preset values in the program whereas a different cam is needed if the times have to be changed with the mechanical solution. The software solution is much easier to implement than the hardware one.

Fig. 22.6 Cam-operated switch

Fig. 22.7 PLC timer system

A microprocessor-based solution could involve a microprocessor with a memory chip and input/output interfaces. The program is then used to switch an output on and then off after some time delay with the time delay being produced by a block of program in which there is a timing loop. This generates a time delay by branching round a loop the number of cycles required to generate the requisite time. Thus, in assembly language we might have:

```
DELAY    LDX      #F424      F424 is number of loops
LOOP     DEX
         BNE      LOOP
         RTS
```

DEX decrements the index register, and this and BNE, branch if not equal, each take 4 clock cycles. The loop thus takes 8 cycles and there will be n such loops until $8n + 3 + 5$ gives the number F424 (LDX takes 3 cycles and RTS takes 5 cycles). In C we could write the program lines using the while function.

An alternative to using a loop is to use a timer module, e.g. 555, with the microprocessor. With the 555 timer the timing intervals are set by external resistors and capacitors. Figure 22.8 shows the timer and the external circuitry needed to give an on-output when triggered, the duration of the on-output being $1.1RC$. Large times need large values of R and C. R is limited to about 1 MΩ, otherwise leakage becomes a problem, and C is limited to about 10 μF if electrolytic capacitors with the problems of leakage and low accuracy are to be avoided. Thus the circuit shown is limited to times less than about 10 s. The lower limit is about $R = 1$ kΩ and $C = 100$ pF, i.e. times of a fraction of a millisecond. For longer times, from 16 ms to days, an alternative timer such as the ZN1034E can be used.

Fig. 22.8   555 timer

Another possibility is to use the timer system in a microcontroller such as MC68HC11. The timer system is based on a 16-bit counter TCNT operating from the system E-clock signal (Fig. 22.9). The system E-clock can be prescaled by setting

Fig. 22.9 Generating 2 MHz internal clock

PR1	PR0	Prescale factor
0	0	1
0	1	4
1	0	8
1	1	16

Fig. 22.10 Prescale factor

Fig. 22.11 Output compare

bits in the timer interrupt mask register 2 (TMSK2), address $1024 (Fig. 22.10). The TCNT register starts at $0000 when the processor is reset and counts continuously until it reaches the maximum count of $FFFF. On the next pulse it overflows and reads $0000 again. When it overflows it sets the timer overflow flag TOF (bit 7 in miscellaneous timer interrupt flag register 2, TFLG2 at address $1025). Thus with a prescale factor of 1 and an E-clock frequency of 2 MHz, overflow occurs after 32.768 ms.

One way of using this for timing is for the TOF flag to be watched by polling. When the flag is set, the program increments its counter. The program then resets the flag, by writing a 1 to bit 7 in the TFLG2 register. Thus the timing operation just consists of the program waiting for the required number of overflag settings.

A better way of timing involves the use of the output-compare function. Port A of the microcontroller can be used for general inputs or outputs or for timing functions. The timer has output pins, OC1, OC2, OC3, OC4 and OC5, with internal registers TOC1, TOC2, TOC3, TOC4 and TOC5. We can use the output-compare function to compare the values in the TOC1 to TOC5 registers with the value in the free-running counter TCNT. This counter starts at 0000 when the CPU is reset and then runs continuously. When a match occurs between a register and the counter then the corresponding OCx flag bit is set and output occurs through the relevant output pin. Figure 22.11 illustrates this. Thus by programming the TOCx registers, so the times at which outputs occur can be set. The output-compare function can generate timing delays with much higher accuracy than the timer overflow flag.

The following program illustrates how output compare can be used to produce a time delay. The longest delay that can be generated in one output-compare operation is 32.7 ms when the E-clock is 2 MHz. In order to generate longer delays, multiple output-compare operations are required. Thus we might have each output-compare operation producing a delay of 25 ms and repeating this 40 times to give a total delay of 1 s.

```
REGBAS   EQU   $1000      Base address of registers
TOC2     EQU   $18        Offset of TOC2 from REGBAS
TCNT     EQU   $0E        Offset of TCNT from REGBAS
TFLG1    EQU   $23        Offset of TFLGI from REGBAS
OC1      EQU   $40        Mask to clear OC1 pin and OC1F flag
CLEAR    EQU   $40        Clear OC2F flag
D25MS    EQU   50000      Number of E-clock cycles to generate a 25 ms delay
NTIMES   EQU   40         Number of output-compare operations needed to give 1 s delay
         ORG   $1000
COUNT    RMB   1          Memory location to keep track of the number of output-compare
                          operations still to be carried out

         ORG   $C000      Starting address of the program
         LDX   #REGBAS
```

```
        LDAA    #OC1            Clear OC1 flag
        STAA    TFLG1,X
        LDAA    #NTIMES         Initialise the output-compare count
        STAA    COUNT
        LDD     TCNT,X
WAIT    ADDD    #D25MS          Add 25 ms delay
        STD     TOC2,X          Start the output-compare operation
        BRCLR   TFLG1,X OC1     Wait until the OC1F flag is set
        LDAA    #OC1            Clear the OC1F flag
        STAA    TFLG1,X
        DEC     COUNT           Decrement the output-compare counter
        BEQ     OTHER           Branch to OTHER if 1 s elapsed
        LDD     TOC2,X          Prepare to start the next compare operation
        BRA     WAIT
OTHER                           The other operations of the program which occur after the 1 s
                                delay
```

22.2.2 Windscreen-wiper motion

Consider a requirement for a device which will oscillate an arm back and forth in an arc like a windscreen wiper. A mechanical solution is shown in Figure 22.12. Rotation of arm 1 causes arm 2 to impart an oscillatory motion to arm 3.

An alternative solution is to use a stepper motor. Figure 22.13 shows how a microprocessor with a PIA, or a microcontroller, might be used with a stepper. The input to the stepper is required to cause it to rotate a number of steps in one direction and then reverse to rotate the same number of steps in the other direction.

Fig. 22.12   Wiper mechanism

Fig. 22.13   Interfacing a stepper

If the stepper is to be in the 'full-step' configuration then the outputs need to be as shown in Table 22.1. Thus to start and rotate the motor in a forward direction involves the sequence A, 9, 5, 6 and then back to the beginning with 1 again. To reverse

Table 22.1 Full-step configuration

Step	Bit 3	Bit 2	Bit 1	Bit 0	Code
1	1	0	1	0	A
2	1	0	0	1	9
3	0	1	0	1	5
4	0	1	1	0	6
1	1	0	1	0	A

Table 22.2 Half-step configuration

Step	Bit 3	Bit 2	Bit 1	Bit 0	Code
1	1	0	1	0	A
2	1	0	0	0	8
3	1	0	0	1	9
4	0	0	0	1	1
5	0	1	0	1	5
6	0	1	0	0	4
7	0	1	1	0	6
8	0	0	1	0	2
1	1	0	1	0	A

we would use the sequence 6, 5, 9, A and then back to begin with 6 again. If 'half-step' configuration is used then the outputs need to be as shown in Table 22.2. Forward motion then involves the sequence A, 8, 9, 1, 5, 4, 6, 2 and then back to A, with reverse requiring 2, 6, 4, 5, 1, 9, 8, A and back to 2.

The basic elements of a program could be:

Advance a step
Jump to time delay routine to give time for the step to be completed
Loop or repeat the above until the requisite number of steps completed in the forward direction
Reverse direction
Repeat the above for the same number of steps in reverse direction

Such a program in C might, for three half-steps forward and three back and following the inclusion of an appropriate header file, have the following elements:

```
main ( )
{
    portB = 0xa;  /*first step*/
    delay ( );  */incorporate delay program for, say, 20 ms*/
    portB = 0x8;  /*second step*/
    delay ( ); */incorporate delay program for 20 ms*/
    port B = 0x9; /*third step*/
    delay ( ); */incorporate delay program for 20 ms*/
    port B = 0x8; /*reverse a step*/
    delay ( ); */incorporate delay program for 20 ms*/
    port B = 0xa; /*reverse a further step*/
    delay ( ); /*incorporate delay program for 20 ms*/
    port B = 0x2; /* reverse back to where motor started*/
    delay ( ); /* incorporate delay program for 20 ms*/
}
```

Where there are many steps involved, a simpler program is to increment a counter with each step and loop until the counter value reaches the required number. Such a program would have the basic form of:

Advance a step
Jump to time delay routine to give time for the step to be completed
Increment the counter
Loop or repeat the above with successive steps until the counter indicates the requisite number of steps completed in the forward direction
Reverse direction
Repeat the above for the same number of steps in reverse direction

Fig. 22.14 Integrated circuit
SAA 1027 for stepper motor

Integrated circuits are available for step motor control and their use can simplify the interfacing and the software. Figure 22.14 shows how such a circuit can be used. All that is then needed is the requisite number of input pulses to the trigger, the motor stepping on the low-to-high transition of a high-low-high pulse. A high on the rotation input causes the motor to step counter-clockwise while a low gives clockwise rotation. Thus we just need one output from the microcontroller for output pulses to the trigger and one output to rotation. An output to set is used to reset the motor back to its original position.

The above has indicated how we might use a stepper motor to give an angular rotation. But how will a stepper motor behave when given a voltage signal input? Can we expect it to rotate directly to the angle concerned with no over shoot and no oscillations before settling down to the required angle? As an illustration of how we can develop a model for the stepper motor system and so predict its behaviour, consider the following simplified analysis (for a more detailed analysis, see *Stepping Motors and their Microprocessor Controls* by T. Kenjo (Oxford University Press 1984).

The system involving a stepper motor being driven by pulses from a microcontroller is an open-loop control system. The permanent magnet stepper motor (see Section 7.7) has a stator with a number of poles, the poles being energised by current being passed through coils wound on them. We can determine a model for how the rotor will rotate when there is a voltage pulse input to it by considering, for simplicity, a stepper with just a pair of poles and treat it in the same manner as the d.c. motor that was analysed in Section 9.3.2. If v is the voltage supplied to the motor pair of coils and v_b the back e.m.f. then:

$$v - v_b = L\frac{di}{dt} + Ri$$

where L is the inductance of the circuit, R the resistance and i the circuit current. We will make the simplifying assumption that the inductance does not significantly change and so treat L as a constant.

The back e.m.f will be proportional to the rate at which the magnetic flux is changing for the pair of coils. This will depend on the angle θ of the rotor relative to the poles concerned. Thus we can write:

$$v_b = -k_b \frac{d}{dt} \cos\theta = k_b \sin\theta \frac{d\theta}{dt}$$

where k_b is a constant. Thus:

$$v - k_b \sin\theta \frac{d\theta}{dt} = L\frac{di}{dt} + Ri$$

Taking the Laplace transform of this equation gives:

$$V(s) - k_b s \sin\theta\, \theta(s) = sL\,I(s) + R\,I(s) = (sL + R)\,I(s)$$

As with the d.c. motor, the current through a pair of coils will generate a torque (the torque on the magnet, i.e. the rotor, being the reaction resulting from the torque exerted on the coils – Newton's third law). The torque is proportional to the product of the flux density at the coil turns and the current through them. The flux density will depend on the angular position of the rotor and thus we can write:

$$T = k_t i \sin\theta$$

where k_t is a constant. This torque will cause an angular acceleration a and since $T = Ja$, where J is the moment of inertia of the rotor:

$$T = J\frac{d^2\theta}{dt^2} = k_t i \sin\theta$$

Taking the Laplace transform of this equation gives:

$$s^2 J\, \theta(s) = k_t \sin\theta\, I(s)$$

and so we can write:

$$V(s) - k_b s \sin\theta\, \theta(s) = (sL + R)(s^2 J\, \theta(s)/k_t \sin\theta)$$

and the transfer function between the input voltage and the resulting angular displacement is:

$$G(s) = \frac{\theta(s)}{V(s)} = \frac{k_t \sin\theta}{J(sL + R)s^2 + k_b k_t s \sin^2\theta}$$

$$= \frac{1}{s} \times \frac{k_t \sin\theta}{JLs^2 + JRs + k_b k_t \sin^2\theta}$$

When there is a voltage impulse supplied to the motor coils, since for a unit impulse $V(s) = 1$:

$$\theta(s) = \frac{1}{s} \times \frac{k_t \sin\theta}{JLs^2 + JRs + k_b k_t \sin^2\theta}$$

$$= \frac{1}{s} \times \frac{(k_t \sin\theta)/JL}{s^2 + (R/L)s + (k_b k_t \sin^2\theta)/JL}$$

The quadratic equation in s is of the form $s^2 + 2\zeta\omega_n s + \omega_n^2$ (see Section 11.3) and thus has a natural frequency ω_n of $\sqrt{(k_b k_t \sin^2\theta/JL)}$ and a damping factor ζ of $(R/L)/2\omega_n$. The rotor will rotate to some angle and gives oscillations about that angle with the oscillations dying away with time.

22.2.3 Bathroom scales

Consider the design of a simple weighing machine, i.e. bathroom scales. The main requirements are that a person can stand on a platform and the weight of that person will be displayed on some form of readout. The weight should be given with reasonable speed and accuracy and be independent of where on the platform the person stands. The following are possible solutions.

One possible solution is to use the weight of the person on the platform to deflect an arrangement of two parallel leaf springs (Fig. 22.15a)). With such an arrangement the deflection is virtually independent of where on the platform the person stands. The deflection can be transformed into movement of a pointer across a scale by using the arrangement shown in Figure 22.15(b). A rack-and-pinion is used to transform the linear motion into a circular motion about a horizontal axis. This is then transformed into a rotation about a vertical axis, and hence movement of a pointer across a scale, by means of a bevel gear.

Another possible solution involves the use of a microprocessor. The platform can be mounted on load cells employing electrical resistance strain gauges. When the person stands on the platform the gauges suffer strain and change resistance. If the gauges are mounted in a four-active-arm Wheatstone bridge then the out-of-balance voltage output from the bridge is a measure of the weight of the person. This can be amplified by a differential operational amplifier. The resulting analogue signal can then be fed through a latched analogue-to-digital converter for inputting to the microprocessor, e.g. the Motorola 6820. Figure 22.16 shows the input interface. There will also be a need to provide a non-erasable memory and this can be provided by an EPROM chip, e.g. Motorola 2716. The output to the display can then be taken through a PIA, e.g. Motorola 6821.

Weight

Leaf springs

(a)

Pointer moving
across a scale

Bevel
gear

Pinion

Rack attached
to centre of
spring

(b)

Fig. 22.15   Bathroom scales

Fig. 22.16   Input interface

However, if a microcontroller is used then memory is present within the single microprocessor chip, and by a suitable choice of microcontroller, e.g. M68HC11, we can obtain analogue-to-digital conversion for the inputs. The system then becomes: strain gauges feeding through an operational amplifier a voltage to the port E (the ADC input) of the microcontroller, with the output passing through suitable drives to output through ports B and C to a decoder and hence a LED display (Fig. 22.17).

The program structure might be:

Initialisation by clearing LED displays and memory

Start
   Is someone on the scales ? If not display 000
   If yes
      input data
      convert weight data into suitable output form
      output to decoder and LED display
      time delay to retain display
   Repeat from start again to get new weight

In considering the design of the mechanical parts of the bathroom scales we need to consider what will happen when some one stands on the scales. We have a spring–damper–mass system comparable to that described in Figure 8.6 (see Section 8.2.2) and so can describe its behaviour by:

$$m\frac{\mathrm{d}^2x}{\mathrm{d}t^2} + c\frac{\mathrm{d}x}{\mathrm{d}t} + kx = F$$

where x is the vertical deflection of the platform when a force F is applied. Taking the Laplace transform gives:

Fig. 22.17 Bathroom scales

$$ms^2\,X(s) + cs\,X(s) + k\,X(s) = F(s)$$

and so the system can be described by a transfer function of the form:

$$G(s) = \frac{X(s)}{F(s)} = \frac{1}{ms^2 + cs + k}$$

We can consider a person of weight W standing on the platform as a step input and so:

$$X(s) = \frac{1}{ms^2 + cs + k} \times \frac{W}{s}$$

The quadratic term is of the form $s^2 + 2\zeta\omega_n s + \omega_n^2$ (see Section 11.3) and thus has a natural frequency ω_n of $\sqrt{(k/m)}$ and a damping factor ζ of $c/(2\sqrt{mk})$.

When a person stands on the scales he/she wants the scales to quickly indicate the weight value and not oscillate for a long time about the value. If the damping was adjusted to be critical it would take too long to reach the value and so the damping needs to be adjusted to allow some oscillations which are rapidly damped away. We might decide that a 2% settling time t_s (see Section 10.4) of, say, 4 s was desirable. Since $t_s = 4/\zeta\omega_n$ then we require $\zeta\omega_n = 1$ and so $\zeta = \sqrt{(m/k)}$. A simple way we can alter the damping is thus by changing the mass.

The above indicates how we can use a mathematical model to predict the behaviour of a system and what factors we can then change to improve its performance.

22.3 Case studies of mechatronic systems

Electronic control systems are very widely used and the following are outlines of examples of such mechatronic systems.

22.3.1 A pick-and-place robot

Figure 22.18 shows the basic form of a pick-and-place robot unit. The robot has three axes about which motion can occur: rotation in a clockwise or counter-clockwise direction of the unit on its base, arm extension or contraction and arm up or down; also the gripper can open or close. These movements can be actuated by the use of pneumatic cylinders operated by solenoid controlled valves with limit switches to indicate when a motion is completed. Thus the clockwise rotation of the unit might result from the piston in a cylinder being extended and the counter-clockwise direction by its retraction. Likewise the upward movement of the arm might result from the piston in a linear cylinder being extended and the downward motion from it retracting; the extension of the arm by the piston in another cylinder extending and its return movement by the piston retracting.

The gripper can be opened or closed by the piston in a linear cylinder extending or retracting, Figure 22.19 showing a basic mechanism that could be used. Figure 22.20 shows how a microcontroller could be used to control the solenoid valves and hence the movements of the robot unit.

Fig. 22.18 Pick-and-place

22.3.2 Car park barriers

As an illustration of the use of a PLC, consider the coin-operated barriers for a car park. The in-barrier is to open when the correct money is inserted in the collection box and the out-barrier is to open when a car is detected at the car park side of the barrier. Figure 22.21 shows the types of valve systems that can be used to lift and lower the pivoted barriers.

Fig. 22.19 One form of gripper

Fig. 22.20   Robot control

Fig. 22.21 System for raising and lowering barrier

When a current flows through the solenoid of valve A, the piston in a cylinder moves upwards and causes the barrier to rotate about its pivot and raise to let a car through. When the current through the solenoid of valve A ceases, the return spring of the valve results in the valve position changing back to its original position. When the current is switched to through the solenoid of valve B the pressure is applied to lower the barrier. Limit switches are used to detect when the barrier is down and also when fully up.

With two of the systems shown in Figure 22.21, one for the entrance barrier and one for the exit barrier, and the connections to PLC inputs and outputs shown in Figure 22.22, the ladder program can be of the form shown in Figure 22.23.

X400 Coin-operated switch at entrance to car park
X401 Entrance barrier up switch
X402 Entrance barrier down switch
X403 Input when car at exit barrier
X404 Exit barrier up switch
X405 Exit barrier down switch

Y430 Solenoid on valve A for entrance barrier
Y431 Solenoid on valve B for entrance barrier
Y432 Solenoid on valve A for exit barrier
Y433 Solenoid on valve B for exit barrier

Fig. 22.22 PLC connections

To lift the entrance barrier, coin-operated input X400 required, Y430 then energises solenoid A

When barrier up input X401 starts timer

After 10 s the internal relay M100 energised

This results in Y431 energising solenoid B of entrance barrier and barrier comes down

To lift the exit barrier, car-operated input X403 required, Y432 then energises solenoid A of exit barrier

When barrier up input X404 starts timer

After 10 s the internal relay M101 energised

This results in Y433 energising solenoid B of exit barrier and barrier comes down

Fig. 22.23   Ladder program

22.3.3 Automatic camera

Figure 22.24 shows the basic features of the Canon EOS model, automatic, auto-focus, reflex cameras. The cameras have interchangeable lenses. There is a main microcontroller M68HC11 in the camera body and another microcontroller in the lens housing, the two communicating with each other when a lens is attached to the camera body. Figure 22.25 shows a block diagram of the system.

Fig. 22.24   Reflex camera

Fig. 22.25   Block diagram of electronic system

When the photographer presses the shutter button to its first position, that of being partially depressed, the main micro-controller calculates the shutter speed and aperture settings from the input from the metering sensor and displays them in the viewfinder and an external LCD display. At the same time, the main microcontroller processes the input from the range sensor and sends signals to the lens microcontroller. This issues signals to drive motors to adjust the focusing of the lens. When the photographer presses the shutter button to its second position, that

Fig. 22.26   Light sensors

of being fully depressed, the main microcontroller issues signals to drive the mirror up, change the aperture to that required, open the shutter for the required exposure time, and then, when the shutter has closed, advance the film ready for the next photograph.

The metering sensor has six light sensors arranged as shown in Figure 22.26. Signal conditioning is used to obtain the average value of C1, C2, C3 and C4; the A, B and average C value are then analysed to find the required exposure value. This, for example, reveals whether the scene is a scene with a relatively constant luminosity or perhaps a close up of a person so that there is a bright central zone surrounded by a dark background. The type of program that is used is:

> if B is equal to A and C minus B is less than 0
>   then exposure set on value of A

> if B is equal to A and C minus B is 0
>   then exposure set on value of C

This information is translated by the microcontroller into an appropriate shutter speed and aperture value. If the camera is operated with the shutter speed preselected by the photographer then only the aperture value is supplied; similarly if the aperture is preselected then only the shutter speed is supplied.

The range sensor has two 48-bit linear arrays of photodetectors. The light from the object, after passing through the camera lens, falls on this array (Fig. 22.27). When the image is in focus the spacing of the images on the detector array is a particular value, the spacing deviating from this when the image is out of focus. The amount of this deviation is used to give an error signal which is fed to the lens microcontroller and used to give an output to adjust the focusing of the lens. An encoder is used to provide feedback of this adjustment so that the microcontroller knows when the focusing has been completed. The program is thus of the form:

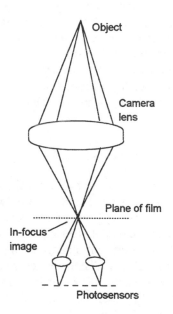

Fig. 22.27   Automatic focusing

> For the main microcontroller
>   send start command to lens microcontroller
>   take input from range sensor
>   calculate lens movement required
>   send lens movement data to lens microcontroller
>   wait for verification of lens movement from microcontroller
>   send in-focus signal to viewfinder display

> For the lens microcontroller
>   wait for start command from main microcontroller
>   determine initial lens position
>   wait for lens movement data from main microcontroller
>   read lens movement data
>   calculate new lens position

while lens is not in new position drive the motor
send verification signal of in-focus to main microcontroller

The diaphragm drive system is a stepper motor which opens or closes a set of diaphragm blades. The focusing involves two forms of drive, the arc form drive and the ultrasonic motor. The arc form drive uses a brushless permanent magnet d.c. motor (see Section 7.5.5), Hall sensors being used to detect the position of the rotor. The drive from the motor is transmitted through gears to move the focusing lens along the optical axis. The ultrasonic motor has a series of piezoelectric elements in the form of a ring (Fig. 22.28). When a current is supplied to a piezoelectric element it expands or contracts according to the polarity of the current. By switching the current to the piezoelectric elements in the appropriate sequence a displacement wave can be made to travel around the piezoelectric ring of elements in either a clockwise or counter-clockwise direction and consequently rotate a rotor which is in contact with its surface, hence driving the focusing element.

Piezoelectric elements

Fig. 22.28 Piezoelectric drive elements

As an illustration of the use of the modelling techniques discussed in earlier chapters of this book, consider this ultrasonic motor. The torque T generated by the motor is required to rotate the motor ring to some angular position θ. The ring is very light and so we neglect its inertia in comparison to the friction between the rings. Assuming the frictional force is proportional to the angular velocity ω, then $T = c\omega = c\,d\theta/dt$, where c is the friction constant. Integration then gives:

$$\theta = \frac{1}{c} \int dt$$

The control system for the ultrasonic motor is of the form shown in Figure 22.29. y_n is the nth input pulse and x_n the nth output pulse. With the microprocessor exercising proportional control gain K, the input to it is $y_n - x_n$ and the output is $K(y_n - x_n)$. This then passes through the DAC to give an analogue output consisting of a number of steps (Fig. 22.29). The motor acts as an integrator and so its output will be $1/c$ times the progressive sum of the areas under the steps (Fig. 22.29). Each step has an area of (DAC change in output for the step) $\times T$. Thus:

$$x_n - x_{n-1} = (\text{DAC output for } x_{n-1})T/c$$

$$= K(y_{n-1} - x_{n-1})T/c$$

Hence:

$$x_n = [1 - (KT/c)]x_{n-1} + (KT/c)y_{n-1}$$

Fig. 22.29   Control system

Suppose we have $K/c = 5$ and a sampling interval of 0.1 s. Thus

$$x_n = 0.5y_{n-1} + 0.5x_{n-1}$$

If there is an input to the control system for the focusing of a sequence of pulses of constant size 1, prior to that there being no input, i.e. $y_0 = 1$, $y_1 = 1$, $y_2 = 1$, ... etc., then:

$x_0 = 0$
$x_1 = 0.5 \times 0 + 0.5 \times 1 = 0.5$
$x_2 = 0.5 \times 0.5 + 0.5 \times 1 = 0.75$
$x_3 = 0.5 \times 0.75 + 0.5 \times 1 = 0.875$
$x_4 = 0.5 \times 0.875 + 0.5 \times 1 = 0.9375$
$x_5 = 0.5 \times 0.9375 + 0.5 \times 1 = 0.96875$
$x_6 = 0.5 \times 0.96875 + 0.5 \times 1 = 0.984375$
$x_7 = 0.5 \times 0.984365 + 0.5 \times 1 = 0.9921875$
and so on

The output thus takes about seven sampling periods, i.e. 0.7 s for the focusing to be achieved. This is too long. Suppose, however we choose values so that $KT/c = 1$. The difference equation then becomes $x_n = y_{n-1}$. Then we have:

$x_0 = 0$
$x_1 = 1$
$x_2 = 1$
$x_3 = 1$
etc.

This means that the output will reach the required position after just one sample. This is a much faster response. By using a high sampling rate a very fast response can be achieved. This form of response is termed a *deadbeat response*.

22.3.4 Car engine management

The modern car is likely to include many electronic control systems involving microcontrollers, the engine control system being one with its aim being to ensure that the engine is operated at its optimum settings. Figure 22.30 shows a generalised block diagram of such a system. The system consists of sensors supplying, after suitable signal conditioning, the input signals to the microcontroller and it providing output signals via drivers to actuate actuators. Figure 22.31 shows some of these elements in relation to an engine, only one cylinder being shown.

The engine speed sensor is an inductive sensor and consists of a coil for which the inductance changes as the teeth of the sensor wheel pass it and so gives an oscillating voltage. The temperature sensor is usually a thermistor. The mass air flow sensor may be a hot wire sensor; as air passes over a heated wire it will be cooled, the amount of cooling depending on the mass rate of flow. The oxygen sensor is generally a closed-end tube made of zirconium oxide with porous platinum electrodes on the inner and outer surfaces. Above about 300°C the sensor becomes permeable to oxygen ions and so gives a voltage between the electrodes.

Fig. 22.30 Engine management system

Fig. 22.31 An engine management system

ISBN 0 – 582 – 25634 – 8

9 7 8 0 5 8 2 2 5 6 3 4 7

Fig. 22.32 Bar code

22.3.5 Bar code reader

The familiar scene at the check-out of a supermarket is of the purchases being passed in front of a light beam or a hand-held wand being passed over the goods so that the bar code can be read and the nature of the purchase and hence its price automatically determined. The code consists of a series of black and white bars of varying widths. For example, there is such a bar code on the back of this book.

Figure 22.32 shows the basic form of the bar code used in the retail trade. The bar code represents a series of numbers. There is a prefix which identifies the coding scheme being used; this is a single digit for the regular *Universal Product Coding* (UPC) scheme used in the United States and two digits for the *European Article Number* (EAN) scheme used in Europe. The UPC uses a 0 prefix for grocery and a 3 for pharmaceuticals. The EAN prefix is from 00 to 09 and is such that the UPC code can be read within the EAN code. This is followed by five digits to represent the manufacturer, each manufacturer having been assigned a unique number. This brings up the centre of the code pattern which is identified by two taller bar patterns. The five-digit number that then follows represents the product. The final number is a check digit which is used to check that the code has been correctly read. A guard pattern of two taller bars at the start and end of the bar pattern is used to frame the bars.

Each number is coded as seven 0 or 1 digits. The codes used on either side of the centre line are different so that the direction of the scan can be determined. To the right the characters have an even number of 1s and so even parity and, for UPC, to the left an odd number of 1s and so odd parity; the EAN coding for the left being a mixture. Table 22.3 shows the UPC and EAN codings, UPC being the left A coding and the EAN using both left A and left B character codes.

Table 22.3 UPC and EAN codings

Decimal number	Left A characters	Left B characters	Right characters
0	0001101	0100111	1110010
1	0011001	0110011	1100110
2	0010011	0011011	1101100
3	0111101	0100001	1000010
4	0100011	0011101	0011100
5	0110001	0111001	0001110
6	0101111	0000101	1010000
7	0111011	0010001	1000100
8	0110111	0001001	1001000
9	0001011	0010111	1110100

1 1 0 1 1 0 0

Fig. 22.33 Bar code for right-hand 2

Each 1 is entered as a dark bar and thus the right-hand character 2 would be represented 1101100 and, with the adjacent dark bars run together, it appears as a double-width dark bar followed by a narrow space and then another double-width dark wide bar followed by a double-width space. This is illustrated in Figure 22.33. The guard pattern at the ends of the code represents 101 and the central band of bars is 01010.

The bar code shown in Figure 22.32 was that for the first edition of this book. It uses the EAN code and has the prefix 97 to identify it as a publication, 80582 to identify the publisher, 25634 to identify the particular book and a check digit of 7. Note that the bar code contains the relevant parts of the ISBN number, this also being a number to identify the publisher and the book concerned.

The procedure for using the check code digit is:

1   Starting at the left, sum all the characters, excluding the check digit, in the odd positions, i.e. first, third, fifth, etc. and then multiply the sum by 3.

2   Starting at the left, sum all the characters in the even positions.

3   Add the results of steps 1 and 2. The check character is the smallest number which when added to this sum produces a multiple of 10.

As an illustration of the use of the check digit, consider the bar code for the book where we have

   9780582256347

For the odd characters we have $9 + 8 + 5 + 2 + 5 + 3 = 32$ and when multiplied by 3 we have 96. For the even characters we have $7 + 0 + 8 + 2 + 6 + 4 = 27$. The sum is 123 and thus the check digit should be 7.

Reading the bar code involves determining the widths of the dark and light bands. This can involve a solid-state laser being used to direct an intense, narrow, beam of light at the code and detecting the reflected light by means of a photocell. Usually with the supermarket version the scanner is fixed and a spinning mirror is used to direct the light across the bar code and so scan all the bars. Signal conditioning involves amplification of the output of the photocell using operational amplifiers and then using an operational amplifier circuit as a comparator in order to give a high, i.e. 1, output when a black bar is scanned and a low, i.e. 0, output when a white space is scanned. This sequence of 0s and 1s can then be an input to, say, a PIA connected to a Motorola 6800 microprocessor. The overall form of the microprocessor program is:

1   Initialisation to clear the various memory locations.

2   Recovering the data from the input. This involves continually testing the input to determine whether it is 0 or 1.

3   Processing the data to obtain the characters in binary format. The input is a serial signal consisting of different duration 0s and 1s depending on the width of the spaces of black bars. The microprocessor system is programmed to find the module time width by dividing the time of scan between the end marker bars by the number of modules, a module being a light or dark band to represent a single 0 or 1. The program can then determine whether a dark or light band is a single digit or more than one and hence interpret the scanner signal.

4   Process the binary outcome into a statement of the item purchased and cost.

Problems and assignments

1   Present outline solutions of possible designs for the following:
    (a) A temperature controller for an oven.
    (b) A mechanism for sorting small, medium and large size objects moving along a conveyor belt so that they each are diverted down different chutes for packaging.
    (c) An x–y plotter (such a machine plots graphs showing how an input to x varies as the input to y changes).

Research assignments

2   Research the anti-lock braking system used in cars and describe the principles of its operation.
3   Research the mechanism used in the dot matrix printer and describe the principles of its operation.
4   Research the control area network (CAN) protocol used with cars.

Design assignments

5   Design a digital thermometer system which will display temperatures between 0 and 99°C. You might like to consider a solution based on the use of a microprocessor with RAM and ROM chips or a microcontroller solution.
6   Design a digital ohmmeter which will give a display of the resistance of a resistor connected between its terminals. You might like to consider basing your solution on the use of a monostable multivibrator, e.g. 74121, which will provide an impulse with a width related to the time constant RC of the circuit connected to it.
7   Design a digital barometer which will display the atmospheric pressure. You might like to base your solution on the use of the MPX2100AP pressure sensor.
8   Design a system which can be used to control the speed of a d.c. motor. You might like to consider using the M68HC11 evaluation board.
9   Design a system involving a PLC for the placing on a conveyor belt of boxes in batches of four.

Appendix A
The Laplace transform

A.1 The Laplace transform

This appendix gives more details of the Laplace transform than appears in Chapter 11. For a more detailed discussion of the Laplace transform, and examples of their use, the reader is referred to *Laplace and z-Transforms* by W. Bolton (Mathematics for Engineers Series, Longman 1994).

Consider a quantity which is a function of time. We can talk of this quantity being in the *time domain* and represent such a function as $f(t)$. In many problems we are only concerned with values of time greater than or equal to 0, i.e. $t \geq 0$. To obtain the Laplace transform of this function we multiply it by e^{-st} and then integrate with respect to time from zero to infinity. Here s is a constant with the unit of 1/time. The result is what we now call the *Laplace transform* and the equation is then said to be in the *s-domain*. Thus the Laplace transform of the function of time $f(t)$, which is written as $\mathcal{L}\{f(t)\}$, is given by

$$\mathcal{L}\{f(t)\} = \int_0^\infty e^{-st} f(t)\, dt$$

The transform is *one-sided* in that values are only considered between 0 and $+\infty$, and not over the full range of time from $-\infty$ to $+\infty$.

We can carry out algebraic manipulations on a quantity in the s-domain, i.e. adding, subtracting, dividing and multiplying in the normal way we do any algebraic quantities. We could not have done this on the original function, assuming it to be in the form of a differential equation, when in the time domain. By this means we can obtain a considerably simplified expression in the s-domain. If we want to see how the quantity varies with time in the time domain then we have to carry out the inverse transformation. This involves finding the time domain function that could have given the simplified s-domain expression.

When in the s-domain a function is usually written, since it is a function of s, as $F(s)$. It is usual to use a capital letter F for the

Laplace transform and a lower-case letter f for the time-varying function $f(t)$. Thus

$$\mathcal{L}\{f(t)\} = F(s)$$

For the inverse operation, when the function of time is obtained from the Laplace transform, we can write

$$f(t) = \mathcal{L}^{-1}\{F(s)\}$$

This equation thus reads as: $f(t)$ is the inverse transform of the Laplace transform $F(s)$.

A.1.1 The Laplace transform from first principles

To illustrate the transformation of a quantity from the time domain into the s-domain, consider a function that has the constant value of 1 for all values of time greater than 0, i.e. $f(t) = 1$ for $t \geq 0$. This describes a *unit-step* function and is shown in Figure ApA.1. The Laplace transform is then

$$\mathcal{L}\{f(t)\} = F(s) = \int_0^\infty 1\,e^{-st}\,dt = -\frac{1}{s}[e^{-st}]_0^\infty$$

Since with $t = \infty$ the value of e is 0 and with $t = 0$ the value of e^{-0} is -1, then

$$F(s) = \frac{1}{s}$$

As another example, the following shows the determination, from first principles, of the Laplace transform of the function e^{at}, where a is a constant. The Laplace transform of $f(t) = e^{at}$ is thus

$$F(s) = \int_0^\infty e^{at}e^{-st}\,dt$$

$$= \int_0^\infty e^{-(s-a)t}\,dt = -\frac{1}{s-a}\left[e^{-(s-a)t}\right]_0^\infty$$

When $t = \infty$ the term in the brackets becomes 0 and when $t = 0$ it becomes -1. Thus

$$F(s) = \frac{1}{s-a}$$

$f(t)$

1

0　　　　t

Fig. ApA.1　Unit-step function

A.2 Unit steps and impulses

Common input functions to systems are the unit step and the impulse. The following indicates how their Laplace transforms are obtained.

A.2.1 The unit-step function

Figure ApA.1 shows a graph of a unit-step function. Such a function, when the step occurs at $t = 0$, has the equation

$f(t) = 1$ for all values of t greater than 0
$f(t) = 0$ for all values of t less than 0

The step function describes an abrupt change in some quantity from zero to a steady value, e.g. the change in the voltage applied to a circuit when it is suddenly switched on.

The unit-step function thus cannot be described by $f(t) = 1$ since this would imply a function that has the constant value of 1 at all values of t, both positive and negative. The unit-step function that switches from 0 to +1 at $t = 0$ is conventionally described by the symbol $u(t)$ or $H(t)$, the H being after the originator O. Heaviside. It is thus sometimes referred to as the *Heaviside function*.

The Laplace transform of this step function is, as derived in the previous section,

$$F(s) = \frac{1}{s}$$

The Laplace transform of a step function of height a is

$$F(s) = \frac{a}{s}$$

A.2.2 Impulse function

Consider a rectangular pulse of size $1/k$ that occurs at time $t = 0$ and which has a pulse width of k, i.e. the area of the pulse is 1. Figure ApA.2 shows such a pulse. The pulse can be described as

$f(t) = \frac{1}{k}$ for $0 \leq t < k$
$f(t) = 0$ for $t > k$

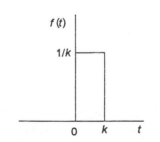

Fig. ApA.2   Rectangular pulse

If we maintain this constant pulse area of 1 and then decrease the width of the pulse (i.e. reduce k), the height increases. Thus, in the limit as $k \to 0$ we end up with just a vertical line at $t = 0$ with the height of the graph going off to infinity. The result is a graph that is zero except at a single point where there is an infinite spike (Fig. ApA.3). Such a graph can be used to represent an impulse. The impulse is said to be a unit impulse because the area enclosed by it is 1. This function is represented by $\delta(t)$, the *unit impulse function* or the *Dirac-delta function*.

The Laplace transform for the unit area rectangular pulse in Figure ApA.2 is given by

Fig. ApA.3   Impulse

$$F(s) = \int_0^\infty f(t)e^{-st}\,dt$$

$$= \int_0^k \frac{1}{k} e^{-st} \, dt + \int_k^\infty 0 \, e^{-st} \, dt$$

$$= \left[-\frac{1}{sk} e^{-st} \right]_0^k$$

$$= -\frac{1}{sk} (e^{-sk} - 1)$$

To obtain the Laplace transform for the unit impulse we need to find the value of the above in the limit as $k \to 0$. We can do this by expanding the exponential term as a series. Thus

$$e^{-sk} = 1 - sk + \frac{(-sk)^2}{2!} + \frac{(-sk)^3}{3!} + \dots$$

and so we can write

$$F(s) = 1 - \frac{sk}{2!} + \frac{(sk)^2}{3!} + \dots$$

Thus in the limit as $k \to 0$ the Laplace transform tends to the value 1.

$$\mathcal{L}\{\delta(t)\} = 1$$

Since the area of the above impulse is 1 we can define the size of such an impulse as being 1. Thus the above equation gives the Laplace transform for a *unit impulse*. An impulse of size a is represented by $a\delta(t)$ and the Laplace transform is

$$\mathcal{L}\{a\delta(t)\} = a$$

A.3 Standard Laplace transforms

In determining the Laplace transforms of functions it is not usually necessary to evaluate integrals since tables are available that give the Laplace transforms of commonly occurring functions. These, when combined with a knowledge of the properties of such transforms (see the next section), enable most commonly encountered problems to be tackled. Table ApA.1 lists some of the more common time functions and their Laplace transforms.

A.3.1 Properties of Laplace transforms

In this section the basic properties of the Laplace transform are outlined. These properties enable the table of standard Laplace transforms to be used in a wide range of situations.

Table ApA.1 Laplace transforms

	Time function $f(t)$	Laplace transform $F(s)$
1	$\delta(t)$, unit impulse	1
2	$\delta(t-T)$, delayed unit impulse	e^{-sT}
3	$u(t)$, a unit step	$\dfrac{1}{s}$
4	$u(t-T)$, a delayed unit step	$\dfrac{e^{-sT}}{s}$
5	t, a unit ramp	$\dfrac{1}{s^2}$
6	t^n, nth-order ramp	$\dfrac{n!}{s^{n+1}}$
7	e^{-at}, exponential decay	$\dfrac{1}{s+a}$
8	$1 - e^{-at}$, exponential growth	$\dfrac{a}{s(s+a)}$
9	$t\,e^{-at}$	$\dfrac{1}{(s+a)^2}$
10	$t^n e^{-at}$	$\dfrac{n!}{(s+a)^{n+1}}$
11	$t - \dfrac{1-e^{-at}}{a}$	$\dfrac{a}{s^2(s+a)}$
12	$e^{-at} - e^{-bt}$	$\dfrac{b-a}{(s+a)(s+b)}$
13	$(1-at)\,e^{-at}$	$\dfrac{s}{(s+a)^2}$
14	$1 - \dfrac{b}{b-a}\,e^{-at} + \dfrac{a}{b-a}\,e^{-bt}$	$\dfrac{ab}{s(s+a)(s+b)}$
15	$\dfrac{e^{-at}}{(b-a)(c-a)} + \dfrac{e^{-bt}}{(c-a)(a-b)} + \dfrac{e^{-ct}}{(a-c)(b-c)}$	$\dfrac{1}{(s+a)(s+b)(s+c)}$
16	$\sin \omega t$, a sine wave	$\dfrac{\omega}{s^2 + \omega^2}$
17	$\cos \omega t$, a cosine wave	$\dfrac{s}{s^2 + \omega^2}$
18	$e^{-at} \sin \omega t$, a damped sine wave	$\dfrac{\omega}{(s+a)^2 + \omega^2}$
19	$e^{-at} \cos \omega t$, a damped cosine wave	$\dfrac{s+a}{(s+a)^2 + \omega^2}$
20	$1 - \cos \omega t$	$\dfrac{\omega^2}{s(s^2 + \omega^2)}$
21	$t \cos \omega t$	$\dfrac{s^2 - \omega^2}{(s^2 + \omega^2)^2}$

(Continued overleaf)

	Time function $f(t)$	Laplace transform $F(s)$
22	$t \sin \omega t$	$\dfrac{2\omega s}{(s^2 + \omega^2)^2}$
23	$\sin(\omega t + \theta)$	$\dfrac{\omega \cos \theta + s \sin \theta}{s^2 + \omega^2}$
24	$\cos(\omega t + \theta)$	$\dfrac{s \cos \theta - \omega \sin \theta}{s^2 + \omega^2}$
25	$e^{-at} \sin(\omega t + \theta)$	$\dfrac{(s+a) \sin \theta + \omega \cos \theta}{(s+a)^2 + \omega^2}$
26	$e^{-at} \cos(\omega t + \theta)$	$\dfrac{(s+a) \cos \theta - \omega \sin \theta}{(s+a)^2 + \omega^2}$
27	$\dfrac{\omega}{\sqrt{1-\zeta^2}} e^{-\zeta \omega t} \sin \omega \sqrt{1 - \zeta^2}\, t$	$\dfrac{\omega^2}{s^2 + 2\zeta \omega s + \omega^2}$
28	$1 - \dfrac{1}{\sqrt{1-\zeta^2}} e^{-\zeta \omega t} \sin\left(\omega \sqrt{1 - \zeta^2}\, t + \phi\right)$, $\cos \phi = \zeta$	$\dfrac{\omega^2}{s(s^2 + 2\zeta \omega s + \omega^2)}$

Note: $f(t) = 0$ for all negative values of t. The $u(t)$ terms have been omitted from most of the time functions and have to be assumed.

Linearity property
If two separate time functions, e.g. $f(t)$ and $g(t)$, have Laplace transforms then the transform of the sum of the time functions is the sum of the two separate Laplace transforms.

$$\mathcal{L}\{af(t) + bg(t)\} = a\mathcal{L}f(t) + b\mathcal{L}g(t)$$

a and b are constants.

Thus, for example, the Laplace transform of $1 + 2t + 4t^2$ is given by the sum of the transforms of the individual terms in the expression. Thus, using items 1, 5 and 6 in Table ApA.1,

$$F(s) = \frac{1}{s} + \frac{2}{s^2} + \frac{8}{s^3}$$

s-Domain shifting property
This property is used to determine the Laplace transform of functions that have an exponential factor and is sometimes referred to as the *first shifting property*. If $F(s) = \mathcal{L}\{f(t)\}$ then

$$\mathcal{L}\{e^{at}f(t)\} = F(s - a)$$

For example, the Laplace transform of $e^{at}t^n$ is, since the Laplace transform of t^n is given by item 6 in Table ApA.1 as $n!/s^{n+1}$, given by

$$\mathcal{L}\{e^{at}t^n\} = \frac{n!}{(s-a)^{n+1}}$$

Time domain shifting property
If a signal is delayed by a time T then its Laplace transform is multiplied by e^{-sT}. If $F(s)$ is the Laplace transform of $f(t)$ then

$$\mathcal{L}\{f(t - T)u(t - T)\} = e^{-sT}F(s)$$

This delaying of a signal by a time T is referred to as the *second shift theorem*.

The time domain shifting property can be applied to all Laplace transforms. Thus for an impulse $\delta(t)$ which is delayed by a time T to give the function $\delta(t - T)$, the Laplace transform of $\delta(t)$, namely 1, is multiplied by e^{-sT} to give $1e^{-sT}$ as the transform for the delayed function.

Periodic functions
For a function $f(t)$ which is a periodic function of period T, the Laplace transform of that function is

$$\mathcal{L}f(t) = \frac{1}{1 - e^{-sT}} F_1(s)$$

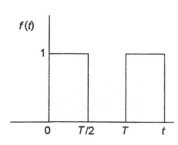

$f(t)$

1

0     $T/2$ T t

Fig. ApA.4   Rectangular pulses

where $F_1(s)$ is the Laplace transform of the function for the first period.

Thus, for example, consider the Laplace transform of a sequence of periodic rectangular pulses of period T, as shown in Figure ApA.4. The Laplace transform of a single rectangular pulse is given by $(1/s)(1 - e^{-sT/2})$. Hence, using the above equation, then the Laplace transform is

$$\frac{1}{1 - e^{-sT}} \times \frac{1}{s}(1 - e^{-sT/2}) = \frac{1}{s(1 + e^{-sT/2})}$$

Initial- and final-value theorems
The initial-value theorem can be stated as: if a function of time $f(t)$ has a Laplace transform $F(s)$ then in the limit as the time tends to zero the value of the function is given by

$$\lim_{t \to 0} f(t) = \lim_{s \to \infty} sF(s)$$

The final-value theorem can be stated as: if a function of time $f(t)$ has a Laplace transform $F(s)$ then in the limit as the time tends to infinity the value of the function is given by

$$\lim_{t \to \infty} f(t) = \lim_{s \to 0} sF(s)$$

Derivatives
The Laplace transform of a derivative of a function $f(t)$ is given by

$$\mathcal{L}\left\{\frac{d}{dt}f(t)\right\} = sF(s) - f(0)$$

where $f(0)$ is the value of the function when $t = 0$. For a second derivative

$$\mathcal{L}\left\{\frac{d^2}{dt^2}f(t)\right\} = s^2F(s) - sf(0) - \frac{d}{dt}f(0)$$

where $df(0)/dt$ is the value of the first derivative at $t = 0$. Examples of the Laplace transforms of derivatives are given in Chapter 11.

Integrals
The Laplace transform of the integral of a function $f(t)$ which has a Laplace transform $F(s)$ is given by

$$\mathcal{L}\left\{\int_0^t f(t)\,dt\right\} = \frac{1}{s}F(s)$$

For example, the Laplace transform of the integral of the function e^{-t} between the limits 0 and t is given by

$$\mathcal{L}\left\{\int_0^t e^{-t}\,dt\right\} = \frac{1}{s}\,\mathcal{L}\{e^{-t}\} = \frac{1}{s(s+1)}$$

A.4 The inverse transform

The inverse Laplace transformation is the conversion of a Laplace transform $F(s)$ into a function of time $f(t)$. This operation can be written as

$$\mathcal{L}^{-1}\{F(s)\} = f(t)$$

The inverse operation can generally be carried out by using Table ApA.1. The linearity property of Laplace transforms means that if we have a transform as the sum of two separate terms then we can take the inverse of each separately and the sum of the two inverse transforms is the required inverse transform.

$$\mathcal{L}^{-1}\{aF(s) + bG(s)\} = a\mathcal{L}^{-1}F(s) + b\mathcal{L}^{-1}G(s)$$

Thus, to illustrate how rearrangement of a function can often put it into the standard form shown in the table, the inverse transform of $3/(2s + 1)$ can be obtained by rearranging it as

$$\frac{3(1/2)}{s + (1/2)}$$

The table (item 7) contains the transform $1/(s + a)$ with the inverse of e^{-at}. Thus the inverse transformation is just this multiplied by the constant $(3/2)$ with $a = (1/2)$, i.e. $(3/2)\,e^{-t/2}$.

As another example, consider the inverse Laplace transform of $(2s + 2)/(s^2 + 1)$. This expression can be rearranged as

$$2\left[\frac{s}{s^2+1} + \frac{1}{s^2+1}\right]$$

The first term in the brackets has the inverse transform of $\cos t$ (item 17 in Table ApA.1) and the second term $\sin t$ (item 16 in Table ApA.1). Thus the inverse transform of the expression is

$$2\cos t + 2\sin t$$

A.4.1 Partial fractions

Often $F(s)$ is a ratio of two polynomials and cannot be readily identified with a standard transform in Table ApA.1. It has to be converted into simple fraction terms before the standard transforms can be used. The process of converting an expression into simple fraction terms is called decomposing into *partial fractions*. This technique can be used provided the degree of the numerator is less than the degree of the denominator. The degree of a polynomial is the highest power of s in the expression. When the degree of the numerator is equal to or higher than that of the denominator, the denominator must be divided into the numerator until the result is the sum of terms with the remainder fractional term having a numerator of lower degree than the denominator.

We can consider there to be basically three types of partial fractions:

1   The denominator contains factors which are only of the form $(s + a)$, $(s + b)$, $(s + c)$, etc. The expression is of the form

$$\frac{f(s)}{(s+a)(s+b)(s+c)}$$

and has the partial fractions of

$$\frac{A}{(s+a)} + \frac{B}{(s+b)} + \frac{C}{(s+c)}$$

2   There are repeated $(s + a)$ factors in the denominator, i.e. the denominator contains powers of such a factor, and the expression is of the form

$$\frac{f(s)}{(s+a)^n}$$

This then has partial fractions of

$$\frac{A}{(s+a)^1} + \frac{B}{(s+a)^2} + \frac{C}{(s+a)^3} + \cdots + \frac{N}{(s+a)^n}$$

3   The denominator contains quadratic factors and the quadratic does not factorise without imaginary terms. For an expression of the form

$$\frac{f(s)}{(as^2+bs+c)(s+d)}$$

the partial fractions are

$$\frac{As+B}{as^2+bs+c}+\frac{C}{s+d}$$

The values of the constants A, B, C, etc. can be found by either making use of the fact that the equality between the expression and the partial fractions must be true for all values of s or that the coefficients of s^n in the expression must equal those of s^n in the partial fraction expansion. The use of the first method is illustrated by the following example where the partial fractions of

$$\frac{3s+4}{(s+1)(s+2)} \text{ are } \frac{A}{s+1}+\frac{B}{s+2}$$

Then, for the expressions to be equal, we must have

$$\frac{3s+4}{(s+1)(s+2)}=\frac{A(s+2)+B(s+1)}{(s+1)(s+2)}.$$

and consequently

$$3s+4=A(s+2)+B(s+1)$$

This must be true for all values of s. The procedure is then to pick values of s that will enable some of the terms involving constants to become zero and so enable other constants to be determined. Thus if we let $s = -2$ then we have

$$3(-2)+4=A(-2+2)+B(-2+1)$$

and so $B = 2$. If we now let $s = -1$ then

$$3(-1)+4=A(-1+2)+B(-1+1)$$

and so $A = 1$. Thus

$$\frac{3s+4}{(s+1)(s+2)}=\frac{1}{s+1}+\frac{2}{s+2}$$

Appendix B
Logic gates

The following are the truth tables and symbols used for logic gates. Different sets of standard circuit symbols have been used with the main form having originated in the United States; an international standard form (IEEE/ANSI) has, however, now been developed which removes the distinctive shape used for a symbol and uses a rectangle with the logic function written inside it. Both formats are given here.

AND gate

Input		
A	B	Output
0	0	0
0	1	0
1	0	0
1	1	1

OR gate

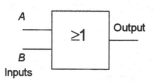

Input		
A	B	Output
0	0	0
0	1	1
1	0	1
1	1	1

NOT gate

Input A	Output
0	1
1	0

NAND gate

This can be considered as an AND gate followed by a NOT gate.

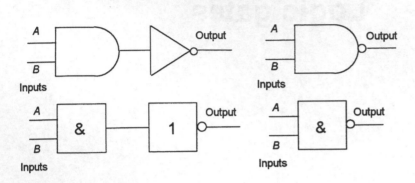

Inputs		
A	B	Output
0	0	1
0	1	1
1	0	1
1	1	0

NOR gate

This can be considered as an OR gate followed by a NOT gate.

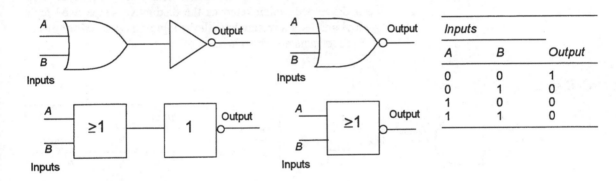

Inputs		
A	B	Output
0	0	1
0	1	0
1	0	0
1	1	0

EXCLUSIVE-OR (XOR) gate

This can be considered as an OR gate with a NOT gate applied to one of its inputs; alternatively it can be considered as an AND gate with a NOT gate applied to one of its inputs.

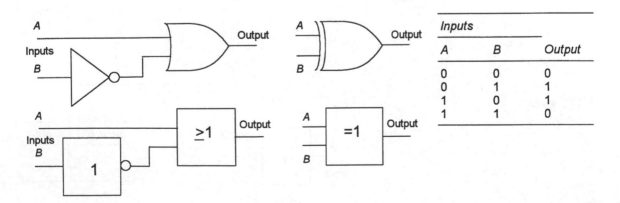

Inputs		
A	B	Output
0	0	0
0	1	1
1	0	1
1	1	0

Appendix C
Instruction sets

The following are the instructions used with the Motorola M68HC11, Intel 8051 and PIC16Cxx microcontrollers.

M68HC11

Instruction	Mnemonic	Instruction	Mnemonic
Loading		*Store registers*	
Load accumulator A	LDAA	Store contents of accumulator A	STAA
Load accumulator B	LDAB	Store contents of accumulator B	STAB
Load double accumulator	LDD	Store contents of double acc.	STD
Load stack pointer	LDS	Store stack pointer	STS
Load index register X	LDX	Store index register X	STX
Load index register Y	LDY	Store index register Y	STY
Pull data from stack and load acc. A	PULA	Push data from acc. A onto stack	PSHA
Pull data from stack and load acc. B	PULB	Push data from acc. B onto stack	PSHB
Pull index register X from stack	PULX	Push index reg. X contents onto stack	PSHX
Pull index register Y from stack	PULY	Push index reg. Y contents onto stack	PSHY
Transfer registers		*Logic*	
Transfer from acc. A to acc. B	TAB	AND with contents of accumulator A	ANDA
Transfer from acc. B to acc. A	TBA	AND with contents of accumulator B	ANDB
From stack pointer to index reg. X	TSX	Exclusive-OR with contents of acc. A	EORA
From stack pointer to index reg. Y	TSY	Exclusive-OR with contents of acc. B	EORB
From index reg. X to stack pointer	TXS	OR with contents of accumulator A	ORAA
From index reg. Y to stack pointer	TYS	OR with contents of accumulator B	ORAB
Exchange double acc. and index reg. X	XGDX	Replace memory with ones complement	COM
Exchange double acc. and index reg. Y	XGDY	Replace acc. A with ones complement	COMA
Decrement/increment		Replace acc. B with ones complement	COMB
Subtract 1 from contents of memory	DEC	*Arithmetic*	
Subtract 1 from contents of acc. A	DECA	Add contents of acc. A to acc. B	ABA
Subtract 1 from contents of acc. B	DECB	Add contents of acc. B to index reg. X	ABX
Subtract 1 from stack pointer	DES	Add contents of acc. B to index reg. Y	ABY
Subtract 1 from index register X	DEX	Add memory to acc. A without carry	ADDA
Subtract 1 from index register Y	DEY	Add memory to acc. B without carry	ADDB
Add 1 to contents of memory	INC	Add mem. to double acc. without carry	ADDD
Add 1 to contents of accumulator A	INCA	Add memory to acc. A with carry	ADCA
Add 1 to contents of accumulator B	INCB	Add memory to acc. B with carry	ADCB
Add 1 to stack pointer	INS	Decimal adjust	DAA
Add 1 to index register X	INX	Subtract contents of acc. B from acc. A	SBA
Add 1 to index register Y	INY	Subtract mem. from acc. A with carry	SBCA

(Continued on next page)

Instruction	Mnemonic	Instruction	Mnemonic
Rotate/shift		Subtract mem. from acc. B with carry	SBCB
Rotate bits in memory left	ROL	Subtract mem. from accumulator A	SUBA
Rotate bits in accumulator A left	ROLA	Subtract mem. from accumulator B	SUBB
Rotate bits in accumulator B left	ROLB	Subtract mem. from double acc.	SUBD
Rotate bits in memory right	ROR	Replace acc. A with twos complement	NEGA
Rotate bits in accumulator A right	RORA	Replace acc. B with twos complement	NEGB
Rotate bits in accumulator B right	RORB	Multiply unsigned acc. A by acc. B	MUL
Arithmetic shift bits in memory left	ASL	Unsigned integer divide D by i. reg. X	IDIV
Arithmetic shift bits in acc. A left	ASLA	Unsigned fractional divide D by i. reg. X	FDIV
Arithmetic shift bits in acc. B left	ASLB	*Conditional branch*	
Arithmetic shift bits in memory right	ASR	Branch if minus	BMI
Arithmetic shift bits in acc. A right	ASRA	Branch if plus	BPL
Arithmetic shift bits in acc. B right	ASRB	Branch if overflow set	BVS
Logical shift bits in memory left	LSL	Branch if overflow clear	BVC
Logical shift bits in acc. A left	LSLA	Branch if less than zero	BLT
Logical shift bits in acc. B left	LSLB	Branch if greater than or equal to zero	BGE
Logical shift bits in acc. D left	LSLD	Branch if less than or equal to zero	BLE
Logical shift bits in memory right	LSR	Branch if greater than zero	BGT
Logical shift bits in acc. A right	LSRA	Branch if equal	BEQ
Logical shift bits in acc. B right	LSRB	Branch if not equal	BNE
Logical shift bits in acc. C right	LSRD	Branch if higher	BHI
Data test with setting of condition codes		Branch if lower or same	BLS
Logical test AND between acc. A & memory	BITA	Branch if higher or same	BHS
Logical test AND between acc. B & memory	BITB	Branch if lower	BLO
Compare accumulator A to accumulator B	CBA	Branch if carry clear	BCC
Compare accumulator A and memory	CMPA	Branch if carry set	BCS
Compare accumulator B and memory	CMPB	*Jump and branch*	
Compare double accumulator with memory	CPD	Jump to address	JMP
Compare index register X with memory	CPX	Jump to subroutine	JSR
Compare index register Y with memory	CPY	Return from subroutine	RTS
Subtract $00 from memory	TST	Branch to subroutine	BSR
Subtract $00 from accumulator A	TSTA	Branch always	BRA
Subtract $00 from accumulator B	TSTB	Branch never	BRN
Interrupt		Branch bits set	BRSET
Clear interrupt mask	CLI	Branch bits clear	BRCLR
Set interrupt mask	SEI	*Condition code*	
Software interrupt	SWI	Clear carry	CLC
Return from interrupt	RTI	Clear overflow	CLV
Wait for interrupt	WAI	Set carry	SEC
Complement and clear		Set overflow	SEV
Clear memory	CLR	Transfer from acc. A to condition code reg.	TAP
Clear A	CLRA	Transfer from condition code reg. to acc. A	TPA
Clear B	CLRB	*Miscellaneous*	
Clear bits in memory	BCLR	No operation	NOP
Set bits in memory	BSET	Stop processing	STOP
		Special test mode	TEST

Note: The number of bits in a register depends on the processor. An 8-bit microprocessor generally has 8-bit registers. Sometimes two of the data registers may be used together to double the number of bits. Such a combined register is referred to as a doubled register.

Intel 8051

Instruction	Mnemonic	Instruction	Mnemonic
Data transfer		*Logical operations*	
Move data to accumulator	MOV A, #data	AND accumulator to direct byte	ANL direct,A
Move register to accumulator	MOV A, Rn	AND immediate data to direct byte	ANL direct, #data
Move direct byte to accumulator	MOV A, direct	AND immediate data to acc.	ANL A, #data
Move indirect RAM to accumulator	MOV A, @Ri	AND direct byte to accumulator	ANL A, direct
Move accumulator to direct byte	MOV direct, A	AND indirect RAM to accumulator	ANL A, @Ri
Move accumulator to external RAM	MOVX @Ri, A	AND register to accumulator	ANL A, Rn
Move accumulator to register	MOV Rn, A	OR accumulator to direct byte	ORL direct, A
Move direct byte to indirect RAM	MOV @Ri, direct	OR immediate data to direct byte	ORL direct, #data
Move immediate data to register	MOV Rn, #data	OR immediate data to accumulator	ORL A, #data
Move direct byte to direct byte	MOV direct, direct	OR direct byte to accumulator	ORL A, direct
Move indirect RAM to direct byte	MOV direct, @Ri	OR indirect RAM to accumulator	ORL A, @Ri
Move register to direct byte	MOV direct, Rn	OR register to accumulator	ORL A, Rn
Move immediate data to direct byte	MOV direct, #data	XOR accumulator to direct byte	XRL direct, A
Move immediate data to indirect RAM	MOV @Ri, #data	XOR immediate data to acc.	XRL direct, #data
Load data pointer with a 16-bit constant	MOV DPTR, #data16	XOR immediate data to acc.	XRL A, #data
Move code byte relative to DPTR to acc.	MOV A, @A+DPTR	XOR direct byte to accumulator	XRL A, direct
Move external RAM, 16-bit addr., to acc.	MOVX A, @DPTR	XOR indirect RAM to accumulator	XRL A, @Ri
Move acc. to external RAM, 16-bit addr.	MOVX @DPTR, A	XOR register to accumulator	XRL A, Rn
Exchange direct byte with accumulator	XCH A, direct	*Addition*	
Exchange indirect RAM with acc.	XCH A, @Ri	Add immediate data to acc.	ADD A, #data
Exchange register with accumulator	XCH A, Rn	Add direct byte to accumulator	ADD A, direct
Push direct byte onto stack	PUSH direct	Add indirect RAM to accumulator	ADD A, @Ri
Pop direct byte from stack	POP direct	Add register to accumulator	ADD A, Rn
Branching		Add immediate data to acc.	
Absolute jump	AJMP addr 11	with carry	ADDC A, #data
Long jump	LJMP addr 16	Add direct byte to acc. with carry	ADDC A, direct
Short jump, relative address	SJMP rel	Add indirect RAM to acc. with carry	ADDC A, @Ri
Jump indirect relative to the DPTR	JMP @A+DPTR	Add register to acc. with carry	ADDC A, Rn
Jump if accumulator zero	JZ rel	*Subtraction*	
Jump if accumulator is not zero	JNZ rel	Subtract immediate data from acc.	
Compare direct byte to acc. and jump if		with borrow	SUBB A, #data
not equal	CJNE A, direct, rel	Subtract direct byte from acc. with	
Compare immediate to acc. and jump if		borrow	SUBB A, 29
not equal	CJNE A, #data, rel	Subtract indirect RAM from acc.	
Compare immediate to register and		with borrow	SUBB A, @Ri
jump if not equal	CJNE Rn, #data, rel	*Multiplication and division*	
Compare immediate to indirect and		Multiply A and B	MUL AB
jump if not equal	CJNE @Ri, #data, rel	Divide A by B	DIV AB
Decrement register and jump if not zero	DJNZ Rn, rel	*Decimal maths operations*	
Decrement direct byte, jump if not zero	DJNZ A, direct, rel	Exchange low-order digit indirect	
Jump if carry is set	JC rel	RAM with accumulator	XCHD A, @Ri
Jump if carry not set	JNC rel	Swap nibbles within the acc.	SWAP A
Jump if direct bit is set	JB bit, rel	Decimal adjust accumulator	DA A
Jump if direct bit is not set	JNB bit, rel	*Increment and decrement*	
Jump if direct bit is set and clear bit	JBC bit, rel	Increment accumulator	INC A
Subroutine call		Increment direct byte	INC direct
Absolute subroutine call	ACALL addr 11	Increment indirect RAM	INC @Ri
Long subroutine call	LCALL addr 16	Increment register	INC Rn
Return from subroutine	RET	Decrement accumulator	DEC A
Return from interrupt	RETI	Decrement direct byte	DEC direct

(Continued on next page)

Instruction	Mnemonic	Instruction	Mnemonic
Bit manipulation		Decrement indirect RAM	DEC @Ri
Clear carry	CLR C	Decrement register	DEC Rn
Clear bit	CLR bit	Increment data pointer	INC DPTR
Set carry but	SETB C	*Clear and complement operations*	
Set bit	SETB bit	Complement accumulator	CPL A
Complement carry	CPL C	Clear accumulator	CLR A
AND bit to carry bit	ANL C,bit	*Rotate operations*	
AND complement of bit to carry bit	ANL C,/bit	Rotate accumulator right	RR A
OR bit to carry bit	ORL C,bit	Rotate accumulator right thro. C	RRC A
OR complement of bit to carry bit	ORL C,/bit	Rotate accumulator left	RL A
Move bit to carry	MOV C,bit	Rotate accumulator left through C	RLC A
Move carry bit to bit	MOV bit,C	*No operation*	
		No operation	NOP

Note: A value preceded by # is a number, #data16 is a 16-bit constant; Rn refers to the contents of a register; @Ri refers to the value in memory where the register points, DPTR is the data pointer, direct is the memory location where data used by an instruction can be found.

PIC16Cxx

Instruction	Mnemonic	Instruction	Mnemonic
Add a number with number in working reg.	addlw number	Move (copy) the number in a file reg. into the working reg.	movf FileReg,w
Add number in working reg. to number in file register and put number in file register	addwf FileReg,f	Move (copy) number into working reg.	movlw number
Add number in working reg. to number in file register and put number in working reg.	addwf FileReg,w	Move (copy) the number in file reg. into the working reg.	movwf FileReg
AND a number with the number in the working reg. and put result in working reg.	andlw number	No operation	nop
AND a number in the working reg. with the number in file reg., and put result in file reg.	andwf FileReg,f	Return from a subroutine and enable global interrupt enable bit	refie
Clear a bit in a file reg., i.e. make it 0	bcf FileReg,bit	Return from a subroutine with a number in the working register	retlw number
Set a bit in a file reg., i.e. make it 1	bsf FileReg,bit	Return from a subroutine	return
Test a bit in a file reg. and skip the next instruction if the bit is 0	btfsc FileReg,bit	Rotate bits in file reg. to the left through the carry bit	rlf FileReg,f
Test a bit in a file reg. and skip the next instruction if the bit is 1	btfss FileReg,bit	Rotate bits in file reg. to the right through the carry bit	rrf FileReg,f
Call a subroutine, after which return to where it left off	call AnySub	Send the PIC to sleep, a low power consumption mode	sleep
Clear, i.e. make 0, the number in file reg.	clrf FileReg	Subtract the number in working reg. from a number	sublw number
Clear , i.e. make 0, the no. in working reg.	clrw	Subtract the no. in working reg. from number in file reg., put result in file reg.	subwf FileReg,f
Clear the number in the watchdog timer	clrwdt	Swap the two halves of the 8 bit no. in a file reg, leaving result in file reg.	swapf FileReg, f
Complement the number in file reg. and leave result in file register	comf FileReg,f	Use the number in working reg. to specify which bits are input or output	tris PORTX
Decrement a file reg., result in file reg.	decf FileReg,f		
Decrement a file reg. and if result zero skip the next instruction	decfsz FileReg,f	XOR a number with number in working register	xorlf number
Go to point in program labelled	gotot label		
Increment file reg. and put result in file reg.	incf FileReg,f	XOR the number in working reg. with number in file reg. and put result in the file reg.	xorwf FileReg,f
OR a number with number in working reg.	iorlw number		
OR the number in working reg. with the number in file reg., put result in file reg.	iorwf FileReg,f		

Note: f is used for the file register, w for the working register and b for bit. The mnemonics indicate the types of operand involved, e.g. movlw indicates the move operation with the lw indicating a literal value, i.e. a number, is involved in the working register w; movwf indicates the move operation when the working register and a file register are involved.

Appendix D
C library functions

The following are some common C library functions. This is not a complete list of all the functions within any one library or a complete list of all the libraries that are likely to be available with any one compiler.

<ctype.h>

isalnum	int isalnum(int ch)	Tests for alphanumeric characters, returning nonzero if its argument is either a letter or a digit or a 0 if it is not alphanumeric.
isalpha	int isalpha(int ch)	Tests for alphabetic characters, returning nonzero if a letter of the alphabet, otherwise 0.
iscntrl	int iscntrl(int ch)	Tests for control character, returning nonzero if between 0 and 0x1F or is equal to 0x7F (DEL), otherwise 0.
isdigit	int isdigit(int ch)	Tests for decimal digit character, returning nonzero if a digit (0 to 9), otherwise 0.
isgraph	int isgraph(int ch)	Tests for a printable character (except space), returning nonzero if printable, otherwise 0.
islower	int islower(int ch)	Tests for lowercase character, returning nonzero if lowercase, otherwise 0.
isprint	int isprint(int ch)	Tests for printable character (including space), returning nonzero if printable, otherwise 0.
ispunct	int ispunct(int ch)	Tests for punctuation character, returning nonzero if a punctuation character, otherwise 0.
isspace	int isspace(int ch)	Tests for space character, returning nonzero if a space, tab, form feed, carriage return or newline character, otherwise 0.
isupper	int isupper(int ch)	Tests for uppercase character, returning nonzero if uppercase, otherwise 0.
isxdigit	int isxdigit(int ch)	Tests for hexadecimal character, returning nonzero is a hexadecimal digit, otherwise 0.

<math.h>

acos	double acos(double arg)	Returns the arc cosine of the argument.
asin	double asin(double arg)	Returns the arc sine of the argument.
atan	double atan(double arg)	Returns the arc tangent of the argument. Requires one argument.
atan2	double atan2(double y, double x)	Returns the arc tangent of y/x.
ceil	double ceil(double num)	Returns the smallest integer that is not less than num.
cos	double cos(double arg)	Returns the cosine of arg. The value of arg must be in radians.
cosh	double cosh(double arg)	Returns the hyperbolic cosine of arg.
exp	double exp(double arg)	Returns e^x where x is arg.
fabs	double fabs(double num)	Returns the absolute value of num.
floor	double floor(double num)	Returns the largest integer not greater than num.
fmod	double fmod(double x, double y)	Returns the flaoting point remainder of x/y.
ldexp	double ldexp(double x, int y)	Returns x times 2^y.
log	double log(double num)	Returns the natural logarithm of num.
log10	double log10(double num)	Returns the base 10 logarithm of num.
pow	double pow(double base, double exp)	Returns base raised to the exp power.
sin	double sin(double arg)	Returns the sine of arg.
sinh	double sinh(double arg)	Returns the hyperbolic sine of arg.

sqrt	double sqrt(double num)	Returns the square root of num.
tan	double tan(double arg)	Returns the tangent of arg.
tanh	double tanh(double arg)	Returns the hyperbolic tangent of arg.

<stdio.h>

getchar	int getchar(void)	Returns the next character typed on the keyboard.
gets	char gets(char *str)	Reads characters entered at the keyboard until a carriage return is read and stores them in the array pointed to by str.
printf	int printf(char *str, ...)	Outputs the string pointed to by str.
puts	int puts(char *str)	Outputs the string pointed to by str.
scanf	int scanf(char *str, ...)	Reads information into the variables pointed to by the arguments following the control string.

<stdlib.h>

abort	void abort(void)	Causes immediate termination of a program.
abs	int abs(int num)	Returns the absolute value of the integer num.
bsearch	void bsearch(const void *key, const void *base, size_t num, size_t size, int(*compare)(const void *,const void *))	Performs a binary search on the sorted array pointed to by base and returns a pointer to the first member that matches the key pointed to by key. The number of the elements in the array is specified by num and the size in bytes of each element by size.
calloc	void *calloc(size_t num, size_t size)	Allocates sufficient memory for an array of num objects of size given by size, returning a pointer to the first byte of the allocated memory.
exit	void exit(int status)	Causes immediate normal termination of a program. The value of the status is passed to the calling process.
free	void free(void *ptr)	Frees the allocated memory pointed to be ptr.
labs	long labs(long num)	Returns the absolute value of the long int num.
malloc	void *malloc(size_t size)	Returns a pointer to the first byte of memory of size given by size that has been allocated.
qsort	void qsort(void *base, size_t num, size_t size, int(*compare)(const void*, const void*)	Sorts the array pointed to be base. The number of elements in the array is given by num and the size in bytes of each element by size.
realloc	void *realloc(void *ptr, size_t size)	Changes the size of the allocated memory pointed to by ptr to that specified by size.

Note: size_t is the type for 'size of' variables and usually represents the size of another parameter or object.

<time.h>

asctime	char *asctime(const struct tm *ptr)	Converts time from a structure form to a character string appropropriate for display, returning a pointer to the string.
clock	clock_t clock(void)	Returns the number of clock cycles that have occurred since the program began execution.
ctime	char *ctime(const time_t *time)	Returns a pointer to a string of the form day month date hours:minutes:seconds year\n\0 given a pointer to the numbers of seconds elapsed since 00:00:00 Greenwich Mean Time.
difftime	double difftime(time_t time 2, time_t time 1)	Returns the difference in seconds between time1 and time 2.
gmtime	struct tm *gmtime(const time_t *time)	Returns a pointer to time converted from long inter form to a structure form.
localtime	struct tm *localtime(const time_t *time)	Returns a pointer to time converted from long inter form to structure form in local time.
time	time_t time(time_t *system)	Returns the current calendar time of the system.

Note: time_t and clock_t are used as the type for 'time of' and 'number of cycles of' variables.

Further information

The following are sources for further information indicated in the text.

Chapter 1

1 Ed. Noltingk, B. E., *Instrumentation Reference Book*, Butterworth-Heinemann 1995
2 Morris, A.S., *Measurement and Instrumentation Principles*, 3rd edition, Newnes 2001
3 Bolton, W., *Newnes Instrumentation and Measurement*, Newnes 1991, 1996, 2000
4 Denton, T., *Automobile Electrical and Electronic Systems*, Arnold 1995

Chapter 2

5 Ed. Noltingk, B. E., *Instrumentation Reference Book*, Butterworth-Heinemann 1995
6 Bolton, W., *Measurement and Instrumentation Systems*, Newnes 1996
7 Bolton, W., *Newnes Instrumentation and Measurement*, Newnes 1991, 1996, 2000
8 Boyle, H. B., *Transducer Handbook*, Newnes 1992

Chapter 3

9 Horrocks, D. H., *Feedback Circuits and Op. Amps*, Chapman and Hall 1990
10 Gray, P. R. and Meyer, R. G., *Analysis and Design of Analog Integrated Circuits*, Wiley 1993
11 Niewiadomski, S., *Filter Handbook*, Heinemann Newnes 1989

Chapter 4

12 Bolton, W., *Newnes Instrumentation and Measurement*, Newnes 1991, 1996, 2000

13   Mazda, F. F., *Electronic Instruments and Measurement Techniques*, CUP 1987
14   Berlin, H. M. and Getz, F. C., *Principles of Electronic Instrumentation and Measurement*, Merrill 1988
15   Wells, L. K. and Travis, J., *LabVIEW for Everyone*, Prentice-Hall 1997
16   Johnson, G. W., *LabVIEW Graphical Programming*, McGraw-Hill 1994
17   Morris, A. S., *Measurement and Calibration for Quality Assurance*, Prentice-Hall 1991

Chapter 5

18   Bolton, W., *Pneumatic and Hydraulic Systems*, Butterworth-Heinemann 1997
19   Pinches, M. J. and Callear, B. J., *Power Pneumatics*, Prentice-Hall 1996
20   Rohner, P. and Smith G., *Pneumatic Control for Industrial Automation*, Wiley 1987, 1990
21   Rohner, P., *Industrial Hydraulic Control*, Wiley 1984, 1986, 1988, 1995

Chapter 6

22   Bolton, W., *Mechanical Science*, Blackwell Scientific Publications 1993, 1998
23   Norton, R. L., *Design of Machinery*, McGraw-Hill 1992

Chapter 7

24   Morris, N., *Advanced Industrial Electronics*, McGraw-Hill 1974
25   Crecraft, D. I., Gorham, D.A. and Sparkes, J. J., *Electronics*, Chapman and Hall 1993
26   Kenjo, T., *Power Electronics for the Microprocessor Age*, OUP 1990
27   Edwards, J. D., *Electrical Machines and Drives*, Macmillan 1991
28   Gray, C. B., *Electrical Machines and Drive Systems*, Longman 1989
29   Gottlieb, I. M., *Electric Motors and Control Techniques*, TAB Books, McGraw-Hill 1994
30   Miller, T. J. E., *Brushless Permanent-magnet and Reluctance Motor Drives*, OUP 1989
31   Kenjo, T., *Stepping Motors and their Microprocessor Controls*, OUP 1984

Chapter 8

32 Shearer, J. L. and Kulakowski, B. T., *Dynamic Modelling and Control of Engineering Systems*, 2nd edition, Prentice Hall 1997

33 Frederick, C., *Modelling and Analysis of Dynamic Systems*, Houghton Mifflin 1993

Chapter 10

34 Fox. H. and Bolton, W., *Mathematics for Engineers and Technologists*, Butterworth-Heinemann 2002

35 Bolton, W., *Laplace and z-Transforms*, Longman 1994

Chapter 11

36 Bolton, W., *Laplace and z-Transforms*, Longman 1994

37 Pärt-Enander, E., Sjöberg, A., Melin, B. and Isaksson, P., *The MATLAB Handbook*, Addison-Wesley 1996

38 Leonard, N. E. and Levine, W. S., *Using MATLAB to Analyze and Design Control Systems*, Addison-Wesley 1995

Chapter 12

39 Bolton, W., *Complex Numbers*, Longman 1994

Chapter 13

40 Bennett, S., *Real-time Computer Control*, Prentice-Hall 1994

41 Åstrom, K. J. and Wittenmark, B., *Adaptive Control*, 2nd edition, Addison-Wesley 1995

Chapter 14

42 Peterson, D., *Audio, Video and Data Telecommunications*, McGraw-Hill 1992

Chapter 15

43 Bolton, W., *Microprocessor Systems*, Longman 2000

44 Cady, F. M., *Software and Hardware Engineering: Motorola M68HC11*, OUP 1997

45 Spasov, P., *Microcontroller Technology: The 68HC11*, Prentice-Hall 1992, 1996

46 Barnett, R. H., *The 8051 Family of Microcontrollers*, Prentice-Hall 1995

47 MacKenzie I. Scott, *The 8051 Microcontroller*, Prentice Hall 1999, 1995, 1992

48 Yeralan, S. and Ahluwalia A., *Programming and Interfacing the 8051 Microcontroller*, Addison-Wesley 1993

49 Morton, J., *PIC: Your Personal Introductory Course*, 2nd edition, Newnes 2001

50 Ibrahim, D., *PIC Basic: Programming and Projects*, Newnes 2001

Chapter 17

51 Aitken, P. and Jones, B. L., *Teach Yourself C Programming in 21 Days*, Sams Publishing 1995
52 Bronson, G., *C for Engineers and Scientists*, West 1993
53 Van Sickle, T., *Programming Microcontrollers in C*, Newnes 2001
54 Zurrell, K., *C Programming for Embedded Systems*, R & D Books 2000
55 Pont M. J., *Embedded C*, Addison Wesley 2002

Chapter 18

56 Cady, F. M., *Software and Hardware Engineering: Motorola M68HC11*, OUP 1997
57 Spasov, P., *Microcontroller Technology: The 68HC11*, Prentice-Hall 1992, 1996
58 Predko, M., *Programming and Customizing the PIC Microcontroller*, McGraw-Hill 1998

Chapter 19

59 Warnock, I. G., *Programmable Controllers, Operation and Application*, Prentice-Hall 1988
60 Bolton, W., *Programmable Logic Controllers*, Newnes 1996, 2003
61 Rohner, P., *Automation with Programmable Logic Controllers*, Macmillan 1996

Chapter 20

62 Freer, C. G., *Data Communication for Engineers*, Macmillan 1992
63 Tomasi, W., *Advanced Electronic Communications Systems*, 5th edition, Prentice-Hall 2000

Chapter 21

64 *M68HC11EVB Evaluation Board User's Manual*, Motorola 1986
65 Cady, F. M., *Software and Hardware Engineering: Motorola M68HC11*, OUP 1997

Appendix A

66 Bolton, W., *Laplace and z-Transforms*, Longman 1994

Answers

The following are answers to numerical problems and brief clues as to possible answers with descriptive problems.

Chapter 1

1. (a) Sensor – mercury, signal conditoner – fine bore stem, display – marks on the stem, (b) sensor – curved tube, signal conditioner – gears, display – pointer moving across a scale
2. See text
3. Comparison/controller – thermostat, correction – perhaps a relay, process – heat, variable – temperature, measurement – a temperature-sensitive device, perhaps a bimetallic strip
4. See Figure A.1

Fig. A.1   Chapter 1, problem 4

5. See text
6. See text
7. For example: water in, rinse, water out, water in, heat water, rinse, water out, water in, rinse, water out
8. Traditional: bulky, limited functions, requires rewinding. Mechatronics: compact, many functions, no rewinding, cheaper
9. Bimetallic element: slow, limited accuracy, simple functions, cheap. Mechatronics: fast, accurate, many functions, getting cheaper

Chapter 2

1. See the text for explanation of the terms
2. −3.9%

3 67.5 s
4 0.73%
5 0.105 Ω
6 Incremental – angle from some datum, not absolute; absolute – unique identification of an angle
7 162
8 (a) ±1.2°, (b) 3.3 mV
9 See text
10 2.8 kPa
11 19.6 kPa
12 −0.89%
13 +1.54°C
14 Yes
15 −9.81 N, −19.62 N, e.g. strain gauges
16 For example: orifice plate with differential pressure cell
17 For example: differential pressure cell
18 For example: LVDT displacement sensor

Chapter 3

1 As Figure 3.2 with $R_2/R_1 = 50$, e.g. $R_1 = 1$ kΩ, $R_2 = 50$ kΩ
2 200 kΩ
3 Figure 3.5 with two inputs. E.g. $V_A = 1$ V, $V_B = 0$ to 100 mV, $R_A = R_2 = 40$ kΩ, $R_B = 1$ kΩ
4 Figure 3.11 with $R_1 = 1$ kΩ and $R_2 = 2.32$ kΩ
5 $V = K\sqrt{I}$
6 100 kΩ
7 80 dB
8 Fuse to safeguard against high current, limiting resistor to reduce currents, diode to rectify a.c., Zener diode circuit for voltage and polarity protection, low pass filter to remove noise and interference, optoisolator to isolate the high voltages from the microprocessor
9 0.059 V
10 5.25×10^{-5} V
11 As given in the problem
12 24.4 mV
13 9
14 0.625 V
15 1, 2, 4, 8
16 12 µs
17 See text
18 Buffer, digital-to-analogue converter, protection

Chapter 4

1 See text
2 See Section 4.1
3 For example: (a) a galvanometric or potentiometric recorder, (b) a moving coil meter, (c) a magnetic tape recorder, (d) a storage oscilloscope

4 Could be four-active-arm bridge, differential operational amplifier, display of a voltmeter. The values of components will depend on the thickness chosen for the steel and the diameter of a load cell. You might choose to mount the tank on three cells

5 Could be as in Figure 3.8 with cold junction compensation by a bridge (see Section 3.5.2). Linearity might be achieved by suitable choice of thermocouple materials

6 Could be thermistors with a sample and hold element followed by an analogue-to-digital converter for each sensor. This would give a digital signal for transmission, so reducing the effects of possible interference. Optoisolators could be used to isolate high voltages/currents, followed by a multiplexer feeding digital meters

7 This is based on Archimedes' principle, the upthrust on the float equals the weight of fluid displaced

8 Could use a LVDT or strain gauges with a Wheatstone bridge

9 For example: (a) Bourdon gauge, (b) thermistors, galvanometric chart recorder, (c) strain-gauged load cells, Wheatstone bridge, differential amplifier, digital voltmeter, (d) tachogenerator, signal conditioning to shape pulses, counter

Chapter 5

1 See Section 5.3
2 See Section 5.3.2
3 See Section 5.4.2
4 See (a) Figure 5.15, (b) Figure 5.8, (c) Figure 5.11, (d) Figure 5.14
5 A+, B+, A–, B–
6 See Figure A.2
7 0.0057 m^2
8 124 mm
9 1.27 MPa, $3.9 \times 10^{-5} \text{ m}^3/\text{s}$
10 (a) $0.05 \text{ m}^3/\text{s}$, (b) $0.10 \text{ m}^3/\text{s}$
11 (a) $0.42 \text{ m}^3/\text{s}$, (b) $0.89 \text{ m}^3/\text{s}$
12 960 mm

Chapter 6

1 (a) A system of elements arranged to transmit motion from one form to another form. (b) A sequence of joints and links to provide a controlled output in response to a supplied input motion

2 See Section 6.3.1
3 (a) 1, (b) 2, (c) 1, (d) 1, (e) 3
4 (a) Pure translation, (b) pure translation, (c) pure rotation, (d) pure rotation, (e) translation plus rotation
5 Quick-return

Fig. A.2 Chapter 5, problem 6

6 Sudden drop in displacement followed by a gradual rise back up again
7 60 mm
8 Heart-shaped with distance from axis of rotation to top of heart 40 mm and to base 100 mm. See Figure 6.11(b)
9 For example: (a) cams on a shaft, (b) quick-return mechanism, (c) eccentric cam, (d) rack-and-pinion, (e) belt drive, (f) bevel gears
10 1/24
11 (a) Hydrodynamic, (b) dry sliding or rolling

Chapter 7

1 It acts as a flip-flop
2 See text and Figure 7.9
3 (a) Series wound, (b) shunt wound
4 (a) D.C. shunt wound, (b) induction or synchonous motor with an inverter, (c) d.c., (d) a.c.
5 See Section 7.5.5
6 See Section 7.7
7 480 pulses/s

Chapter 8

1 (a) $m\dfrac{d^2x}{dt^2} + c\dfrac{dx}{dt} = F$, (b) $m\dfrac{d^2x}{dt^2} + c\dfrac{dx}{dt} + (k_1 + k_2)x = F$

2 As in Figure 8.6

3 $c\dfrac{d\theta_i}{dt} = c\dfrac{d\theta_o}{dt} + k\theta_o$

4 Two torsional springs in series with a moment of inertia block,

$$T = I\dfrac{d^2\theta}{dt^2} + k_1(\theta_1 - \theta_2) = m\dfrac{d^2\theta}{dt^2} + \dfrac{k_1 k_2}{k_1 + k_2}\theta_1$$

5 $v = v_R + \dfrac{1}{RC}\int v_R\,dt$

6 $v = \dfrac{L}{R}\dfrac{dv_R}{dt} + \dfrac{1}{CR}\int v_R\,dt + v_R$

7 $v = R_1 C\dfrac{dv_C}{dt} + \left(\dfrac{R_1}{R_2} + 1\right)v_C$

8 $RA_2\dfrac{dh_2}{dt} + h_2\rho g = h_1$

9 $RC\dfrac{dT}{dt} + T = T_r$

Charged capacitor discharging through a resistor

10 $RC\dfrac{dT_1}{dt} = Rq - 2T_1 + T_2 + T_3$,

$RC\dfrac{dT_2}{dt} = T_1 - 2T_2 + T_3$

11 $pA = m\dfrac{d^2x}{dt^2} + R\dfrac{dx}{dt} + \dfrac{1}{c}x$,

R = resistance to stem movement, c = capacitance of spring

12 $T = \left(\dfrac{I_1}{n} + n\right)\dfrac{d^2\theta}{dt^2} + \left(\dfrac{c_1}{n} + nc_2\right)\dfrac{d\theta}{dt} + \left(\dfrac{k_1}{n} + nk_2\right)\theta$

Chapter 9

1 $\Delta F = (2kx_o)\,\Delta x$

2 $\Delta E = (a + 2bT_o)\,\Delta T$

3 $\Delta T = (mgL)\,\Delta\theta$

4 $\dfrac{IR}{k_1 k_2}\dfrac{d\omega}{dt} + \omega = \dfrac{1}{k_2}v$

5 $(L_a + L_L)\dfrac{di_a}{dt} + (R_a + R_L)i_a - k_1\dfrac{d\theta}{dt} = 0$,

$I\dfrac{d^2\theta}{dt^2} + B\dfrac{d\theta}{dt} + k_2 i_a = T$

6 Same as armature-controlled motor

7 Inductance in series with resistance and current source, rod with mass, damping and moving against a spring

Chapter 10

1 $4\dfrac{dx}{dt} + x = 6y$

2 (a) 59.9°C, (b) 71.9°C

3 (a) $i = \dfrac{V}{R}(1 - e^{-Rt/L})$, (b) L/R, (c) V/R

4 (a) Continuous oscillations, (b) under-damped, (c) critically damped, (d) over-damped

5 (a) 4 Hz, (b) 1.25, (c) $i = I\left(\frac{1}{3}e^{-8t} - \frac{4}{3}e^{-2t} + 1\right)$

6 (a) 5 Hz, (b) 1.0, (c) $x = (-32 + 6t)e^{-5t} + 6$

7 (a) 9.5%, (b) 0.020 s

8 (a) 4 Hz, (b) 0.625, (c) 1.45 Hz, (d) 0.5 s, (e) 8.1%, (f) 1.4 s

9 (a) 0.59, (b) 0.87

10 2.4

11 0.09

12 3.93 rad/s, 0.63 Hz

Chapter 11

1 (a) $\dfrac{1}{As + \rho g/R}$, (b) $\dfrac{1}{ms^2 + cs + k}$, (c) $\dfrac{1}{LCs^2 + RCs + 1}$

2 (a) 3 s, (b) 0.67 s

3 (a) $1 + e^{-2t}$, (b) $2 + 2\,e^{-5t}$

4 (a) Over-damped, (b) under-damped, (c) critically damped, (d) under-damped

5 $t\,e^{-3t}$

6 $2\,e^{-4t} - 2\,e^{-3t}$

7 (a) $\dfrac{4s}{s^2(s+1) + 4}$, (b) $\dfrac{2(s+2)}{(s+1)(s+2) + 2}$,

(c) $\dfrac{4}{(s+2)(s+3) + 20}$, (d) $\dfrac{2}{s(s+2) + 20}$

8 $5/(s + 53)$

9 $5s/(s^2 + s + 10)$

10 $2/(3s + 1)$

Chapter 12

1 (a) $\dfrac{5}{\sqrt{\omega^2 + 4}}$, $\dfrac{\omega}{2}$, (b) $\dfrac{2}{\sqrt{\omega^4 + \omega^2}}$, $\dfrac{1}{\omega}$,

(c) $\dfrac{1}{\sqrt{4\omega^6 - 3\omega^4 + 3\omega^2 + 1}}$, $\dfrac{\omega(3 - 2\omega^2)}{1 - 3\omega^2}$

2 $0.56 \sin (5t - 38°)$

3 $1.18 \sin (2t + 25°)$

4 (a) (i) ∞, 90°, (ii) 0.44, 450°, (iii) 0.12, 26.6°, (iv) 0, 0°,
(b) (i) 1, 0°, (ii) 0.32, −71.6°, (iii) 0.16, −80.5°, (iv) 0, −90°

5 See Figure A.3

6 (a) $1/s$, (b) $3.2/(1 + s)$, (c) $2.0/(s^2 + 2\zeta s + 1)$,
(d) $3.2/[(1 + s)(0.01s^2 + 0.2\zeta s + 1)]$

Fig. A.3   Chapter 12, problem 5

Chapter 13

1 See Section 13.2
2 (a) 8 minutes, (b) 20 minutes
3 (a) 12 s, (b) 24 s
4 20%, 5
5 62.5%, 0.63%
6 (a) 51.0%, (b) 51.0%, (c) 9%, (d) 49.5%
7 (a) 54%, (b) 66%
8 (a) 55%, (b) 65.25%, (c) 76%
9 See the text. In particular P – offset, PI and PID no offset
10 3, 666 s, 100 s
11 3, 100 s, 25 s
12 See Sections (a) 13.12.1, (b) 13.12.2, (c) 13.12.3
13 1.6
14 First-order response with time constant c/K_p

Chapter 14

1 255
2 (a) 11, (b) 529
3 (a) 1A7, (b) 211
4 (a) 781, (b) 157
5 (a) 1010 0110, (b) 1101 1101
6 (a) 0, (b) 1
7 (a) 8, (b) 12

Fig. A.4 Chapter 14, problem 17

8 For example: (a) ticket selected AND correct money in, correct money decided by OR gates analysis among possibilities
   (b) AND with safety guards, lubricant, coolant, work piece, power, etc. all operating or in place

9 (a) $A \cdot (B + C)$, (b) $(A + B) \cdot (C + D)$, (c) $\overline{A} + B$, (d) $\overline{A} \cdot B$

10 (a) $Q = (A \cdot B + C \cdot D) \cdot E$, (b) $Q = (A \cdot B + B) \cdot C$

11

A	B	C	Q
0	0	0	0
0	0	1	0
0	1	0	0
0	1	1	1
1	0	0	0
1	0	1	1
1	1	0	0
1	1	1	1

12 (a) $Q = C \cdot (A + D)$, (b) $Q = A \cdot B$, (c) $Q = A \cdot \overline{B} \cdot C + C \cdot D$

13 As given in the problem

14 (a) $Q = A + B$, (b) $Q = C + \overline{A} \cdot C$

15 (a) $Q = \overline{A} \cdot \overline{B} + \overline{A} \cdot \overline{C}$, (b) $Q = A \cdot B \cdot D + A \cdot B \cdot \overline{C} + \overline{C} \cdot D$

16 Four input AND gates with two NOT gates if correct combination is 1, 1, 0, 0:
   $$Q = A \cdot B \cdot \overline{C} \cdot \overline{D}$$

17 See Figure A.4

18 As in the text, Section 14.7.1, for cross-coupled NOR gates

Chapter 15

1 See Section 15.2
2 256
3 $64 \text{ K} \times 8$
4 See Section 15.3
5 See Figure 15.10 and associated text
6 (a) E, (b) C, (c) D, (d) B
7 256
8 (a) 0, (b) 1
9 See Section 15.3.1, item 6
10 See Section 15.3.2, item 5
11 High to reset pin
12 (a) IF A
         THEN
            BEGIN B
            END B
         ELSE
            BEGIN C
            END C
         ENDIF A

(b) WHILE A
   BEGIN B
   END B
  ENDWHILE A

Chapter 16

1 (a) 89, (b) 99
2 No address has to be specified since the address is implied by the mnemonic
3 (a) CLRA, (b) STAA, (c) LDAA, (d) CBA, (e) LDX
4 (a) LDAA $20, (b) DECA, (c) CLR $0020, (d) ADDA $0020
5 (a) Store accumulator B value at address 0035
  (b) Load accumulator A will data F2
  (c) Clear the carry flag
  (d) Add 1 to value in accumulator A
  (e) Compare C5 to value in accumulator A
  (f) Clear address 2000
  (g) Jump to address given by index register plus 05
6 (a)

```
DATA1      EQU      $0050
DATA2      EQU      $0060
DIFF       EQU      $0070
           ORG      $0010
           LDAA     DATA1      ; Get minuend
           SUBA     DATA2      ; Subtract subtrahend
           STAA     DIFF       ; Store difference
           SWI                 ; Program end
```

(b)

```
MULT1      EQU      $0020
MULT2      EQU      $0021
PROD       EQU      $0022
           ORG      $0010
           CLR      PROD       ; Clear product
                               ; address
           LDAB     MULT1      ; Get first number
SUM        LDAA     MULT2      ; Get multiplicand
           ADDA     PROD       ; Add multiplicand
           STAA     PROD       ; Store result
           DECB                ; Decrement acc. B
           BNE      SUM        ; Branch if adding not
                               ; complete
           WAI                 ; Program end
```

(c)

```
FIRST      EQU      $0020
           ORG      $0000
           CLRA                ; Clear accumulator
           LDX      #0
MORE       STAA     $20,X
           INX                 ; Increment index reg.
           INCA                ; Increment
                               ; accumulator
           CMPA     #$10       ; Compare with
                               ; number 10
           BNE      MORE       ; Branch if not zero
```

```
                      WAI                        ; Program end
       (d)
                      ORG        $0100
                      LDX        #$2000          ; Set pointer
       LOOP           LDA A      $00,X           ; Load data
                      STA A      $50,X           ; Store data
                      INX                        ; Increment index
                                                 ; register
                      CPX        $3000           ; Compare
                      BNE        LOOP            ; Branch
                      SWI                        ; Program end

     7 YY             EQU        $??             ; Value chosen to give
                                                 ; required time delay
       SAVEX          EQU        $0100
                      ORG        $0010
                      STA        SAVEX           ; Save accumulator A
                      LDAA       YY              ; Load accumulator A
       LOOP           DECA                       ; Decrement acc. A
                      BNE        LOOP            ; Branch if not zero
                      LDA        SAVEX           ; Restore accumulator
                      RTS                        ; Return to calling
                                                 ; program

     8                LDA        $2000           ; Read input data
                      AND A      #$01            ; Mask off all bits but
                                                 ; bit 0
                      BEQ        $03             ; If switch low, branch
                                                 ; over JMP which is 3
                                                 ; program lines
                      JMP        $3000           ; If switch high no
                                                 ; branch and so
                                                 ; execute JMP
                      Continue
```

Chapter 17

1 (a) The variable counter is an integer, (b) the variable num is
  assigned the value 10, (c) the word name will be displayed,
  (d) the display is Number 12, (e) include the file stdio.h
2 (a) Calls up the library necessary for the printf( ) function,
  (b) indicates the beginning and end of a group of statements,
  (c) starts a new line, (d) problem 3
3 The number is 12
4 # include <stdio.h>

```
  int main(void);
  {
    int len, width;
    printf("Enter length: ");
    scanf("%d", &len);
    printf("Enter width: ");
    scanf("%d", &width);
```

```
printf("Area is %d", lens * width);
return 0;
{
```

5 Similar to program given in Section 17.3, item 4
6 Divides first number by second number unless second is 0

Chapter 18

1 See Section 18.3
2 See Section 18.3. A parallel interface has the same number of input/output lines as the microprocessor. A serial interface has just a single input/output line
3 See Section 18.2
4 See Section 18.4
5 See Section 18.4 and Figure 18.13
6 See Section 18.4.1
7 See Section 18.3.3. Polling involves the interrogation of all peripherals at frequent intervals, even when some are not activated. It is thus wasteful of time. Interrupts are only initiated when a peripheral requests it and so is more efficient
8 CRA 00110100, CRB 00101111
9 As the program in 18.4.2 with LDAA #$03 replaced by LDAA #$34 and LDAA #$34 replaced by LDAA #$2F
10 As the program in Section 18.4.2 followed by
READ LDAA $2000   ; Read port A
Perhaps after some delay program there may then be
BRA READ

Fig. A.5   Chapter 19, problem 7

Chapter 19

1 (a) AND, (b) OR
2 (a) Figure 19.9(b), (b) Figure 19.10(b), (c) a latch circuit, Figure 19.16, with Input 1 the start and Input 2 the stop switches
3 0 LD X400, 1 LD Y430, 2 ORB, 3 ANI X401, 4 OUT Y430
4 0 LD X400, 1 OR Y430, 3 OUT Y430, 4 OUT T450, 5 K 50; delay-on timer
5 0 LD X400, 1 OR Y430, 2 ANI M100, 3 OUT Y430, 4 LD X401, 5 OUT M100; reset latch
6 As in Figure 19.28 with Timer 1 having $K = 1$ for 1 s and Timer 2 with $K = 20$ for 20 s
7 Figure A.5
8 Figure A.6
9 Figure A.7
10 Figure A.8
11 An output would come on, as before, but switch off when the next input occurs
12 See Section 19.10

Fig. A.6   Chapter 19, problem 8

Fig. A.7 Chapter 19, problem 9

Fig. A.8 Chapter 19, problem 10

Fig. A.10 Chapter 19, problem 14

Fig. A.9 Chapter 19, problem 13

13 Two latch circuits, as in Figure A.9
14 Fig. A.10

Chapter 20

1 See Section 20.2
2 See Section 20.3
3 Bus
4 Broad band
5 See Section 20.5.1
6 See Section 20.4
7 See Section 20.3.1
8 NRFD to PD0, DAV to STRA and IRQ, NDAC to STRB, data to Port C
9 TTL to RS-232C signal level conversion
10 See Section 20.6.4

Chapter 21

1 See Section 21.1
2 See Section 21.2
3 See Section 21.2

4 See Section 21.5.3 for programmable checks and checksum
   and Section 21.2 for watchdog timer
5 See Section 21.5.3

Chapter 22

1 Possible solutions might be: (a) thermocouple, cold junction
   compensation, amplifier, ADC, PIA, microprocessor, DAC,
   thryristor unit to control the oven heating element, (b) light
   beam sensors, PLC, solenoid operated delivery chute
   deflectors, (c) closed-loop control with, for movement in each
   direction, a d.c. motor as actuator for movement of pen,
   microprocessor as comparator and controller, and feedback
   from an optical encoder.

Research assignments

The following are brief indications of the type of information
which might be contained in an answer.

2 A typical ABS system has sensors, inductance types, sensing
   the speeds of each of the car wheels, signal conditioning to
   convert the sensor signals into 5 V pulses, a microcontroller
   with a program to calculate the wheels' speed and rate of
   deceleration during braking so that when a set limit is
   exceeded the microcontroller gives an output to solenoid
   valves in the hydraulic modulator unit to either prevent an
   increase in braking force or if necessary reduce it.
3 The carriage motor moves the printer head sideways while
   the print head prints the characters. After printing a line the
   paper feed motor advances the paper. The print head consists
   of solenoids driving pins, typically a row of nine, to impact
   on an ink ribbon. A microcontroller can be used to control
   the outputs. For more details, see *Microcontroller
   Technology: The 68HC11* by P. Spasov (Prentice-Hall 1992,
   1996).
4 The CAN bus operates with signals which have a start bit,
   followed by the name which indicates the message
   destination and its priority, followed by control bits, followed
   by the data being sent, followed by CRC bits, followed by
   confirmation of reception bits, and concluding with end bits.

Design assignments

The following are brief indications of possible solutions.

5 A digital thermometer using a microprocessor might have a
   temperature sensor such as LM35, an ADC, a ROM chip
   such as the Motorola MCM6830 or Intel 8355, a RAM chip
   such as the Motorola MCM6810 or Intel 8156, a microprocessor such as Motorola M6800 or Intel 8085A and a
   driver with LED display. With a microcontroller such as the

Motorola MC68HC11 or Intel 8051 there might be just the temperature sensor, with perhaps signal conditioning, the microcontroller and the driver with LED display.

6 A digital ohmmeter might involve a monostable multivibrator which provides an impulse with a duration of $0.7RC$. A range of different fixed capacitors could be used to provide different resistance ranges. The time interval might then be determined using a microcontroller or a microprocessor plus memory, and then directed through a suitable driver to an LED display.

7 This might involve a pressure sensor, e.g. the semiconductor transducer Motorola MPX2100AP, signal conditioning to convert the small differential signal from the sensor to the appropriate level, e.g. an instrumentation amplifier using operational amplifiers, a microcontroller, e.g. MC68HC11, an LCD driver, e.g. MC145453, and a four-digit LCD display.

8 This could be tackled by using the M68HC11EVM board with a PWM output to the motor. Where feedback is wanted an optical encoder might be used.

9 The arrangement might be for each box to be loaded by current being supplied to a solenoid valve to operate a pneumatic cylinder to move a flap and allow a box down a chute. The box remains in the chute which is closed by a flap. Its presence is detected by a sensor which then indicates the next box can be allowed into the chute. This continues until four boxes are counted as being in the chute. The flap at the end of the chute might then be activated by another solenoid valve operating a cylinder and so allowing the boxes onto the belt. The arrival of the boxes on the belt might be indicated by a sensor mounted on the end of the chute. This can then allow the entire process to be repeated.

Index